Geographical Information Systems
Systems
Trends and Technologies

Geographical Information Systems
Trends and Technologies

Editor

Elaheh Pourabbas

National Research Council
Institute of Systems Analysis and Computer Science "Antonio Ruberti"
Rome, Italy

CRC Press
Taylor & Francis Group
Boca Raton London New York

CRC Press is an imprint of the
Taylor & Francis Group, an **informa** business

A SCIENCE PUBLISHERS BOOK

Visit the Taylor & Francis Web site at
http://www.taylorandfrancis.com

CRC Press Web site at
http://www.crcpress.com

Science Publishers Web site at
http://www.scipub.net

Preface

Geographical Information Systems (GIS) since its inception in the late 1960s have seen an increasing rate of theoretical, technological and organizational development. Developments in each decade of the last 50 years highlight particular innovations in this field. The mid 1960s witnessed the initial development of GIS in combining spatially referenced data, spatial data models and data visualization. The early 1970s witnessed the ability of computer mapping in automatic map drafting and using data format. In the 1980s, computer mapping capabilities have been merged with traditional database management systems capabilities to generate spatial database management systems. Accordingly, the ability to select, sort, extract, classify and display geographic data on the basis of complex topological and statistical criteria was available to users. The 1990s saw map analysis and modeling advances in GIS, and these systems became real management information tools as computing power increased. During this decade, the Open GIS Consortium, aimed at developing publicly available geoprocessing specifications, was founded. Since 2000, with the advent of Web 2.0, mobile, and wireless technologies, GIS have been moving towards an era in which the power of such systems is continuously increasing in multiple facets consisting of computing, visualizing, mining, reasoning data. The latest changes in technologies and trends have brought new challenges and opportunities in GIS domain. Specifically, mobile and internet devices, Cloud computing, NoSQL databases, Semantic Web, Web services offer new ways of accessing, analyzing, and elaborating geospatial information in both real-world and virtual spaces.

This book presents the latest developments, trends and technologies in GIS. It addresses the following areas: Big data, Cloud computing, NoSQL Geographic Databases, Web services, Geoprocessing-Scientific workflow, Mobile GIS, Spatial Data Warehousing-OLAP, and Semantic similarity. The general aim of this book is to offer a comprehensive overview of the methodological aspects of such technologies and their applications in GIS, and also to outline the major challenges and future perspectives.

The book comprises ten chapters. Chapters 1 to 4 focus on the efficient storage, indexing and querying of huge amount of distributed geographic data. In particular, Chapter 1 provides an overview of the various technologies necessary to build a functional array analytics system for geospatial data. Chapter 2 focuses on spatial similarity join techniques, and describes the design and implementation of *MapReduce-based Similarity Join* algorithms for large geographical data. Chapter 3 describes spatial indexing techniques in the cloud environments and presents an analysis of data modelling and the basic operations of cloud data managements such as HBase and Cassandra. Chapter 4 provides an overview of the main NoSQL systems that cope with geographical data. Chapters 5 to 7 address efficient geographic data exchanging and geographic data sharing over a network through Web services as well as efficient geographic data processing supported by scientific workflow to facilitate a vast range of scientific activities such as simulation, data modeling and visualization. Chapter 5, specifically, focuses on combining of traditional web services with geographical web services. Chapter 6 discusses frameworks to enhance the semantics in GIS Web Services, such as the Web services WMS (Web Map Service) and WFS (Web Feature Service). Chapter 7 discusses Geo-enabled Model Web from the perspective of geoprocessing web services and geoscientific workflows, and explores opportunities and implications by jointly applying these technologies to various scenarios. Chapter 8 presents an overview of the mobile-oriented geospatial applications and explores the problems of combining GIS and Semantic Web techniques within a smartphone working off-line. Chapter 9 explores the existing research related to Spatial DWs, spatial ETL, and SOLAP, and investigates various stages of the development process that use available free software. Finally, Chapter 10 presents an overview of methods for measuring semantic similarity of geographic data based on weighted ontology.

The primary target audience for the book includes researchers, application specialists and developers, practitioners and students who work or have an interest in the area of integrating GIS with the latest technologies.

Acknowledgments

The editor wishes to acknowledge the contributions of authors from various institutions, who wrote many of the chapters in this book. I am grateful to all authors for their conscientious cooperation. The editor would like to thank a number of colleagues who have critically read various parts of the manuscript and offered many valuable comments. In particular, special thanks go to Kenneth M. Anderson, Yvan Bédard, Jonathan D. Blower, Alain Bouju, Carlos Granell, Leticia Gomez, Shuichi Kurabayashi , Chen Li, Wei Luo, Sattam Mubark Alsubaiee, Adam Iwariak, Elias Ioup, Gabriele Pozzani, Ahmet Sayar, Stefano de Sabbata, Rafael Jesús Segura Sánchez, Kurt Stockinger, Aida Valls, Guanfeng Wang, Ling-Yi Wei, Lena Wiese, Huayi Wu, Kesheng Wu, Peng Yue, Peisheng Zhao. The editor also wishes to acknowledge the support of the Institute of Systems Analysis and Computer Science (IASI) "Antonio Ruberti"—National Research Council of Italy.

Contents

CHAPTER 1

Preparing Array Analytics for the Data Tsunami

Peter Baumann,[a,] * *Jinsongdi Yu,*[a,c] *Dimitar Misev,*[a]
Kinga Lipskoch,[a] *Alan Beccati,*[a] *Piero Campalani*[a] and
Michael Owonibi[b]

Introduction

Array data, commonly also known as "grid" or "raster" data, represent a large part of today's operational data, including 1D timeseries, 2D remote sensing imagery, 3D x/y/t image timeseries and x/y/z geophysical data, 4D x/y/z/t ocean data, as well as 5D grid meteo simulations, which are 3D spatial cubes over two time axes—today's analysis and yesterday's one-day forecast have the same validity time but different forecast times (Domenico et al. 2006). Being one particular class of the general category of space/time varying data it forms a good study field for further scientific progress into additional data access and analytics.

Management and analytics of scientific data are a central challenge today, with the growth doubling every year (Szalay and Gray 2006). Sensors are becoming more and more powerful and ubiquitous; resolution is

[a] Jacobs University Bremen, Campus Ring 1, 28759 Bremen, Germany.
  Emails: p.baumann@jacobs-university.de; j.yu@jacobs-university.de;
  d.misev@jacobs-university.de; k.lipskoch@jacobs-university.de;
  a.beccati@jacobs-university.de; p.campalani@jacobs-university.de
[b] Institut für Informatik, Friedrich-Schiller-University, Humboldtstraße 11 07743 Jena Germany.
  Email: michael.owonibi@uni-jena.de
[c] Fuzhou University, 523 Gongye Rd, Gulou, Fuzhou, Fujian, China.
  Email:yyx350@126.com
* Corresponding author

constantly increasing while cost and sensor size are going down. The result is data growth at a rate exceeding Moore's law (doubles approximately every two years) according to the leading researches in the field, which becomes even more impressive as the trend predicted by Moore's law is slowly coming to an end (Hilbert and López 2011). In other words, hardware development alone cannot keep up in providing the expected resources for real-time and near real-time analysis of such Big Data (Laney 2001; Snow 2012; IBM 2013). It is just as important, if not more, to advance the development of software that scales to huge volumes of data in all aspects of working with it—from collecting and storing the data, to searching it, processing and analyzing it, to visualizing the results.

Traditionally, raster data repositories are implemented in a one-file-per-image manner where images often serve as "dead" backdrops which can be displayed, but not get analyzed further at the data repository. In terms of data management and provisioning, file-based solutions are based on particular models (typically induced by the particular design choices of the data exchange format used) and often actually lacking a clearly stated, informationally coherent model for different formats, e.g., GeoTiff and NetCDF do not know each other unless format rules are introduced under a coherent information model. Consequently, every extension or add-ons with new structures is causing severe implementation, performance, and interoperability problems. Therefore, it requires tremendous effort to provide flexible and real-time answers in this case.

Array databases provide a more convenient way to process large arrays. For example, queries can be quicker than file-based systems because of the potential for optimization and for organizing data physically around the kinds of queries that one expects. Database technology adds substantial advantages by offering scalable, flexible storage and retrieval/manipulation on arrays of (conceptually) unlimited size, but have to be extended and adapted in several ways, leading to the goal of the recently emerged research field of *Array Analytics* which combines insights from databases, programming languages, high-performance computing, and further areas with the domain expertise of Earth, Space, Life, and Social sciences and engineering.

Effective analytics on pixel level is more than simply a matter of shuffling high volumes or optimizing the search performance by introducing a database system to store file locations in the database; it heralds an era of finding insights in historical and emerging information in a more agile and adaptive way, and of answering questions that were previously unconsidered because they were thought to be intractable. This paradigm shift is fundamentally impacting all geo-scientific disciplines, such as Cryospheric, Atmospheric, Solid Earth, and Ocean research. The promise is a better understanding of the Earth System, e.g., global warming and climate changes.

This chapter, therefore, focuses on Array Analytics approaches and technologies. The state of the art in Array Analytics will be discussed, together with research examples[1] for preparing the analytics in the era of Big Data.

Array Analytics History

Support for massive arrays in the dimension of Petabytes has long been neglected by the database community; Figure 1 gives a schematic overview of important representatives. PICDMS (Chock and Cardenas 1983) considered arrays, but only in a very specialized manner (as stacks of 2D raster maps) and without dedicated architecture. EXTRA-EXCESS (Vandenberg 1991) was an early attempt to incorporate arrays into a database type system.

Since a few years, Array Database research has gained impetus, following a first Array Database Workshop (Baumann et al. 2011). The most advanced ones are picked and presented with their relevant characteristics as in Table 1.

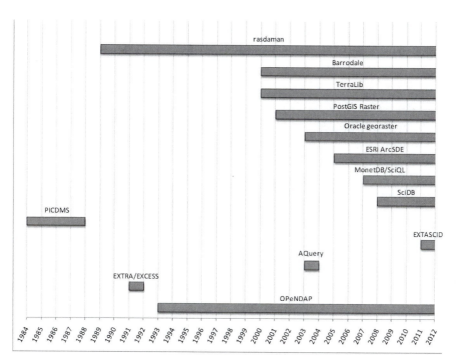

Fig. 1. The time line and timely duration of systems manipulating array data.

---

[1] The study is based on ample real-life examples, many of which are freely accessible over the Web, see: http://www.rasdaman.org

Table 1. Characteristics of systems manipulating array data.

	Algebra	Query Language	Dimensions	Tiling	Query Optimization	SQL Integration
rasdaman	yes	yes	unlimited	arbitrary	yes	yes
Barrodale[2]	no	yes	4D	yes	no	yes
TerraLib[3]	yes	yes	unlimited	regular	yes	yes
PostGIS Raster[4]	yes	map-algebra	2D	arbitrary	no	yes
Oracle georaster[5]	no	PL/SQL	2D	yes	no	yes
MonetDB/SciQL[6]	yes	yes	unlimited	regular	no	yes
SciDB[7]	yes	yes	unlimited	yes	yes	no
PICDMS	no	yes	2D	no	no	no
EXTASCID	no	no	unlimited	no	no	no
OpeNDAP[8]	no	no	unlimited	regular	no	no

The most important fundamental and functional characteristics concerning broadness and high level of applicability, such as being based on an algebra, defining a query language, being able to handle multidimensional data, applying specific tiling strategies, optimizing queries and integrating SQL into the system, are summarized. The systems which are considered to be representative are contained in the table by highlighting their characteristics. Hence, a comparison of the listed systems with respect to these characteristics can be directly derived from the table.

In Fig. 1 the time line and the timely duration of different systems handling array data are presented. Some of the systems are the same ones from the characteristics table and some other systems are also mentioned. The additional systems are not reference systems, but they contributed to one or other aspect of handling raster data.

The ArcSDE (Spatial Database Engine)[9] technology of ESRI (Environmental Systems Research Institute) manages spatial data in a relational database management system and enables it to be accessed by ArcGIS clients. It provides a framework to support long transactions, which facilitates the versioned editing environment in multiuser geodatabases. Its raster capabilities are constrained to 2D.

A Query (Lerner and Shasha 2003) is an algebra and a query language, which supports order as a first class algebraic concept from-the-ground-up.

---

[2] http://www.barrodale.com/bcs-grid-datablade/, "last seen on 31 March 2013"
[3] http://www.terralib.org/, "last seen on 31 March 2013"
[4] http://postgis.net/, "last seen on 31 March 2013"
[5] http://docs.oracle.com/html/B10827_01/geor_intro.htm, "last seen on 31 March 2013"
[6] http://www.scilens.org/, "last seen on 31 March 2013"
[7] http://www.scidb.org/, "last seen on 31 March 2013"
[8] http://opendap.org/, "last seen on 31 March 2013"
[9] http://www.esri.com/software/arcgis/arcsde, "last seen on 31 March 2013"

Order-related query transformations can be considered in its context. No operational implementation is known, though.

SciQL is an array extension to the SQL: 2003 based query language of the MonetDB column store DBMS. It integrates the semantics of set, sequence, and array. With minimal modifications to the SQL syntax, it allows for expressing array operations within the common language framework. The drawback for scientists is that it is mainly targeted for experienced SQL programmers. Also, SciQL still lacks an array storage manager—currently, arrays are emulated by relations, which leads to inefficiencies.

The shared-nothing parallel database system SciDB was designed for dense array processing. It handles multi-dimensional, nested arrays within array cells. SciDB is novel in that it has no built-in operators, but operators which are all user-defined functions. SciDB is supposed to be more than an Array DBMS, a complete system for managing and analyzing scientific data. Therefore, it is supposed to include a complex versioning system, *in situ* data processing, array and cell provenance, and support for uncertain cell values (Rusu and Cheng 2013).

EXTASCID (Cheng and Rusu 2012) is a parallel system for handling efficient analysis of large-scale scientific data. It manages massive heterogeneous data by partitioning data, executing at parallel, and having a relational and array data model. EXTASCID is architecture independent employing multi-threading and shared-nothing parallelism.

A special case in the table is OPeNDAP—it is not a database system but offers dedicated array access and processing functionality. Other non-database solutions for manipulating raster data are offered within MATLAB,[10] the programming language R,[11] Orfeo toolbox,[12] etc.

The first Array DBMS based on a rigid formalization and comprising a fully fledged query language (Baumann 1994; 1999) backed by the highly effective optimizations (Baumann et al. 2009) is rasdaman, short for "raster data manager".[13] It will be detailed further down below, as it is the Reference Implementation[14] for the OGC WCS 2.0 geoservice standards suite.

Array Models

In this section, array data models, which have undergone a rigid formalization, are informally presented. Important candidates are *Array Algebra*, *AML*, *AQL*, and *RAM*. Although recently more relevant models

---

[10] http://www.mathworks.com/products/matlab/, "last seen on 31 March 2013"
[11] http://www.r-project.org/, "last seen on 31 March 2013"
[12] http://www.orfeo-toolbox.org/otb/, "last seen on 31 March 2013"
[13] http://www.rasdaman.org, "last seen on 31 March 2013"
[14] http://www.esri.com/software/arcgis/arcsde, "last seen on 31 March 2013"

have emerged, such as PostGIS Raster, SciQL, SciDB, and EXTASCID, these still lack a rigorous semantics definition so that they have sufficiently been treated in the previous section.

As it turns out, AML, AQL and RAM can be mapped to *Array Algebra*, therefore this common representation eases the comparison by highlighting the common functionality and differences between the four models (Baumann and Holsten 2012).

Array Algebra

An algebraic approach for array modeling is being offered by the *Array Algebra* (Baumann 1994; 1999). Originally, it has been developed for studying image processing and computer graphics such as (Carson and McGinnis 1986; ISO 1994; Felger et al. 1990). The Image Algebra of the Air Force Armament Laboratory (AFATL) (Ritter et al. 1994) has proven to be a valuable starting point for database-centric domain-independent array processing. The traditional application domains of Array Algebra in the past consisted of sensor, image, simulation, and statistics data services. The latest emphasis is on large-scale Earth Science (Gutierrez and Baumann 2007) and Life Science (Roland et al. 2001) data.

Array Algebra is implemented by the *rasdaman* Array Database management system and its query language *rasql* (Baumann 2013). The Web Coverage Processing Service (WCPS), (Baumann 2008) geo raster query language standard issued by the Open Geospatial Consortium (OGC) in 2008 is based on the concepts introduced by Array Algebra.

The core model specifies an array to be a mapping from integer coordinates—the *spatial domain*—to some array cell data type. The model consists of three core operators: *array constructor, condense operator,* and *array sorter.* The condense operator reduces an array to a scalar value by using some aggregation function. The array sorter proceeds along a given dimension by reordering the corresponding hyperslices. All other operations, such as common statistics and image processing operators like matrix multiplication, image convolution, and Fast Fourier Transform, can be expressed as combinations of these operations.

AML

The *Array Manipulation Language* (*AML*) is another algebra-based, high-level language which allows querying data and defining new arrays in terms of existing ones (Marathe and Salem 1997; 1999). The targeted application domain of the model was image databases especially for remote sensing, although the authors point out that their model allows customization to multiple application domains.

AML uses infinite vector of integers for representing array data. Special characteristics of the model are bit patterns. The core operators of AML are defined with the help of bit patterns. AML has three core operators: *subsample, merge* and *apply*. While the *subsample* operator eliminates cells in an array, the *merge* operator combines two arrays and the *apply* operator applies a given function to an array.

AQL

The third array model presented in this chapter is *AQL*, which relies on Nested Relational Calculus for Arrays (NRCA) as extension of the Nested Relational Calculus (NRC) (Libkin et al. 1996; Machlin 2007). The objective of AQL was to support the application domains of the NetCDF data exchange format and particularly scientific array data motivated by the fact to be able to algorithmically generate and manipulate array data.

NRC contains objects like products and sets. The value set can be generalized into an uninterpreted base type (i.e., black box). NRCA adds to NRC natural numbers, constants, basic arithmetics, an index set generator, and a summation construct. By this, arrays can be algorithmically generated and manipulated.

NRCA introduces four operators: *subscripting* for returning the content of a cell and *length* of a dimension, which both operate on arrays, and *array tabulation* and *indexing*, which contribute to the generation of arrays.

RAM

As the last array model presented here, the Random-access memory (RAM) was designed as extension to the neo-relational database management system MonetDB (Cornacchia et al. 2008). Contrarily to AQL, this array model is completely separated from the relational query formalism. Both the motivation and the corresponding application area lay in multimedia analysis (van Ballegooij 2004; van Ballegooij et al. 2005; Cornacchia et al. 2004). However, the model was meant for OLAP only and does not claim to be suitable for geo services; SciQL is a successor which aims at incorporating such application domains.

RAM offers a generic array construction operator similar to the array tabulation of AQL. RAM introduces the concept of *aligned arrays* which are defined as arrays with identical shape representing related data. Other six operators are: *const, grid, map, apply, choice* and *aggregate* for filling a new array with a constant value, applying a given function to aligned elements in a set of arrays, applying a given array to aligned elements in a set of index-arrays, choosing cell values from two input arrays and applying the aggregation function along the first given axes of an array.

Comparison

By mapping the array models to Array Algebra (Baumann and Holsten 2012), the following common properties and differences were established. All four array models share the concept of arrays as functions over a domain. But, typing is possible only within Array Algebra. Compared to AML, AQL and RAM, Array Algebra allows not only non-negative but also negative indices allowing hyperobjects to be located anywhere with respect to the coordinate reference system. Array Algebra, AQL and RAM all have a generic array generator. Only AML does not have such a constructor. Array Algebra additionally introduces the sorting operator. With respect to the relational embedding, RAM maps arrays to relations which are accessible by SQL. AQL is embedded into the nested relational calculus and in the case of AML, the relational embedding is not discussed (Marathe and Salem 1997; 1999). Contrarily, Array Algebra does not make any assumptions about an embedding data model, it provides typed arrays as basis for an array sub-model of some other data model. In fact, the rasdaman API and implementation relies on the object-oriented model of the Object Database Management Group (ODMG) standard (Catell and Cattell 2000), and the first implementation has been on the object-oriented DBMS, O_2, while ports to various relational systems have been added only later.

Array DBMS

Array DBMS provides flexible, scalable storage and retrieval services specifically for array data. As in practice, all facets of database design need to be adapted for embedding array model into some overall data model, either as an analogy to tables or an additional attribute type. There are many open research questions, addressing storage management, query language design and formalization, query optimization, parallelization and distributed processing, and scalability issues in general. A sample array manipulation query[15] is as below:

```
for c in (OIL_SPILL_MONTEREY_BAY)
    return
    encode ((((c[t(22)] > 0) and (c[t(12)] = 0)), "png")
```

Fast access is the prime task of the array storage manager. Traditional database tuples and objects tend to fit well into a single database page—a unit of disk access on server, typically 4 KB, while array objects can easily

---

[15] The demo is based on the Monterey Bay oil spill monitoring's image timeseries, to map areas which are contaminated on day 22 and are not contaminated on the day 10 of the spill, accessible over the Web, see: http://www.earthlook.org

span over several storage media. To establish quick access patterns, arrays are partitioned in tiles or chunks of convenient size which are basic access units during query evaluation (Baumann 1994). Additional geo indexes assist the tiles to act in a well-performing manner.

Array query languages, like SQL, give declarative access to arrays. These queries are parsed, optimized, and executed to create, manipulate, search, and delete arrays in flexible ways. The parser receives the query and generates the operation tree. Then, algebraic optimization rules are applied to the query tree where applicable. Without considering the parallelism, the execution addresses tiles sequentially. The tile-by-tile processing strategy leads to an architecture allowing servers to process arrays orders of magnitude far beyond the main memory.

Extensions are made for achieving scalability. Normally, scalability on a single machine is guided by parameters like the number of processor cores and the amount of main memory. The trends in processor development are towards increasing the number of cores in one chip, rather than increasing the power of a single core. By processing each tile on separate nodes or cores, parallel processing becomes a critical development paradigm for scalable software, allowing full utilization of these new processor architectures. At a certain point, the hardware resources of a single machine will not be enough to handle all tasks. It becomes necessary to distribute the data and workload to further machines (nodes) according to some strategy. Multiple machines present new challenges, however, like limited connection speed between nodes, optimizing data distribution, minimizing data movement between nodes. This drives database development towards distributed and cloud computing. When data duplication would be inevitable with the standard storage management mechanisms, the *in situ* processing capability is an alternative way of adding value to legacy systems and preserving scalability.

Parallel processing

Parallel databases seek to improve performance by parallelizing all steps involved in the query evaluation whenever possible. Parallel processing in the context of Array DBMS specifically means *data parallelism* (Hahn et al. 2002), which focuses on data distribution across different computing nodes for parallel evaluation. Pipeline parallelism which is widely used in RDBMS is not particularly suitable, as the granularity of query evaluation is much larger with array tiles and a pipeline buffer would quickly overflow the main memory (Hahn et al. 2002).

Parallel DBMS typically exploit one of the following three architectures. In *shared-memory* systems, CPUs are interconnected and have access to a common memory region. CPUs in a *shared-disk* architecture have access to a common disk space, but maintain their private main memory space. Finally, in a *shared-nothing* architecture CPUs do not share any memory or disk, and communicate with each other via a network connection.

The shared-nothing architecture is widely regarded as the most scalable approach providing linear speed-up and scale-up to the number of CPUs and disks in proportion to data volume. Shared-memory and shared-disk systems are in general limited because of interference, i.e., performance improvement slows down as more and more CPUs are added due to memory and network contention. SciDB, for example, has adopted a shared-nothing architecture from the start (Brown 2010). A SciDB instance can be deployed on a network of computers, each with its own disk, which are logically unified through a central system catalog. The rasdaman system, on the other hand, supports shared-nothing, and additionally a variant of shared-disk to accommodate existing massive archive infrastructures; currently, this is being evaluated by data centers like NASA and ESA.

Parallelization techniques include inter-query and intra-query parallelization. Inter-query parallelization dispatches incoming queries in completeness, thereby achieving good load balancing in particular when many smaller queries come in. Intra-query parallelization splits incoming queries and executes them in parallel on different nodes; obviously, this needs to take query "objects" into consideration when large, complex queries arrive. Therefore, intra-query parallelization is further subdivided into intra- and inter-object parallelization. An "object" in this case is an array, as opposed to single tuples in conventional databases. This calls for different methods as the size of an array is typically orders of magnitude greater. Intra-array parallelism splits and distributes array data as finer granules based on the physical model of the data. As arrays get partitioned for storage anyway, this partitioning may serve as a unit of distribution as well. This is individually tailored for each array operation, and the query tree is further adapted in such a way that data transfer within the tree as well as between processes is minimized. Inter-array parallelism distributes a single query to a set of peers holding their own data. This provides significant speedup especially for very complex and expensive, long-running queries.

Rasdaman implements both intra-query (Hahn et al. 2002) and inter-query (Hahn et al. 2002) parallelism—hence, a shared-nothing architecture is one of the paradigms that can be employed. Rasdaman can further utilize just-in-time (JIT) compilation of queries into machine code which is dynamically linked into the server and executed by the CPU as well as the GPU (Baumann et al. 2009). While multi-threaded JIT code already incurs advantages by utilizing the available cores, this improves even more in presence of GPUs which are particularly optimized for image processing operations as they provide hundreds of cores with built-in scheduling, making them very cost-efficient for exploiting hardware parallelism. In fact,

GPU-based queries have turned out to be limited only by bus bandwidth, not by operational complexity, for a large class of queries (Baumann et al. 2009).

Oracle Spatial GeoRaster makes use of the parallel processing framework available in the Oracle DBMS, but further implements parallelization specific to raster processing operations, since Oracle's generic framework itself is optimized for row-based data (Xie et al. 2012). Parallelization of raster processing starts by splitting the raster object into several subsets depending on the degree of parallelism, or user input. Then the Oracle parallel execution framework splits the task into subtasks that will each process one subset independently of the others, and distributes them to separate subprocesses for execution. Benchmarks have shown that this results in significant performance speedup in comparison to a non-parallelized Oracle implementation. This approach resembles tile-based/chunk-based processing similar to the other architectures presented, except for the additional overhead of managing heavy-weight operating system processes. It should be noted, though, that Oracle GeoRaster does not come with a general-purpose raster query language but supports only sequential execution of predefined functions.

Cloud processing

Cloud computing is an internet based computing model that enables easy and dynamic provisioning, configuring, and de-provisioning of a pool of resources (such as compute, storage, data, network, applications, services) that are used as needed to satisfy a workload requirement. According to the US National Institute of Standards and Technology (NIST), the five main characteristics of cloud computing include (Mell and Grance 2011):

- *on-demand self-service:* resource provisioning by the user without human interaction at the server;
- *broad network access:* resource accessibility to different devices from anywhere over the network;
- *resource pooling:* resources are shared among multiple consumers in a non-dedicated manner;
- *rapid elasticity:* scaling out/in of resources in proportion commensurate with the user's demand;
- *measured service:* metering and measuring of resource usage.

Similarly, NIST defines cloud deployment models which can follow either a private (used within a single organization), a community (used within a community of consumers with shared concerns), a public (use open to the public), or a hybrid cloud computing model (any combination of the aforementioned variants).

Typically, the motivation for using cloud computing includes:

- *Scalability:* it enables smooth scaling in many dimensions;
- *High Availability:* it provides highly available services;
- *Flexibility and reliability:* it supports redundant, self-recovering (fault-tolerant) programming models that allow services to recover from many unavoidable hardware/software failures;
- *Un-predictable resource usage support:* it supports application whose resource usage cannot be fully determined a-priori;
- *Transparency:* it virtualizes data processing and workflow execution;
- *Thin client support:* all forms of processing and data storage are done in the cloud, therefore, it offers support for any kind of client, more especially, thin client;
- *Service orientation:* everything in the cloud is offered as a service; therefore, there is possibility of composing more complex services from already existing services;
- *Cost efficiency:* it offers low cost of entry and high return on investment.

Cloud geoprocessing applications are getting more complex, compute-intensive, and data-intensive. Furthermore, the growth of the web and sensor-based devices has resulted in an increase in the number of users of these applications, their requirement with respect to functionality and Service Level Agreement (SLA), and the sizes and distribution of the data processed. However, the infinite scale of computing resources is idealized a potential solution.

At the moment, several research efforts have gone into running geospatial applications within the cloud. The remaining part of this section discuss the approaches used with the context of the NIST classification of cloud computing service models which includes Software as a Service (SaaS), Platform as a Service (PaaS), and Infrastructure as a Service (IaaS).

Software as a Service (SaaS)

In this cloud model, application software (together with any other needed software, operating system and hardware) is provided to the user, typically through the some thin client interface (web browser or some other service oriented interfaces). Common examples of these services include Google Apps, Web-based email services, Microsoft Live Mesh, Google Earth, Map, Gmail, and Docs, OGC Web services. The mechanisms ensuring availability and scalability of this service can vary widely. For instance, some services can be scaled directly as below:

- Deploying them on a PaaS or IaaS infrastructure (Baranski et al. 2010; Schäffer et al. 2010).
- Running several instances of the service transparently, especially during peak request periods.
- Deploying the services on distributed computing frameworks such as Hadoop.[16]
- Decomposing the service request into a workflow (either dynamically or otherwise) and orchestrating the workflow as a composite service.

In some deployment, the SaaS developer has to take care of the task decomposition, scheduling and execution logic, e.g., Hadoop-based SaaS. Owonibi (2012) has demonstrated the dynamic decomposition and orchestration of geoprocessing query requests based on computation intensiveness of the query, data location, and different server capabilities among a group of collaborating servers.

Platform as a Service (PaaS)

In this service model, users are provided with a computing platform for application design, development, testing and deployment. This typically includes the operating system, databases, and application development and execution environment. Common examples of this service include AWS Elastic Beanstalk,[17] Oracle PaaS Platform,[18] Windows Azure[19] and Google Apps Engine[20] (GAE), etc. Microsoft Azure and GAE are used to describe how this model can be coupled with on-line geoprocessing.

Typically, PaaS consists of an application server (inside which the developed web application is deployed), a computing environment for executed non-web-based applications, a data persistence layer, and cloud management functionalities, and all are offered as services. Data persistence layer in GAE consists of the App Engine Datastore (which is a schema-less object datastore implemented on BigTable[21] and Google File System[22] for

---

[16] The Apache Hadoop software library is a framework that allows for the distributed processing of large data sets across clusters of computers using simple programming models. See http://hadoop.apache.org/, "last seen on 31 March 2013"

[17] AWS Elastic Beanstalk is an even easier way to quickly deploy and manage applications in the Amazon Web Services Cloud, http://aws.amazon.com/elasticbeanstalk/, "last seen on 31 March 2013"

[18] http://www.oracle.com/us/solutions/cloud/platform/overview/index.html, "last seen on 31 March 2013"

[19] Microsoft's solutions for application deployment in their cloud. See http://www.windowsazure.com/en-us/, "last seen on 31 March 2013"

[20] Used to deploy application in Google's cloud. https://developers.google.com/appengine/, "last seen on 31 March 2013"

[21] BigTable is a distributed storage system for managing structured data that is designed to scale to a very large size: petabytes of data across thousands of commodity servers.

[22] A scalable distributed file system for large distributed data-intensive applications

providing robust, scalable storage), BlobStore (for serving data objects, called *blobs,* that are much larger than the size allowed for objects in the Datastore service), and MySQL database for storing relational data. Similarly, Azure provides scalable services for storing blobs, big table (which provides a structured way of storing non-relational data), and relational data. The cloud management functionalities can be used to configure scalability, performance, fault tolerance, replication, and availability of the developed application. Once the configuration is set, the cloud manages the system and the main issues which the developer has to deal with is the migration of applications to the cloud which effectively means adapting the code to the PaaS architecture. To demonstrate geoprocessing application deployment in a PaaS cloud Wang et al. (2009) created an application for creating spatial indexes which was deployed in GAE, Blower (2010) deployed OGC Web Map Service in GAE, where a WPS-based spatial buffer analysis was implemented in GAE, and Gong et al. (2010) provided a WPS-based service for computation of terrain slope on the Microsoft Azure platform.

Notably, this conglomerate of different techniques still does not provide a semantically adequate level for storing and processing arrays—BlobStore remains at the level of standard relational databases offering only 1D Blobs for nD arrays, whereas Array Databases can accommodate the multidimensional nature, thereby being more expressive, efficient, and also more flexible: new algorithms can be formulated as queries by the users and they do not require additional server-side implementation.

Another criticism of the PaaS-based cloud deployment is that one cannot control the distribution of the geoprocessing task. The application logic is not aware of task decomposition and scheduling, hence, each service request is executed by a processor, and therefore, there is a limit on how fast the request can be executed. Capacities of the processor can be increased though, and the data read from the underlying data storage can be parallelized, and other service requests can be sent to other service instances, however, this will mean very little for a computation-intensive application. Therefore, there is the argument for using a framework which allows one to better control task distribution—a natural argument for the use of IaaS which is discussed below.

Infrastructure as a Service (IaaS)

In IaaS, computing hardware including memory, processors, storage, and networking hardware, is offered as service. Typically, the computing resources are virtualized from a large hardware resource pool in a data center. Common examples include Amazon Elastic Compute Cloud (EC2) and Simple Storage Service (S3), Azure Services Platform, DynDNS, Google Compute Engine, and HP Cloud. Users can dynamically provision the virtualized hardware resources and configure them. One advantage of IaaS over the other models is that the cloud user has more control over the distribution logic of tasks necessary to complete either a single or multiple service requests. Servers can be dynamically provisioned at peak request times, and de-provisioned thereafter. Similarly, for long running processes, servers can be provisioned provided the task scheduling and distribution logic for the task is known. In addition to EC2 and S3 cloud platforms, Amazon also offers Amazon Elastic MapReduce (EMR) which uses the Hadoop framework (Hadoop Distributed File System (HDFS),[23] MapReduce,[24] Pig,[25] Hive,[26] etc.) for running big data storage and analytics. To test the utilization of IaaS cloud for Geosciences applications, Huang et al. (2010) deployed the Global Earth Observation System of Systems (GEOSS) Clearinghouse clearing metadata catalog service in the Amazon EC2. Similarly, Baranski et al. (2010) proposed a pay-per-use revenue model for geoprocessing services in the Cloud to support future business models of geoprocessing in the cloud.

In situ processing

In situ data processing denotes the ability to access data directly "in-place", without having to import it to the database beforehand (Alagiannis et al. 2012). That means that complex, ad-hoc analytics can be easily performed on external data sources, avoiding pre-loading into the database and all overhead incurred by this. In certain situations, *in situ* processing would be preferred and perhaps is the only way to work with data, even though in some cases it might be slower than in-database processing due to a lack of adaptability to I/O access patterns and internal optimization.

In situ processing is most useful when working with existing, legacy data archives where data is already stored in a certain structure, and many services are built to assume this structure. Modifying the data archive is not an option and importing it into a database leads to unnecessary data duplication. *In situ* processing is non-invasive, so a database with such

---

[23] Similar to Google File System, HDFS is a scalable, fault tolerant, distributed file system designed to run on commodity hardware

[24] MapReduce is a software framework and programming model for easily writing applications which process vast amounts of data in-parallel on large clusters of commodity hardware in a reliable, fault-tolerant manner

[25] Pig is a platform for analyzing large data sets that consist of a high-level language for expressing analysis programs on data stored on Hadoop

[26] Hive is a data warehouse system for Hadoop that facilitates easy data summarization, ad-hoc queries, and the analysis of large datasets stored in Hadoop compatible file systems

capabilities is the perfect candidate for this scenario. The data is registered with the database, and typically only selection queries will be allowed, whereas updates are left to mechanisms already in place in the data repository. For example, rasdaman (Baumann 1994) implements *in situ* processing mainly for this purpose. Registering data for *in situ* processing in rasdaman is possible with a simple extension of the regular insert statement (Baumann 2013):

insert into *collName* referencing
(*typeName*)
filePath [*domain*],

...,
filePath [*domain*]

For example, registering a grayscale TIFF image test.tif of size 1000x1000 would be done with:

insert into *GreyColl* referencing (*GreyImage*) "/path/test.tif" [0:999,0:999]

In situ may also be preferred by regular desktop users, especially when they often work with GIS tools that do not have connectors for the particular DBMS, and work best and fastest with the filesystem. "Locking" the data in a database would require exporting it first, before making use of it with such tools. By registering the data *in situ* however, it can be easily accessed by other software, and database queries can still be performed on it. Support for *in situ* in PostGIS Raster is mainly motivated by this use-case, as PostGIS Raster focuses on 2D GIS raster data.

The main disadvantage of working with data *in situ* is the lack of adaptability to the variety aspect of Big Data. In the case of image timeseries analysis this becomes particularly evident. Images are inserted into the archive slice by slice in time, and this is how they are usually stored. In contrast, knowing that temporal queries are important in an Array DBMS can rearrange incoming data into time "columns" that give access to a particular location's timeseries in one disk. Hence, the advantage of circumventing data copying has to be balanced against a possible loss of performance in query evaluation. Generally, ingesting data into a database usually involves translating the data into an internal format, optimized towards a certain pattern of queries that are most commonly exercised on the particular dataset. The data is typically broken and stored as tiles (also called chunks); a detailed view on how this is done in Rasdaman is given in Baumann (2012). Combined with an appropriate tiling and indexing strategy, e.g., a redundant tile schema or an index on the *in situ* data, fast access to interested tiles can be enabled. *In situ* evaluation circumvents these mechanisms, as the original data will not be modified.

Conclusion and Outlook

Earth sciences are among the major landscapes inundated by today's Data Deluge. Arrays tend to be Big, with single objects frequently ranging into Petabyte and soon Exabyte sizes; for example, the Low-Frequency Array for radio astronomy (LOFAR) distributed radio telescope consists of hundreds of antenna stations all over Europe, each of which delivers 3 GB/sec sustained to the central computing center. Array databases aim at offering flexible, scalable storage and retrieval on this information category. Databases add substantial advantages with their extensibility and scalability mechanisms, leading to Array Analytics. Several more emerging technologies are on the table, such as parallelism, cloud paradigms, *in situ* processing, etc., which are presented as potential solutions to the data tsunami challenge. Although "no one size fits all" (Stonebraker and Cetintemel 2005) is a commonly heard response nowadays, research efforts on seamless integration of emerging analytics technologies receive high attention.

Acknowledgments

The authors acknowledge the EU-FP7 EarthServer (http://www.earthserver.eu) and SCIDIP-ES (http://www.scidip-es.eu) projects, which provided the research support.

References

Alagiannis, I., R. Borovica, M. Branco, S. Idreos and A. Ailamaki. 2012. Efficient Query Execution on Raw Data Files, SIGMOD '12, Scottsdale, USA. 241–252.

Baranski, B., T. Deelmann and B. Schäffer. 2010. Pay-Per-Use Revenue Models for Geoprocessing Services in the Cloud. Proceedings of the First International Workshop on Pervasive Web Mapping, Geoprocessing and Services (WebMGS 2010), Como, Italy.

Baumann, P. 1994. On the management of multi-dimensional discrete data, VLDB Journal, Special Issue on Spatial Database Systems. 3(4): 401–444.

Baumann, P. 1999. A database Array Algebra for spatio-temporal data and beyond. Proceedings 4th International Workshop on Next Generation Information Technologies and Systems (NGITS '99), Zikhron-Yaakov, Israel. 1649: 76–93.

Baumann, P. (ed.). 2008. Web Coverage Processing Service (WCPS) Implementation Specification, OGC 08–068. 75 pp.

Baumann, P., C. Jucovschi and S. Stancu-Mara. 2009. Efficient Map Portrayal Using a General-Purpose Query Language. Proceedings of the 20th International Conference on Database and Expert Systems Applications (DEXA '09), Linz, Austria. 153–163.

Baumann, P., B. Howe, K. Orsborn, S. Stefanova (Eds.). 2011. Proceedings of the 2011 EDBT/ICDT Workshop on Array Databases, Uppsala, Sweden. ACM 2011 ISBN 978-1-4503-0614-0.

Baumann, P. and S. Holsten. 2012. A comparative analysis of array models for databases, International Journal of Database Theory and Application. 5(1): 89–120.

Baumann, P. 2013. rasdaman query language guide, rasdaman GmbH, 8.4 edition. 91pp.

Brown, P.G. 2010. Overview of sciDB: large scale array storage, processing and analysis. Proceedings of the 2010 ACM SIGMOD International Conference on Management of data (SIGMOD '10), Indianapolis, Indiana, USA. 963–968.

Blower, J.D. 2010. GIS in the Cloud: Implementing a Web Map Service on Google App Engine. Proceedings of the First International Conference on Computing for Geospatial Research and Application, Washington, DC, USA. 1–4.

Carson, G. and E. McGinnis. 1986. The Reference Model for Computer Graphics, IEEE Computer Graphics and Applications. 6(8): 17–23.

Catell, R. and R.G.G. Cattell. 2000. The Object Data Standard, ODMG 3.0. 288 pp.

Cheng, Y. and F. Rusu. 2012. EXTASCID: An Extensible System for the Analysis of Scientific Data, Poster XLDB Stanford, California, USA.

Chock, M. and A.F. Cardenas. 1983. Database structure and manipulation capabilities of a Picture Database Management System (PICDMS), IEEE Transactions on Pattern Analysis & Machine Intelligence. 484–492.

Cornacchia, R., A. van Ballegooij and A.P. de Vries. 2004. A case study on array query optimization, CVDB '04: Proceedings of the 1st international workshop on Computer vision meets databases, New York, NY, USA. 3–10.

Cornacchia, R., S. Heman, M. Zukowski, A. de Vries and P. Boncz. 2008. Flexible and efficient IR using array databases, VLDB Journal. 7(1): 151–168.

Domenico, B., S. Nativi, J. Caron, L. Bigagli and E.R. Davis. 2006. A standards-based, web services gateway to netCDF datasets. Proceedings of AMS—22nd IIPS Conference, Atlanta, Georgia.

Felger, W., M. Frühauf, M. Göbel, R. Gnatz and G. Hofmann. 1990. Towards a reference model for scientific visualization systems. Proceedings of Eurographics Workshop on Visualization in Scientific Computing, Clamart, France. 63–74.

Gong, J., P. Yue and H. Zhou. 2010. Geoprocessing in the Microsoft Cloud Computing Platform—Azure, Special Joint Symposium of ISPRS Technical Commission IV and AutoCarto 2010 with ASPRS/CaGIS 2010 Specialty Conference (Geospatial Data and Geovisualization: Environment, Security, and Society), Florida, USA.

Gutierrez, A.G. and P. Baumann. 2007. Modeling fundamental geo-raster operations with Array Algebra, Workshops Proceedings of the 7th IEEE International Conference on Data Mining (ICDM 2007), Omaha, Nebraska, USA. 607–612.

Hahn, K., B. Reiner, G. Höfling and P. Baumann. 2002. Parallel query support for multidimensional data: Inter-object parallelism. Database and Expert Systems Applications. Springer. 820–830.

Huang, Q., C. Yang, D. Nebert, K. Liu and H. Wu. 2010. Cloud Computing for Geosciences: Deployment of GEOSS Clearinghouse on Amazon's EC2. Proceedings of the ACM SIGSPATIAL International Workshop on High Performance and Distributed Geographic Information Systems, HPDGIS, San Jose, CA, USA. 35–38.

Hilbert, M. and P. López. 2011. The World's Technological Capacity to Store, Communicate, and Compute Information. Science. 332(6025): 60–65.

IBM. 2013. What is big data? http://www-01.ibm.com/software/data/bigdata/, last seen on 2013-mar-31.

ISO. 1994. Information technology—Computer graphics and image processing—Image Processing and Interchange (IPI)—Functional specification—Part 2: Programmer's imaging kernel system application programme interface. ISO/IEC 12087-2:1994. 865 pp.

Laney, D. 2001. 3D Data Management: Controlling Data Volume, Velocity and Variety. MetaGroup. http://blogs.gartner.com/doug-laney/files/2012/01/ad949-3D-Data-Management-Controlling-Data-Volume-Velocity-and-Variety.pdf, last seen on 2013-mar-31.

Lerner, A. and D. Shasha. 2003. A Query: Query Language for Ordered Data, Optimization Techniques, and Experiments, Proceedings of the 29th VLDB Conference, New York, USA. 345–356.

Libkin, L., R. Machlin and L. Wong. 1996. A query language for multidimensional arrays: Design, implementation, and optimization techniques. 228–239.

Machlin, R. 2007. Index-based multidimensional array queries: safety and equivalence. In: L. Libkin (ed.). PODS, ACM. 175–184.

Marathe, A.P. and K. Salem. 1997. A language for manipulating arrays, Proceedings of VLDB, San Francisco, CA, USA. 46–55.

Marathe, A.P. and K. Salem. 1999. Query processing techniques for arrays, In SIGMOD '99: Proceedings of the 1999 ACM SIGMOD international conference on Management of data, ACM; New York, NY, USA. 323–334.

Mell, P. and T. Grance (eds.). 2011. The NIST Definition of Cloud Computing, Special Publication 800–145. National Institute of Standards and Technology, U.S. Department of Commerce.

Owonibi, M. 2012. Dynamic Resource-Aware Decomposition of Geoprocessing Services Based on Declarative Request Languages, PhD Thesis, Jacobs University, Bremen, Germany. 116 pp.

Ritter, G., J. Wilson and J. Davidson. 1994. Image algebra: An overview, Computer Vision, Graphics, and Image Processing. 49(1): 297–336.

Roland, P., G. Svensson, T. Lindeberg, T. Risch, P. Baumann, A. Dehmel, J. Frederiksson, H. Halldorson, L. Forsberg, J. Young and K. Zilles. 2001. A database generator for human brain imaging, Trends in Neurosciences. 24(10): 562–564.

Rusu, F. and Y. Cheng. 2013. A Survey on Array Storage, Query Languages, and Systems, CoRR 1302.0103. 1–44.

Szalay, A. and J. Gray. 2006. 2020 Computing: Science in an exponential world, Nature. 440: 413–414.

Schäffer, B., B. Baranski and T. Foerster. 2010. Towards Spatial Data Infrastructures in the Clouds. In Geospatial Thinking: Proceedings of the Thirteenth AGILE International Conference on Geographic Information Science, Guimarães, Portugal, Lecture Notes in Geoinformation and Cartography, Springer 399–418.

Snow, D. 2012. Adding a 4th V to BIG Data—Veracity. Available online: http://dsnowondb2.blogspot.de/2012/07/adding-4th-v-to-big-data-veracity.html, last seen on 2013-mar-31.

Stonebraker, M. and U. Cetintemel. 2005. One Size Fits All: An Idea Whose Time Has Come and Gone. Proceedings of 21st Intl. Conf. on Data Engineering, ICDE 2005, Tokyo, Japan. 5–8.

van Ballegooij, A. 2004. Ram: A multidimensional array dbms. In: W. Lindner, M. Mesiti, C. Türker, Y. Tzitzikas and A. Vakali (eds.). EDBT Workshops, Heraklion, Crete, Greece, LNCS, Springer. 3268: 154–165.

van Ballegooij, A., R. Cornacchia, A. de Vries and M. Kersten. 2005. Distribution rules for array database queries, Proceedings of the International Conference on Database and Expert Systems Application (DEXA), Copenhagen, Denmark, Springer. 3588: 55–64.

Vandenberg, S. and D.J. DeWitt. 1991. Algebraic support for complex objects with arrays, identity, and inheritance, Proceedings of the 1991 ACM SIGMOD international conference on Management of data, New York, USA. 158–167.

Wang, Y., S. Wang and D. Zhou. 2009. Retrieving and Indexing Spatial Data in the Cloud Computing Environment. Proceedings of the 1st International Conference on Cloud Computing (CloudCom 2009), Lecture Notes in Computer Science (LNCS), Springer Beijing, China. 22–331.

Xie, Q.J., Z.Z. Zhang and S. Ravada. 2012. In-Database Raster Analytics: Map Algebra and Parallel Processing in Oracle Spatial GeoRaster, Int. Arch. Photogramm. Remote Sens. Spatial Inf. Sci., XXXIX-B4. 91–96.

CHAPTER 2

Similarity Join for Big Geographic Data

Yasin N. Silva,* Jason M. Reed, Lisa M. Tsosie and
Timothy A. Matti

Introduction

Similarity Join is one of the most useful data processing and analysis operations for geographic data. It retrieves all data pairs whose distances are smaller than a predefined threshold ε. Multiple application scenarios need to perform this operation over large amounts of data. Internet companies, for instance, collect massive amounts of information on their customers such as their geographic location and interests. They can use similarity queries to provide enhanced services to their customers; for example, a movie theatre website could recommend neighboring theatres and restaurants in the customer's town. MapReduce, a framework for processing very large datasets using large computer clusters, constitutes an answer to the requirements of processing massive amounts of data in a highly scalable and distributed fashion (Dean and Ghemawat 2004). MapReduce-based systems are composed of large clusters of commodity machines and are often dynamically scalable, i.e., cluster nodes can be added or removed based on the workload. The MapReduce framework quickly processes massive datasets by splitting them into independent chunks that are processed in a highly parallel fashion.

Multiple Similarity Join algorithms and implementation techniques have been proposed. They range from approaches for only internal memory or external memory data to techniques that make use of database operators

Arizona State University, 4701 W. Thunderbird Road, Glendale, AZ 85306, USA.
  Emails: ysilva@asu.edu; jmreed3@asu.edu; lmtsosi1@asu.edu; tmatti@asu.edu
* Corresponding author

to answer Similarity Joins. Unfortunately, however, there has not been much work on the study of MapReduce-based Similarity Join techniques for geographic data. This chapter addresses this problem by focusing on the study, design and implementation techniques of a MapReduce-based Similarity Join algorithm that can be used with geographic data and distance functions. The main contributions of our work are:

- We describe MRSimJoin, an efficient MapReduce Similarity Join algorithm. MRSimJoin extends the single-node QuickJoin algorithm (Jacox and Samet 2008) by adapting it to the distributed MapReduce framework (Dean and Ghemawat 2004) and integrating grouping, sorting and parallelization techniques.
- The proposed algorithm is general enough to be used with any dataset that lies in a metric space. Our focus is on the study of this operation with geographic data, e.g., longitude-latitude pairs, and geographical distance functions, e.g., Euclidean Distance on a plane where a Spherical Earth was projected.
- We present guidelines to implement the algorithm in Hadoop, a highly used open-source MapReduced-based system.
- We thoroughly evaluate the performance and scalability properties of the implemented operation with synthetic and real-world geographic datasets. We show that MRSimJoin performs significantly better than an adaptation of the state-of-the-art MapReduce Theta-Join algorithm (Okcan and Riedewald 2011). MRSimJoin scales very well when important parameters like epsilon, data size, and number of cluster nodes increase.

A generic description of the MRSimJoin algorithm was presented previously (Silva et al. 2012). This chapter, however, includes the following contributions not included in the previous paper: (1) a focus on the study of Similarity Joins in the context of geographical data and distance functions, (2) completely new experimental evaluation of the performance and scalability of MRSimJoin with geographic data (synthetic and real-world datasets), and (3) MRSimJoin's pseudocode, algorithmic details and descriptions not presented previously.

The remaining part of this chapter is organized as follows. First, we present the related work. Then, the MRSimJoin algorithm is described in detail. Next, we present the guidelines to implement MRSimJoin in Hadoop and the performance evaluation of our implementation. Finally, we present the conclusions and future research directions.

Related Work

Most of the previous work on Similarity Join, an operation that retrieves all pairs whose distances are smaller than a predefined threshold ε, has considered the case of non-distributed solutions and proposed techniques to implement it primarily as standalone operations. Among the most relevant implementations, we find approaches that rely on the use of pre-built indices, e.g., eD-index (Dohnal et al. 2003b) and D-index (Dohnal et al. 2003a). These techniques strive to partition the data while clustering together the similar objects. Several non-index-based techniques have also been proposed to solve the Similarity Join problem, e.g., EGO (Bohm et al. 2001), GESS (Dittrich and Seeger 2001) and QuickJoin (Jacox and Samet 2008). The Quickjoin algorithm proposed by Jacox and Samet (2008), which has been shown to outperform EGO and GESS, recursively partitions the data until the subsets are small enough to be efficiently processed using a nested loop join. MRSimJoin, the approach presented in this chapter, builds on Quickjoin's method to partition the data. However, the focus of our work is the design and implementation of a distributed solution. The differences with the work by Jacox and Samet (2008) are: (1) MRSimJoin extends the single-node QuickJoin algorithm by adapting it to the distributed MapReduce framework and integrating parallelization techniques (grouping, sorting) to physically partition and distribute the data, (2) we consider geographical data types and distance functions, and (3) our experimental section evaluates the effect of the number of pivots and number of nodes on performance.

Also of importance is the work on Similarity Join techniques in the context of database systems. Some work has focused on the implementation of Similarity Joins using standard database operators (Chaudhuri et al. 2006; Gravano et al. 2001). These techniques are applicable only to string or set-based data. The general approach pre-processes the data and query, e.g., decomposes data and query strings into sets of grams (substrings of a string that are used as its signature), and stores the results of this stage on separate relational tables. Then, the result of the Similarity Join can be obtained using standard SQL statements. More recently, Similarity Joins have been proposed and studied as first-class database operators (Silva et al. 2010; Silva and Pearson 2012; Silva et al. 2013a; Silva et al. 2013b). This work proposes techniques to implement and optimize Similarity Joins inside database query engines.

The MapReduce framework was introduced by Dean and Ghemawat (2004). The Map-Reduce-Merge variant, as proposed by Yang et al. (2007), extends the MapReduce framework with a *merge* phase after the reduce stage to facilitate the implementation of operations like join. Map-Join-Reduce, as proposed by Jiang et al. (2011), is another MapReduce variant that adds a *join* stage before the reduce stage. In this approach, mappers read from input relations, the output of mappers is distributed to joiners where the actual join task takes place, and the output of joiners is processed by the reducers.

Most of the previous work on MapReduce-based Joins considers the case of equi-joins. The two main types of MapReduce-based joins are Map-side joins and Reduce-side joins. Among the Map-side joins we have Map-Merge and Broadcast Join. The Map-Merge approach has two steps: in the first one, input relations are partitioned and sorted, and in the second one, mappers merge the intermediate results (White 2010). The Broadcast Join approach considers the case where one of the relations is small enough to be sent to all mappers and maintained in memory (Blanas et al. 2010; Chen 2010). The overall execution time is reduced by avoiding sorting and distributing on both input relations. Repartition join is the most representative instance of Reduce-side joins (White 2010). In this approach, the mappers augment each record with a label that identifies its relation. All the records with the same join attribute value are sent to the same reducer. Reducers in turn produce the join pairs.

Recently, MRThetaJoin, a MapReduce-based approach was proposed to implement Theta-joins (Okcan and Riedewald 2011). This previous work proposed a randomized algorithm that requires some basic statistics (input cardinality). The approach proposes a model that partitions the input relations using a matrix that considers all the combinations of records that would be required to answer a cross product. The matrix cells are then assigned to reducers in a way that minimizes job completion time. A memory-aware variant is also proposed for the common scenario where partitions do not fit in memory. This previous work represents the state-of-the-art approach to answer arbitrary joins in MapReduce. In this chapter, we compare MRSimJoin with an adaptation of the memory-aware algorithm to answer Similarity Joins and show that MRSimJoin performs significantly better.

The work by Vernica et al. (2010), studied the problem of Similarity Joins in the context of cloud systems. This work, however, focuses on the study of a different and more specialized type of Similarity Join (Set-Similarity Join) which constrains its applicability to set-based data. The differences between this work and ours are: (1) we consider the case of the most extensively used type of Similarity Join (distance range join), and (2) our approach can be used with data that lies in any metric space.

A comparison of the MapReduce framework and parallel databases was presented by Pavlo et al. 2009. Multiple parallel join algorithms have been proposed in the context of parallel databases (Kitsuregawa and Ogawa 1990; Schneider and DeWitt 1989). The work by Kitesuregawa and Ogawa 1990 presents a comparison of several hash-based and sort-merge-based parallel join algorithms. A hash-based algorithm to address the case of data skew is presented by Schneider and DeWitt (1989). The algorithm dynamically allocates partitions to the processing units with the goal of assigning the same data volume to each unit.

Also related is the work on spatial join. This operation combines two datasets based on some spatial relationship between the objects (usually based on the intersection or containment of objects). Some work focuses on solving this problem in a single-node database (e.g., Patel and DeWitt 1996). More relevant to this chapter is the work on parallel spatial joins that propose algorithms to implement this operation on parallel spatial databases (e.g., Luo et al. 2002; Patel and DeWitt 2000). Some techniques to solve the spatial joins can be extended to support similarity joins (e.g., Luo et al. 2002). A key difference of this work with the one in this chapter, is that we focus on the similarity join problem on the highly distributed and highly fault-tolerant MapReduce framework, which is widely considered as one of the key frameworks for processing big data.

Similarity Joins Using MapReduce

This section presents preliminary information (geographic data and distance functions, and MapReduce) and then presents MRSimJoin.

Geographic Data and Distance Functions

In this work, we assume that the geographic data we need to process uses the common geographic coordinates: latitude (φ) and longitude (λ). These coordinates allow us to specify any point on the earth using a pair of numbers. Several geographical distance functions have been defined in terms of latitude and longitude to compute the distance between two points on earth, e.g., Euclidean distance on a plane where a Spherical or Ellipsoidal Earth was projected, Great-circle distance, Tunnel distance, etc. The algorithm presented in this section can be used with any (metric) distance function. Our presentation, however, considers the case of Euclidean distance on a plane where a Spherical Earth was projected using equirectangular projection (Snyder 1993).This distance function is fast to compute, is reasonably accurate for small distances, and is defined as follows. Given two points $r_1=(\varphi_1,\lambda_1)$ and $r_2=(\varphi_2,\lambda_2)$, the geographical distance between r_1 and r_2, as measured along the surface of the earth, is given by the following expression:

$$geoDist(r_1, r_2) = R\sqrt{(\Delta_\varphi)^2 + (\cos(\varphi_m)\Delta_\lambda)^2},$$

where $\Delta_\varphi = \varphi_2 - \varphi_1$, $\Delta_\lambda = \lambda_2 - \lambda_1$, $\varphi_m = \frac{\varphi_1 + \varphi_2}{2}$, R is the radius of the earth, Δ_φ and Δ_λ are in radians, and *geoDist* is in the same unit as R.

A Quick Introduction to MapReduce

MapReduce is one of the main software frameworks for distributed processing (Dean and Ghemawat 2004). This framework is able to process massive amounts of data and works by dividing the processing task into two phases: *map* and *reduce*, for which the user provides two functions named *map* and *reduce*. These functions have key-value pairs as inputs and outputs which have the following general form:

map: $(k1, v1) \rightarrow \text{list}(k2, v2)$

reduce: $\big(k2, \text{list}(v2)\big) \rightarrow \text{list}(k3, v3)$

Note that the input and output types of each function can be different. However, the input of the *reduce* function should use the same types as the output of the *map* function.

The execution of a MapReduce job works as follows. The framework splits the input dataset into independent data chunks that are processed by multiple independent *map* tasks in a parallel manner. Each *map* call is given a pair $(k1,v1)$ and produces a list of $(k2,v2)$ pairs. The output of the *map* calls is known as the intermediate output. The intermediate data is transferred to the *reduce* nodes by a process known as the *shuffle*. Each *reduce* node is assigned a different subset of the intermediate key space; these subsets are referred as *partitions*. The framework guarantees that all the intermediate records with the same intermediate key ($k2$) are sent to the same reducer node. At each *reduce* node, all the received intermediate records are sorted and grouped. Each formed group will be processed in a single *reduce* call. Multiple *reduce* tasks are also executed in a parallel fashion. Each *reduce* call receives a pair $(k2,\text{list}(v2))$ and produces as output a list of $(k3,v3)$ pairs.

The processes of transferring the *map* outputs to the *reduce* nodes, sorting the records at each destination node, and grouping these records are driven by the *partition*, *sortCompare* and *groupCompare* functions, respectively. These functions have the following form:

partition: $k2 \rightarrow \text{partitionNumber}$
sortCompare: $(k2_1, k2_2) \rightarrow \{-1,0,1\}$
groupCompare: $(k2_1, k2_2) \rightarrow \{-1,0,1\}$

The default implementation of the *partition* function receives an intermediate key ($k2$) as input and generates a partition number based on a hash value for $k2$. The default *sortCompare* and *groupCompare* functions directly compare two intermediate keys ($k2_1$, $k2_2$) and return -1 ($k2_1 < k2_2$), 0 ($k2_1 = k2_2$), or $+1$ ($k2_1 > k2_2$). The result of using the default comparator functions is that all the intermediate records in a *reduce* node are sorted by the intermediate key and a group is formed for each different value of the intermediate key. Custom partitioner and comparator functions can be provided to replace the default functions.

The input and output data are usually stored in a distributed file system (DFS). The MapReduce framework takes care of scheduling tasks, monitoring them and re-executing them in case of failures.

The MRSimJoin Algorithm

The Similarity Join (SJ) operation between two datasets R and S is defined as follows: $R \bowtie_{\theta_\varepsilon(r,s)} S = \{\langle r,s \rangle | \; \theta_\varepsilon(r,s), \; r \in R, \; s \in S\}$, where $\theta_\varepsilon(r, s)$ represents the Similarity Join predicate, i.e., $dist(r, s) \le \varepsilon$. A sample SJ query would be to get the pairs of restaurants (R) and movie theatres (S) that are close to each other (within a certain distance ε).

The MRSimJoin algorithm presented in this section identifies all the pairs (links) that belong to the result of the Similarity Join operation. In general, the input data can be given in one or multiple distributed files. Each input data file contains a sequence of key-value records of the form (*id*, (*id, elem*)) where *id* contains two components, the id of the dataset or relation this record belongs to (*id.relID*) and the id of the record in the relation (*id.uniqueKey*), and *elem* is a latitude-longitude pair. Note that the *id* component in the value of an input record is the same *id* component in the key of that record.

MRSimJoin iteratively partitions the input data into smaller subsets until each subset is small enough to be efficiently processed by a single-node SJ routine. The overall process is divided into a sequence of rounds, where the initial round partitions the input data while any subsequent round divides the data of a previously generated partition. Since each round corresponds to a MapReduce job, the input and output of each job is read from/written to the distributed file system. The output of a round includes: (1) result links for the small partitions that were processed in a single-node, and (2) intermediate data for the partitions that will require further partitioning. Note that the DFS automatically divides and distributes the intermediate data. The execution of a single MRSimJoin round is illustrated in Fig. 1. This figure shows that the partitioning and generation of results or intermediate data are performed in parallel by multiple nodes. All the nodes partition the data using the same set of pivots that are previously sent to each node. MRSimJoin executes the required rounds until all the input and intermediate data is processed.

Data partitioning is performed using a set of K pivots, which are a random subset of the data records to be partitioned. We use random selection since this method was found to be efficient by Jacox and Samet

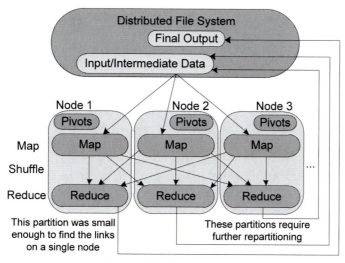

Fig. 1. An MRSimJoin round.

(2008). The process generates two types of partitions: *base partitions* and *window-pair partitions*. The records closest to a given pivot are contained in a base partition, while the records in the boundary between two base partitions are contained in a window-pair partition. Generally, the window-pair records should be a superset of the records whose distance to the hyperplane that separates the base partitions is at most ε. Since this hyperplane does not always explicitly exist in a metric space, it is implicit and known as a *generalized hyperplane* (GHP).The distance of a record t to the GHP between two partitions with pivots P_0 and P_1 cannot always be computed exactly, so a lower bound is used (Hjaltason and Samet 2003): $gen\_hyperplane\_dist(t, P_0, P_1) = (dist(t, P_0) - dist(t, P_1))/2$. This distance can be replaced with an exact distance if this can be computed.

Processing the window-pair partitions guarantees the identification of the links between records that belong to different base partitions. A round that further repartitions, a base partition or the initial input data is referred to as a *base partition round*, a round that repartitions a window-pair partition is referred to as a *window-pair partition round*. At the logical level, the data partitioning in MRSimJoin is similar to the one in the Quickjoin algorithm (Jacox and Samet 2008). The core difference in MRSimJoin, however, is that the partitioning of the data, the generation of the result links, and the storage of intermediate results are performed in a fully distributed and parallel manner.

Consider the restaurant-theatre scenario described at the beginning of this section when dividing a geographical area into partitions. Figure 2 represents a map where the restaurants and movie theatres are located, and illustrates how the region (base partition) is partitioned. In this case, the result of the Similarity Join operation (pairs of restaurants and theatres) on the dataset T is the union of the pairs in $P0$ and $P1$, and the pairs in $P0\_P1$ where one element belongs to window A and the other one to window B. We refer to this last type of pair as *window links*. Figure 3 represents the repartitioning of the window-pair partition $P0\_P1$ of Fig. 2. In this case, the set of window links in $P0\_P1$ is the union of the window links in $Q0$, $Q1$, $Q0\_Q1\{1\}$ and $Q0\_Q1\{2\}$. Note that other window-pair combinations, e.g., $\{E, C\}$, $\{D, F\}$, $\{E, D\}$, and $\{C, F\}$, do not form window-pair partitions because the links that can be identified in these combinations are identified processing other partitions. Specifically, the links in $\{E, C\}$ and $\{D, F\}$ are

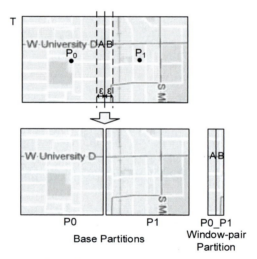

Fig. 2. Partitioning a base partition.

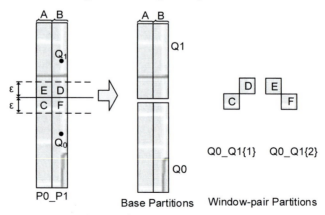

Fig. 3. Partitioning a window-pair partition.

found processing the original base partitions $P0$ and $P1$, respectively; and the window links in $\{E, D\}$ and $\{C, F\}$ are found in partitions $Q0$ and $Q1$, respectively. The examples in Figs. 2 and 3 show that base and window-pair partitions need to be processed differently. MRSimJoin should output links in a base partition and window-links in a window-pair partition. Also, the re-partitioning of the base and window-pair partitions generate different numbers of new window-pair partitions.

The Main Algorithm

The main routine of MRSimJoin is presented in Algorithm 1. The routine uses an intermediate directory (line 1) to store the partitions that will need further repartitioning. The names of intermediate directories that store base partitions have the following format:

$\langle outDir\rangle$/intermediate/B_$\langle roundNum\rangle$_$\langle p\rangle$

The names of intermediate directories storing window-pair partitions have the following format:

$\langle outDir\rangle$/intermediate/W_$\langle roundNum\rangle$_$[\langle uAttr\rangle]$_$\langle p1\rangle$_$\langle p2\rangle$

In these expressions, $p, p1$ and $p2$ are pivot indices. $uAttr$ is an attribute that ensures unique directory names for the partitions generated in a round and is needed only when a window-pair is repartitioned. The details of how $uAttr$ is specified are presented in the Window-pair Partition Round subsection.

Each iteration of the while loop (lines 3 to 22) corresponds to one round and executes a MapReduce job. In each round, the initial input data or a previously generated partition is repartitioned. The Similarity Join output of a partition small enough to be processed in a single node is obtained by running a single-node SJ algorithm. Larger partitions are stored as intermediate data for further processing. For each round, the main routine sets the values of the job input directory (lines 4 to 8) and randomly selects *numPivots* pivots from this directory (line 12). Note that function *GetNextIntermPartitionDir* returns the directory of one of the intermediate partitions that needs further processing. Next, the routine executes a base partition job (line 14) or a window-pair partition job (line 16) based on the type of the job input directory. The routine *MRJob* sets up a MapReduce job that will use the provided *map, reduce, partition* and *compare* functions. The *partition* function will be used to replace the default *partition* function. The *compare* function will be used to replace the default *sortCompare* and *groupCompare* functions. *MRJob* also makes sure that the provided atomic parameters, i.e., *outDir, numPiv, eps* and *memT*, are available at every node that will be used in the MapReduce job and that the *pivots* are available at each node that will execute *map* tasks. If a round is processing a previously

---

Algorithm 1: *MRSimJoin(inDir, outDir, numPiv, eps, memT)*

Input: *inDir* (input directory with the records of datasets *R* and *S*),
  outDir (output directory), *numPiv* (number of pivots), *eps* (epsilon),
  memT (memory threshold)

Output: *outDir* contains all the results of the Similarity Join operation
  $R \bowtie_{\theta_\varepsilon(r,s)} S$

1: *intermDir* ← *outDir* + "/intermediate"
2: *roundNum* ← 0
3: **while** true **do**
4:   **if** *roundNum* = 0 **then**
5:       *job_inDir* ← *inDir*
6:   **else**
7:       *job_inDir* ← *GetNextIntermPartitionDir(intermDir)*
8:   **end if**
9:   **if** *job_inDir* = null **then**
10:      break
11: **end if**
12: *pivots* ← *GeneratePivots(job_inDir, numPiv)*
13: **if** *is BaseRound(job_inDir)* **then**
14:      *MRJob(Map_base, Reduce_base, Partition_base, Compare_base,*
        job_inDir, outDir, pivots, numPiv, eps, memT, roundNum)
15: **else**
16:      *MRJob(Map_windowPair, Reduce_windowPair,*
        Partition_windowPair, Compare_windowPair, job_inDir,
        outDir, pivots, numPiv, eps, memT, roundNum)
17: **end if**
18: *roundNum*++
19: **if** *roundNum* >0 **then**
20:      *RenameFromIntermToProcessed(job_inDir)*
21: **end if**
22: **end while**

---

generated partition, after the MapReduce job finishes, the main routine renames the job input directory to relocate it under the processed directories (line 20).

Figure 4 shows an example of the rounds that are executed by the main routine. Each node MR_N represents a MapReduce job. This figure also shows the partitions generated by each job. Light gray partitions are small partitions that are processed running the single-node SJ routine. Dark gray partitions are partitions that require additional repartitioning. A sample sequence of rounds can be: MR_1, MR_2, MR_3, MR_4, MR_5 and MR_6. The original input data is always processed in the first round. Since the links of any partition can be obtained independently, the routine will generate a correct result independently of the order of rounds.

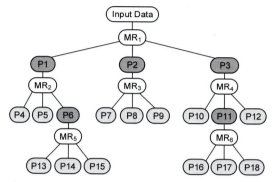

Fig. 4. Example of MRSimJoin rounds and partitions.

Base Partition Round

A base partition round processes the initial input data or a base partition previously generated by a base partition round. The goal of a base partition MapReduce job is to partition its input data and produce: the result links for partitions that are small enough to be processed in a single node, and intermediate data for partitions that require further processing. The main routine sets up each base partition MapReduce job to use *Map_base* and *Reduce_base* as the *map* and *reduce* functions, respectively. Additionally, the default *partition* function is replaced by *Partition_base* and the default *sortCompare* and *groupCompare* functions are replaced by *Compare_base*.

    Map_base, the map function for the base partition rounds, is presented in Algorithm 2. The format of the input key-value pairs, i.e., $k1, v1$, is: $k1 = id$, $v1 = (id, elem)$, and the format of the intermediate key-value pairs, i.e., $k2, v2$, is: $k2 = (part, win)$, $v2 = (id, elem, part)$, where *part* identifies the base partition of the intermediate record and *win* the window partition of the record. We use the value -1 when a given parameter is not applicable or will not be needed in the future. The MapReduce framework divides the job input data into chunks and creates *map* tasks in multiple nodes to process them. Each *map* task is called multiple times and each call executes the *Map_base* function for a given record $(id, (id, elem))$ of the input data. The *Map_base* function identifies the closest pivot p to *elem* (line 1). The function then outputs one intermediate key-value pair of the form $((p,-1), (id, elem,-1))$ for the base partition that *elem* belongs to (line 2) and one pair of the form $((p, i), (id, elem, p))$ for each window-pair partition (corresponding to pivots

---

Algorithm 2: *Map_base*()

Input: ($k1$, $v1$). $k1 = id$, $v1 = (id, elem)$
Output: list($k2$, $v2$). $k2 = (part, win)$, $v2 = (id, elem, part)$
1: $p \leftarrow GetClosestPivotIndx$ (*elem, pivots*)
2: output (($p, -1$), (*id, elem*, -1))
3: **for** $i = 0 \rightarrow numPiv - 1$ **do**
4:    **if** $i \neq p$ **then**
5:       **if** (*dist*(*elem, pivots*[i])$-$*dist*(*elem, pivots*[p]))$/2 \leq eps$ **then**
6:          output ((p, i), (*id, elem, p*))
7:       **end if**
8:    **end if**
9: **end for**

---

p and i) that *elem* belongs to (lines 3 to 9). Figure 5 shows an example of the intermediate key-value pairs generated by *Map_base*. Region T contains all the key-value pairs of the job input data. Different subsets of this region are processed by different *map* tasks on possibly different nodes.

The overall result of the *map* phase is independent of the number or distribution of the *map* tasks. In this example, they will always generate the key-value pairs shown in partitions $P0$, $P1$ and $P0\_P1$. Each input record generates an intermediate key-value pair corresponding to its associated base partition ($P0$ or $P1$). Additionally, each record in the windows between the two base partitions, e.g., $id5$ and $id6$, generates a key-value pair corresponding to the window-pair partition $P0\_P1$.

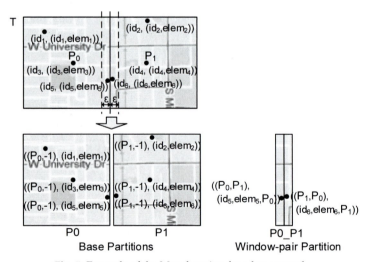

Fig. 5. Example of the Map function for a base round.

The MapReduce framework partitions the intermediate data generated by map tasks. This partitioning is performed calling the *Partition_base* function presented in Algorithm 3. *Partition_base* receives an intermediate key, i.e., $k2 = (part,win)$, as input and generates the corresponding partition number. C1–C3 are constant prime numbers and *NUMPARTITIONS* is the maximum number of partitions set by the MapReduce framework. The partition number for an intermediate key that corresponds to a base partition is computed using a hash function on *part* (line 2). When the key corresponds to a window-pair partition, the partition number is computed using a hash function on *min* $(part,win)$ and $max(part,win)$ (line 6). This last hash function will generate the same partition number for all intermediate records of a window-pair partition independently of the specific window they belong to. In the scenario of Fig. 5, *Partition_base* will generate the same partition number, i.e. $(P0 \times C1)$ mod *NUMPARTITIONS*, for all the intermediate keys that correspond to partition P0. Similarly, the function will generate the same partition number, i.e., $(P0 \times C2 + P1 \times C3)$ mod *NUMPARTITIONS*, for all the intermediate keys that correspond to partition P0_P1.

Algorithm 3: *Partition_base()*

Input: $k2$. $k2 = (part,win)$
Output: $k2$'s partition number
1: **if** $win = -1$ **then** // base partition
2:   $partition \leftarrow (part \times C1)$ mod *NUMPARTITIONS*
3: **else** // window-pair partition
4:   $minVal \leftarrow min(part,win)$
5:   $maxVal \leftarrow max(part,win)$
6:   $partition \leftarrow (minVal \times C2 + maxVal \times C3)$ mod *NUMPARTITIONS*
7: **end if**

After identifying the partition numbers of intermediate records, the shuffle phase of the MapReduce job sends the intermediate records to their corresponding reduce nodes. The intermediate records received at each reduce node are sorted and grouped using the *Compare_base* function presented in Algorithm 4. The main goal of the *Compare_base* function is to group the intermediate records that belong to the same partition. The function establishes the order of partitions shown in Fig. 6a. Base partitions have a lower order than window-pair partitions. Multiple base partitions are ordered based on their pivot indices. Multiple window-pair partitions are ordered based on the two associated pivot indices of each partition.

Compare_base receives as input two intermediate record keys, i.e., $k2_1$, $k2_2$, and returns 0 (when $k2_1$ and $k2_2$ belong to the same group), −1 (when $k2_1$ has lower order than $k2_2$), or +1 (when $k2_1$ has higher order than $k2_2$). The algorithm considers all the possible combinations of the intermediate

---

Algorithm 4: *Compare_base()*

Input: $k2_1$, $k2_2$. $k2_1 = (part_1, win_1)$, $k2_2 = (part_2, win_2)$
Output: 0 ($k2_1$ and $k2_2$ belong to the same group), -1 (group number of $k2_1$<group number of $k2_2$), or $+1$ (group number $k2_1$>group number of $k2_2$)

1: **if** $(win_1 = -1)$ $\wedge(win_2 = -1)$ **then** // basePart-basePart
2:      **if** $(part_1 = part_2)$ **then**
3:          return 0
4:      **else**
5:          return $(part_1 < part_2)$? $-1 : +1$
6:      **end if**
7: **else if** $(win_1 = -1)$ $\wedge(win_2 \neq -1)$ **then** // basePart-winPart
8:      return -1
9: **else if** $(win_1 \neq -1)$ $\wedge(win_2 = -1)$ **then** // winPart-basePart
10:      return +1
11: **else** // $(win_1 \neq -1)$ $\wedge(win_2 \neq -1)$, winPart-winPart
12:      $min1 \leftarrow min(part_1, win_1)$
13:      $max1 \leftarrow max(part_1, win_1)$
14:      $min2 \leftarrow min(part_2, win_2)$
15:      $max2 \leftarrow max(part_2, win_2)$
16:      **if** $(min_1 = min_2)$ $\wedge$ $(max_1 = max_2)$ **then** // elements belong to the
                                                                // same window-pair
17:          return 0
18:      **else** // elements do not belong to the same window-pair
19:          **if** $min_1 = min_2$ **then**
20:              return $(max_1 < max_2)$? $-1 : +1$
21:          **else**
22:              return $(min_1 < min_2)$? $-1 : +1$
23:          **end if**
24:      **end if**
25: **end if**

---

keys. When both keys belong to base partitions, the algorithm orders them, based on their pivot indices (lines 1 to 6). When one key belongs to a base partition and the other one to a window-pair partition, the algorithm orders them giving the base key a lower order than the window-pair key (lines 7 to 10). Finally, if both keys belong to window-pair partitions, the algorithm orders them based on the pair (minimum pivot index, maximum pivot index) using lexicographical order (lines 11 to 25). The *min* and *max* functions are used to group together all the intermediate records of a window-pair independently of the specific window they belong to Fig. 6b shows the order of partitions generated by *Compare_base* for the scenario with two pivots presented in Fig. 5. Figure 6c shows the order of partitions for the case of a base round with three pivots.

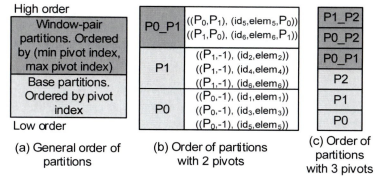

Fig. 6. Group formation in a base round.

After generating the groups in a reduce node, the MapReduce framework calls the *reduce* function *Reduce_base* once for each group. This function is presented in Algorithm 5. The function receives as input the key-value pair (*k2*, *v2List*). *k2* is the intermediate key of one of the records of the group being processed and *v2List* is the list of values of all the records of the group. If the size of the list is small enough to be processed in a single node, the algorithm calls a single-node Similarity Join routine, i.e., *InMemorySimJoin*, to get the links in the current partition (lines 1 to 2). Otherwise all the records of the group are stored in an intermediate directory for further partitioning. If the current group is a base partition, then the

---

Algorithm 5: *Reduce_base()*

Input: (*k2*, *v2List*). *k2* = (*kPart*, *kWin*), *v2List* = list(*id*, *elem*, *part*)
Output: SJ matches or intermediate data. Intermediate data = list(*k3*, *v3*).
  k3 = *id*, *v3* = (*id*, *elem*[, *part*])
1: **if** *sizeInBytes(v2List)* $\leq$ *memT* **then**
2:    *InMemorySimJoin(v2List, outDir, eps)*
3: **else**
4:    **if** *kWin* = -1 **then**
5:       **for** each element *e* of *v2List* **do**
6:          output (*e.id*, (*e.id*, *e.elem*)) to *outDir*/intermediate/B_*(roundNum)_(kPart)*
7:       **end for**
8:    **else**
9:       **for** each element *e* of *v2List* **do**
10:          output (*e.id*, (*e.id*, *e.elem*, *e.part*)) to *outDir*/intermediate/W_*(roundNum)_0_(kPart)_(kWin)*
11:       **end for**
12:    **end if**
13: **end if**

algorithm stores its records in a directory that will be processed in a future base partition round (lines 4 to 7). Likewise, the records of a window-pair partition are stored in a directory that will be processed in a future window-pair partition round (lines 8 to 12). In the latter case, the last part of the directory name includes the indices of the two pivots associated with the window-pair partition. These values will be used in the algorithms of the window-pair rounds. In the scenario represented in Fig. 5, the MapReduce framework calls the *Reduce_base* function for each partition of Fig. 6b: *P*0, *P*1 and *P*0_*P*1.

Window-pair Partition Round

A window-pair partition round processes a window-pair partition generated by a base round or any partition generated by a window-pair round. Similarly to base partition rounds, window-pair partition rounds generate result links and intermediate data. However, the links generated are window links, i.e., links between records of different previous partitions. A window-pair round uses the functions *Map_windowPair*, *Reduce_windowPair*, *Partition_windowPair* and *Compare_windowPair* in a similar way their counterparts are used in a base partition round. This section explains these functions highlighting the differences.

Map_windowPair, the map function for the window-pair partition rounds, is presented in Algorithm 6. In this case, the format of the input key-value pair, i.e., *k*1, *v*1, is: *k*1 = *id*, *v*1 = (*id, elem, prevPart*), and the format of the intermediate key-value pairs, i.e., *k*2, *v*2, is: *k*2=(*part, win, prevPart*), *v*2=(*id, elem, part, prevPart*). The function is similar to *Map_base*. The difference is in the format of the intermediate records. *Map_windowPair* outputs one

---

Algorithm 6: *Map_windowPair*()

Input: (*k*1, *v*1). *k*1 = *id*, *v*1 = (*id, elem, prevPart*)
Output: list(*k*2, *v*2). *k*2 = (*part,win, prevPart*), *v*2 = (*id, elem, part, prevPart*)

1: $p \leftarrow GetClosestPivotIndex (elem, pivots)$
2: output $((p, -1, -1), (id, elem, -1, prevPart))$
3: **for** $i = 0 \rightarrow numPiv - 1$ **do**
4:    **if** $i \neq p$ **then**
5:       **if** $(dist(elem, pivots[i]) - dist(elem, pivots[p]))/2 \leq eps$ **then**
6:          output $((p, i, prevPart), (id, elem, p, prevPart))$
7:       **end if**
8:    **end if**
9: **end for**

---

intermediate key-value pair of the form $((p, -1, -1), (id, elem, -1, prevPart))$ for the base partition with pivot p that *elem* belongs to (line 2) and one key-value pair of the form $((p, i, prevPart), (id, elem, p, prevPart))$ for each window-pair partition (corresponding to pivots p and i) that *elem* belongs to (lines 3 to 9). Figure 7 shows an example of the intermediate key-value pairs generated by *Map_windowPair*.

The MapReduce framework partitions the intermediate data using the *Partition_windowPair* function presented in Algorithm 7. *Partition_windowPair* receives an intermediate key, i.e., $k2 = (part, win, prevPart)$, as input and generates the corresponding partition number. The generation of the partition number is similar to the process in *Partition_base*. The difference is that *Partition_windowPair* distinguishes between the two window-pair partitions of any pair of pivots. The correct identification of the specific window-pair a record belongs to is obtained using the information of the previous partition of the record (lines 6 to 10).

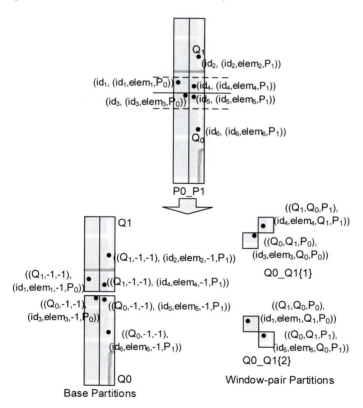

Fig. 7. Example of the Map function for a window-pair round.

Algorithm 7: *Partition_windowPair()*

Input: $k2, W_1, W_2$. $k2 = (part, win, prevPart)$, W_1 and W_2 are the last two components of the job input directory name *job_inDir*

Output: $k2$'s partition number

1: **if** $win = -1$ **then** // base partition
2:     $partition \leftarrow (part \times C4)$ mod *NUMPARTITIONS*
3: **else** // window-pair partition
4:     $minVal \leftarrow min(part, win)$
5:     $maxVal \leftarrow max(part, win)$
6:     **if** $(part > win \wedge prevPart = W1) \vee (part < win \wedge prevPart = W2)$ **then**
7:         $partition \leftarrow (minVal \times C5 + maxVal \times C6)$ mod *NUMPARTITIONS*
8:     **else** // $(part > win \wedge prevPart = W2) \vee (part < win \wedge prevPart = W1)$
9:         $partition \leftarrow (minVal \times C7 + maxVal \times C8)$ mod *NUMPARTITIONS*
10:     **end if**
11: **end if**

In the scenario of Fig. 7, *Partition_windowPair* will generate the same partition number, i.e. $(Q0 \times C4)$ mod *NUMPARTITIONS*, for all the intermediate keys that correspond to partition $Q0$. Similarly, the function will generate the same partition number, i.e. $(Q0 \times C7 + Q1 \times C8)$ mod *NUMPARTITIONS*, for all the intermediate keys that correspond to partition $Q0\_Q1\{1\}$. The partition number of the records in $Q0\_Q1\{1\}$ is generated in line 9 while the partition number of the records in $Q0\_Q1\{2\}$ is generated in line 7. We use the numbers 1 and 2 at the end of the window-pair partitions' names to differentiate between them. We reference this number as the window-pair *sequence*.

After generating the partition numbers of intermediate records, the records are sent to their corresponding reduce nodes. In a window-pair partition round, the records received at each reduce node are sorted and grouped using the *Compare_windowPair* function presented in Algorithm 8. This function groups all the records that belong to the same partition establishing the order of partitions shown in Fig. 8a.

Compare_windowPair receives as input two intermediate record keys, i.e., $k2_1, k2_2$, and returns 0, -1 or $+1$ depending on the order of the associated partitions. The algorithm considers all the possible combinations of the intermediate keys. All the cases are processed as in *Compare_base* with the exception of the case where both keys belong to window-pair partitions. In this case, *Compare_windowPair* orders them based on the tuple (minimum pivot index, maximum pivot index, window-pair sequence) using lexicographical order (lines 11 to 55).

Several sub-cases are considered. When the keys do not belong to the windows between the same pair of pivots, they are ordered based on (minimum pivot index, maximum pivot index) (lines 14 to 20). Otherwise,

---

Algorithm 8 Part 1: *Compare_windowPair()*

Input: $k2_1$, $k2_2$, W_1, W_2. $k2_1 = (part_1, win_1, prevPart_1)$, $k2_2 = (part_2, win_2,$
$prevPart_2)$, W_1 and W_2 are the last two components of the job input
dir. name *job_inDir*

Output: 0 ($k2_1$ and $k2_2$ belong to the same group), -1 (group number of
$k2_1$<group number of $k2_2$), or $+1$ (group number $k2_1$>group
number of $k2_2$)

1: **if** $(win_1 = -1) \wedge (win_2 = -1)$ **then** // baseP-baseP
2:    **if** $(part_1 = part_2)$ **then**
3:       return 0
4:    **else**
5:       return $(part_1 < part_2)? -1 : +1$
6:    **end if**
7: **else if** $(win_1 = -1) \wedge (win_2 \neq -1)$ **then** // baseP-winP
8:    return -1
9: **else if** $(win_1 \neq -1) \wedge (win_2 = -1)$ **then** // winP-baseP
10:    return $+1$
11: **else** // $(win_1 \neq -1) \wedge (win_2 \neq -1)$, winP-winP
12:    min_1, $max_1 \leftarrow min(part_1, win1)$, $max(part_1, win_1)$
13:    min_2, $max_2 \leftarrow min(part_2, win_2)$, $max(part_2, win_2)$
14:    **if** $\neg ((min_1 = min_2) \wedge (max_1 = max_2))$ **then**
15:       **if** $min_1 = min_2$ **then**
16:          return $(max_1 < max_2)? -1 : +1$
17:       **else**
18:          return $(min_1 < min_2)? -1 : +1$
19:       **end if**
20:    **end if**
21:    **if** $(part_1 = part_2) \wedge (prevPart_1 = prevPart_2)$ **then**
         // = partitions, = old partitions
22:       return 0
23:    **end if**
24:    **if** $(part_1 = part_2)$ **then** // = partitions, $\neq$ old partitions
25:       **if** $part_1 < win_1$ **then** // $part_2 < win_2$
26:          **if** $(prevPart_1 = W_1) \wedge (prevPart_2 = W_2)$ **then**
27:             return -1
28:          **else** // $(prevPart_1 = W_2) \wedge (prevPart_2 = W_1)$
29:             return $+1$
30:          **end if**
31:       **else** // $part_1 > win_1 \wedge part_2 > win_2$
32:          **if** $(prevPart_1 = W_1) \wedge (prevPart_2 = W_2)$ **then**
33:             return $+1$

Algorithm 8 Part 2: *Compare_windowPair()*

34:        **else** // $(prevPart_1 = W_2) \wedge (prevPart_2 = W_1)$
35:            return -1
36:        **end if**
37:      **end if**
38:    **end if**
39: **if** $(prevPart_1 = prevPart_2)$ **then** // $\neq$ partitions, = old partitions
40:      **if** $prevPart_1 = win_1$ **then** // $prevPart_2 = win_1$
41:        **if** $part_1 < win_1$ **then** // $part_2 > win_2$
42:            return -1
43:        **else** // $(part_1 > win_1) \wedge (part_2 < win_2)$
44:            return + 1
45:        **end if**
46:      **else** // $prevPart_1 = win_2 \wedge prevPart_2 = win_2$
47:        **if** $part_1 < win_1$ **then** // $part_2 > win_2$
48:            return +1
49:        **else** // $(part_1 > win_1) \wedge (part_2 < win_2)$
50:            return - 1
51:        **end if**
52:      **end if**
53:    **end if**
54:    return 0 // $\neq$ partitions, $\neq$ old partitions
55: **end if**

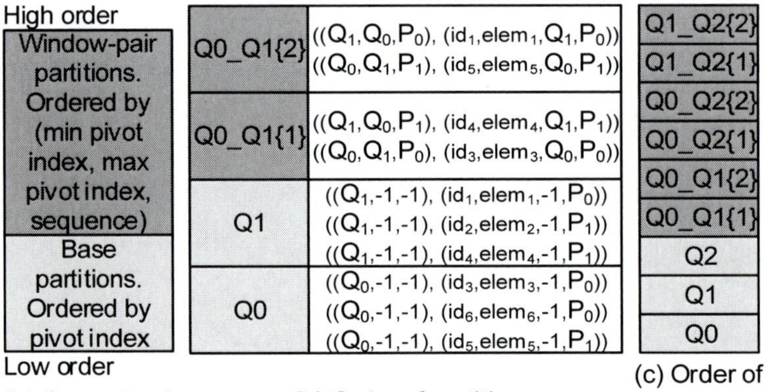 High order			
Window-pair partitions. Ordered by (min pivot index, max pivot index, sequence)	Q0_Q1{2}	$((Q_1,Q_0,P_0), (id_1,elem_1,Q_1,P_0))$ $((Q_0,Q_1,P_1), (id_5,elem_5,Q_0,P_1))$	Q1_Q2{2}
			Q1_Q2{1}
			Q0_Q2{2}
	Q0_Q1{1}	$((Q_1,Q_0,P_1), (id_4,elem_4,Q_1,P_1))$ $((Q_0,Q_1,P_0), (id_3,elem_3,Q_0,P_0))$	Q0_Q2{1}
			Q0_Q1{2}
	Q1	$((Q_1,-1,-1), (id_1,elem_1,-1,P_0))$ $((Q_1,-1,-1), (id_2,elem_2,-1,P_1))$ $((Q_1,-1,-1), (id_4,elem_4,-1,P_1))$	Q0_Q1{1}
			Q2
Base partitions. Ordered by pivot index	Q0	$((Q_0,-1,-1), (id_3,elem_3,-1,P_0))$ $((Q_0,-1,-1), (id_6,elem_6,-1,P_0))$ $((Q_0,-1,-1), (id_5,elem_5,-1,P_1))$	Q1
			Q0
Low order			
(a) General order of partitions	(b) Order of partitions with 2 pivots		(c) Order of partitions with 3 pivots

Fig. 8. Group formation in a window-pair round.

the algorithm considers the following cases: (1) when the keys belong to the same partition and same old partition, they have the same order (lines 21 to 23), (2) when the keys belong to the same partition but different old partition, they are ordered by their window-pair sequences (lines 24 to 38), (3) when the keys belong to different partitions but the same old partition, they are also ordered by their window-pair sequences (lines 39 to 53), and (4) when the keys belong to different partitions and different old partitions, they have the same order (line 54). Figure 8b shows the order of partitions generated by *Compare_windowPair* for the scenario with two pivots presented in Fig. 7. Figure 8c shows the order of partitions for the case of three pivots.

After generating the groups in a reduce node, the MapReduce framework calls the *reduce* function *Reduce_windowPair* once for each group. This function is presented in Algorithm 9 and receives as input a key-value pair (*k2, v2List*). The goal of this function is to generate the window links of the partitions that are small enough to be processed in a single node and to store the data of larger partitions for further repartitioning. If all the records in *v2List* belong to the same old partition, there is no possibility of generating window links and thus the function terminates immediately (lines 1 to 3). If the size of the list is small enough to be processed in a single node, the algorithm calls the function *InMemorySimJoinWin* to get the window links in the current partition (lines 4 to 5). Otherwise, all the records of the group are stored in an intermediate directory for further partitioning (lines 6 to 10). Both the intermediate records and the directory name propagate the information of the previous partitions. This information will enable the correct generation of window links in subsequent rounds.

Algorithm 9 *Reduce_windowPair()*

Input: (*k2, v2List*), W_1, W_2. *k2* = (*kPart, kWin, kPrevPart*), *v2List* = list(*id, elem, part, prevPart*), W_1 and W_2 are the last two components of the job input directory name *job_inDir*

Output: SJ matches or intermediate data. Intermediate data = list(*k3, v3*). *k3* = *id*, *v3* = (*id, elem, part*)

1: **if** all the elements of *v2List* have the same value of *prevPart* **then**
2:     return
3: **end if**
4: **if** *sizeInBytes(v2List)* ≤ *memT* **then**
5:     *InMemorySimJoinWin(v2List, outDir, eps)*
6: **else**
7:     **for** each element *e* of *v2List* **do**
8:         output (*e.id*, (*e.id, e.elem, e.prevPart*)) to
          outDir/intermediate/W_*(roundNum)*_*(k2)*_*(W₁)*_*(W₂)*
9:     **end for**
10: **end if**

Note that, in the case of *Reduce_windowPair*, all partitions that are stored for further processing are set to be repartitioned by a future window-pair partition round. This is the case because the links generated in a window-pair round or in any of its partitions should always be window links. In the scenario represented in Fig. 7, the MapReduce framework calls the *Reduce_windowPair* function for each partition of Fig. 8b: Q0, Q1, Q0_Q1{1} and Q0_Q1{2}. Observe that the value of *uAttr* in the output directory name is *k2*. This component ensures unique directory names. Assuming that the values of *k2* of Q0_Q1{1} and Q0_Q1{2} belong to their bottom windows, the values of *uAttr* are: Q0: $(Q_0,-1,-1)$, Q1: $(Q_1,-1,-1)$, Q0_Q1{1}: (Q_0,Q_1,P_0), and Q0_Q1{2}: (Q_0,Q_1,P_1).

Enhancements for Geographical Distance

Since the MRSimJoin solution presented in the MRSimJoin Algorithm subsection is based on the generalized hyperplane distance, it could be used with any dataset that lies in a metric space. The solution, however, could be enhanced in cases where the distance from a record to the hyperplane between two partitions can be computed exactly (Jacox and Samet 2008). In the case of the geographical distance geoDist defined in the Geographic Data and Distance Functions subsection (Euclidean distance on a plane where a Spherical Earth was projected), the exact distance from a record t to the hyperplane that separates the partitions of two pivots P_0 and P_1 is given by:

$$hDist(t, P_0, P_1) = (geoDist(t, P_0)^2 - geoDist(t, P_1)^2)/(2 \times geoDist(P_0, P_1)).$$

To use this distance, the GHP distance should be replaced by *hDist* in line 5 of *Map_base* and also in line 5 of *Map_windowPair*.

Implementation in Hadoop

The presented MRSimJoin algorithms are generic enough to be implemented in any MapReduce framework. This section presents a few additional guidelines for its implementation on the popular Hadoop MapReduce framework (Apache Hadoop 2013).

Distribution of atomic parameters. One of the tasks of the *MRJob* function, called in the main MRSimJoin routine, is to make sure that the provided atomic parameters, i.e., *outDir, numPiv, eps* and *memT*, are available at every node that will be used in the MapReduce job. In Hadoop, this can be done using the job configuration *jobConf* object and its methods *set* and *get*.

Distribution of pivots. *MRJob* also sends the list of pivots to every node that will execute a *map* task. In Hadoop this can be done using the

DistributedCache, a facility that allows the efficient distribution of application specific, large, read-only files.

Renaming directories. The main MRSimJoin routine renames a directory to flag it as already processed. This can be done using the *rename* method of Hadoop's *FileSystem* class. The method will change the directory path in Hadoop's distributed file system without physically moving its data.

Single-node Similarity Join. *InMemorySimJoin* and *InMemorySimJoin-Win* represent single-node algorithms to get the links and window links in a given dataset, respectively. We have implemented these functions using the Quickjoin algorithm (Jacox and Samet 2008).

Performance Evaluation

We implemented MRSimJoin using the Hadoop 0.20.2 MapReduce framework. In this section we evaluate its performance with synthetic and real-world geographic data.

Test Configuration

We performed the experiments using a Hadoop cluster running on the Amazon Elastic Compute Cloud (Amazon EC2). Unless otherwise stated, we used a cluster of 10 nodes (1 master + 9 worker nodes) with the following specifications: 15 GB of memory, 4 virtual cores with 2 EC2 Compute Units each, 1,690 GB of local instance storage, and 64-bit platform. We set the block size of the distributed systems to 64 MB and the total number of reducers to: $0.95 \times$ ⟨no. worker nodes⟩ × ⟨max reduce tasks per node⟩. We use the following datasets:

- **SynthData** This is a synthetic geographic dataset (longitude-latitude pairs). The dataset for scale factor 1 (SF1) contains 2 million records (86.9MB). The range of latitude and longitude values of the SF1 dataset are [25, 50] and [65, 125], respectively.
- **GeoNames** This dataset contains longitude-latitude pairs extracted from the GeoNames database (GeoNames 2013). These records represent the location of various US geographical features. The SF1 dataset contains 2,023,687 records (52.1 MB) with latitude and longitude ranges of [25, 50] and [65, 125], respectively.

The datasets for SF greater than 1 were generated in such a way that the number of links of any SJ operation in SFN are N times the number of links of the operation in SF1. Specifically, the datasets for higher SFs were obtained adding shifted copies of the SF1 dataset such that the separation between the region of new records and the region of previous records is greater than the maximum value of ε used in our tests. Half of the records of each dataset belong to R and the remaining ones to S. We use the geographical distance *geoDist* with both datasets. The available memory to perform the in-memory SJ algorithm was 32 MB.

We compare the performance of MRSimJoin with MRThetaJoin, an adaptation of the memory-aware 1-Bucket-Theta algorithm (Okcan and Riedewald 2011) that uses the single-node QuickJoin algorithm (Jacox and Samet 2008) in the reduce function.

Performance Evaluation with Synthetic Data

Increasing Scale Factor. Figure 9 compares the way MRSimJoin and MRThetaJoin scale when the data size increases (SF1-SF10). This experiment uses a value of epsilon of 2 miles. The core result in this figure is that MRSimJoin performs and scales significantly better than MRThetaJoin. The execution time of MRThetaJoin grows from being 1.7 times the one of MRSimJoin for SF=1 to 8.4 times for SF=10. The execution time of MRThetaJoin is significantly higher than that of MRSimJoin because the total size of all the partitions of MRThetaJoin is significantly larger than that of MRSimJoin.

Increasing Epsilon. Figure 10 shows how the execution time of MRSimJoin and MRThetaJoin increase when epsilon increases (1–5 miles). These tests use SF1. The performance of MRSimJoin is better than the one of MRThetaJoin for all the values of epsilon. Specifically, the execution time of MRSimJoin is 59.6% of the one of MRThetaJoin for epsilon = 1 mile, and 64.2% for epsilon = 5 miles. We can also observe that, in general, the execution time of both algorithms grows slowly when epsilon grows. The increase in execution time is due to a higher number of distance computations in both algorithms and slightly larger sizes of window-pair partitions in the case of MRSimJoin.

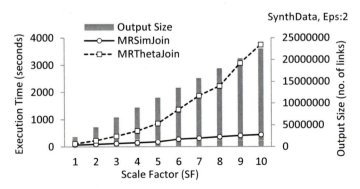

Fig. 9. Increasing SF–SynthData.

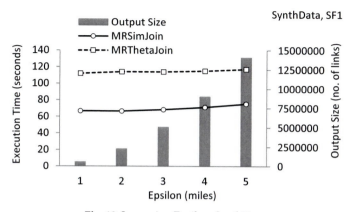

Fig. 10. Increasing Epsilon–SynthData.

Performance Evaluation with GeoNames

Increasing Scale Factor. Figure 11 compares the way MRSimJoin and MRThetaJoin scale when the data size increases (SF1–SF10). The results for GeoNames are similar to the ones we found for the case of synthetic data. Specifically, the execution time of MRThetaJoin grows from being 1.4 times of that of MRSimJoin for SF1 to 8.6 times for SF10.

Increasing Epsilon. Figure 12 shows how the execution times of MRSimJoin and MRThetaJoin increase when epsilon increases. As in the case of SynthData, the performance of MRSimJoin is better than the one of MRThetaJoin for all evaluated values of epsilon. The execution time of MRSimJoin is 64.2% of the one of MRThetaJoin for epsilon = 1 mile, and 89.8% for epsilon = 5 miles.

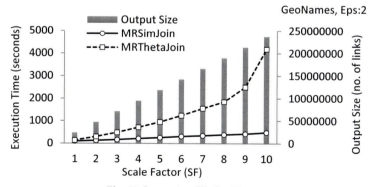

Fig. 11. Increasing SF–GeoNames.

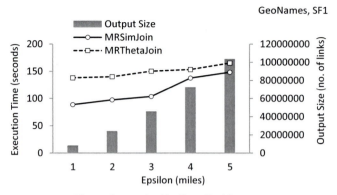

Fig. 12. Increasing Epsilon–GeoNames.

Increasing Number of Nodes and Scale Factor. One of the goals of MapReduce-based operations is to scale efficiently when the number of nodes and the data size increase proportionally. Ideally, the execution time should remain constant. Figure 13 shows the execution time of MRSimJoin and MRThetaJoin when the data size and the number of nodes increase from (SF1, 2 nodes) to (SF5, 10 nodes). MRSimJoin follows the ideal execution time much more closely than MRThetaJoin. MRSimJoin's execution time for (SF5, 10) is only 2.1 times the one for (SF1, 2) while MRThetaJoin's execution time for (SF5, 10) is 4.5 times the one for (SF1, 2). Moreover, the execution time of MRThetaJoin grows from being 1.8 times the one of MRSimJoin for (SF1, 2) to 3.8 times for (SF5, 10).

Increasing Number of Pivots. The execution time and number of rounds for MRSimJoin, as the number of pivots increases, is presented in Fig. 14. The figure shows that a smaller number of pivots generates a higher number of rounds. We also observe that, in general, the execution time decreases when the number of rounds decreases. We found that in most of the experiments presented in this section, the best execution time is achieved using a single round. To compute the number of pivots (P) that will generate a single round for relatively smaller values of epsilon, i.e., smaller than 25%, we can use the fact that the space needed for the in-memory QuickJoin algorithm is about twice the size of the input data (Jacox and Samet 2008). Thus, to ensure that the average MRSimJoin base partition (and window-pair partition) can be solved using the in-memory QuickJoin, we need: $P = 2 \times k \times D/T$, where D is the total input size (282 MB), T is the available memory for QuickJoin (32 MB), and k is a factor that compensates the effect of data skewness on the size of partitions (we used $k = 4$ since the real dataset is considerably skewed). Using this expression, the value of P for this experiment is 70. This value of P generates a single round and an execution time that is only 6% higher than the best execution time (obtained with $P = 105$).

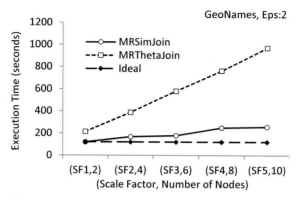

Fig. 13. Increasing Number of Nodes and SF–GeoNames.

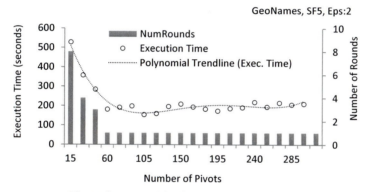

Fig. 14. Increasing Number of Pivots–GeoNames.

Conclusions and Future Work

MapReduce-based systems have become a crucial component to efficiently process and analyze the large amounts of geographical data currently available in many commercial and scientific organizations. The Similarity Join is recognized as one of the most useful data analysis operations and has been used in many application scenarios. While multiple implementation techniques have been proposed for the Similarity Join, very little work has addressed the study of MapReduce-based Similarity Joins for geographical data. This chapter focuses on the study, design, and implementation techniques of MRSimJoin, a MapReduce-based Similarity Join algorithm that can be used with geographical data and distance functions. MRSimJoin iteratively partitions the data until the partitions are small enough to be efficiently processed in a single node. Each iteration executes a MapReduce job that processes the generated partitions in parallel. We implemented

MRSimJoin using the Hadoop MapReduce framework. An extensive performance evaluation of MRSimJoin with synthetic and real-world geographic data shows that it scales very well when important parameters like epsilon, data size, and number of nodes increase. Furthermore, we show that MRSimJoin performs significantly better than an adaptation of the state-of-the-art MapReduce-based algorithm to answer arbitrary joins.

Our paths for future work include the study of: (1) other similarity-aware operators, e.g., kNN Join and kDistance Join, for MapReduce-based systems, (2) indexing techniques that can be exploited to implement Similarity Join operations, and (3) cloud queries with multiple similarity-based operators.

References

Apache Hadoop. 2013. http://hadoop.apache.org/.

Blanas, S., J.M. Patel, V. Ercegovac, J. Rao, E.J. Shekita and Y. Tian. 2010. A comparison of join algorithms for log processing in mapreduce. In ACM SIGMOD '10, USA.

Böhm, C., B. Braunmüller, F. Krebs and H.-P. Kriegel. 2001. Epsilon grid order: an algorithm for the similarity join on massive high-dimensional data. In ACM SIGMOD '01, USA.

Chaudhuri, S., V. Ganti and R. Kaushik. 2006. A primitive operator for similarity joins in data cleaning. In ICDE '06, USA.

Chen, S. 2010. Cheetah: a high performance, custom data warehouse on top of mapreduce. In VLDB '10, Singapore.

Dean, J. and S. Ghemawat. 2004. Mapreduce: simplified data processing on large clusters. In OSDI '04, USA.

Dittrich, J.-P. and B. Seeger. 2001. Gess: a scalable similarity-join algorithm for mining large data sets in high dimensional spaces. In ACM SIGKDD '01, USA.

Dohnal, V., C. Gennaro, P. Savino and P. Zezula. 2003a. Similarity join in metric spaces. In ECIR '03, Italy.

Dohnal, V., C. Gennaro and P. Zezula. 2003b. Similarity join in metric spaces using ed-index. In DEXA '03, Czech Republic.

GeoNames. 2013. http://www.geonames.org/about.html.

Gravano, L., P.G. Ipeirotis, H.V. Jagadish, N. Koudas, S. Muthukrishnan and D. Srivastava. 2001. Approximate string joins in a database (almost) for free. In VLDB '01, Italy.

Hjaltason, G.R. and H. Samet. 2003. Index-driven similarity search in metric spaces. ACM Trans. Database Syst. 28(4): 517–580.

Jacox, E.H. and H. Samet. 2008. Metric space similarity joins. ACM Trans. Database Syst. 33(2): 7:1–7:38.

Jiang, D., A.K.H. Tung and G. Chen. 2011. Map-join-reduce: Toward scalable and efficient data analysis on large clusters. IEEE Trans. on Knowl. And Data Eng. 23(9): 1299–1311.

Kitsuregawa, M. and Y. Ogawa. 1990. Bucket spreading parallel hash: a new, robust, parallel hash join method for data skew in the super database computer (sdc). In VLDB '90, Australia.

Luo, G., J.F. Naughton and C.J. Ellmann. 2002. A non-blocking parallel spatial join algorithm. In ICDE '02, USA.

Okcan, A. and M. Riedewald. 2011. Processing theta-joins using mapreduce. In ACM SIGMOD '11, Greece.

Patel, J.M. and D.J. DeWitt. 1996. Partition based spatial-merge join. In ACM SIGMOD '96, Canada.

Patel, J.M. and D.J. DeWitt. 2000. Clone join and shadow join: two parallel spatial join algorithms. In GIS '00, USA.

Pavlo, A., E. Paulson, A. Rasin, D.J. Abadi, D.J. DeWitt, S. Madden and M. Stonebraker. 2009. A comparison of approaches to large-scale data analysis. In ACM SIGMOD '09, USA.

Schneider, D.A. and D.J. DeWitt. 1989. A performance evaluation of four parallel join algorithms in a shared-nothing multiprocessor environment. In ACM SIGMOD '89, USA.

Silva, Y.N., W.G. Aref and M.H. Ali. 2010. The similarity join database operator. In ICDE '10, USA.

Silva, Y.N., W.G. Aref, P.-A. Larson, S. Pearson and M.H. Ali. 2013a. Similarity Queries: Their Conceptual Evaluation, Transformations and Processing. VLDB Journal. 22(3): 395–420.

Silva, Y.N. and S. Pearson. 2012. Exploiting Database Similarity Joins for Metric Spaces. In VLDB '12, Turkey.

Silva, Y.N., S. Pearson and J.A. Cheney. 2013b. Database Similarity Join for Metric Spaces. In SISAP '13, Spain.

Silva, Y.N., J.M. Reed and L.M. Tsosie. 2012. MapReduce-based Similarity Join for Metric Spaces. In the VLDB International Workshop on Cloud Intelligence '12, Turkey.

Snyder, J. 1993. Flattening the earth: two thousand years of map projections. Chicago: University of Chicago Press.

Vernica, R., M.J. Carey and C. Li. 2010. Efficient parallel set-similarity joins using mapreduce. In ACM SIGMOD '10, USA.

White, T. 2010. Hadoop: The Definitive Guide. Yahoo! Press.

Yang, H.-c., A. Dasdan, R.-L. Hsiao and D.S. Parker. 2007. Map-reduce-merge: simplified relational data processing on large clusters. In ACM SIGMOD '07, China.

CHAPTER 3

Spatial Index Schemes for Cloud Environments

Wen-Chih Peng,[a], Ling-Yin Wei,[a] Ya-Ting Hsu,[a] Yi-Chin Pan[a]*
and *Wang-Chien Lee[b]*

Introduction

In recent years, mobile devices such as smart phones and tablet computers have become popular part of our daily life. Simultaneously, with the increasing prevalence of the Global Positioning System (GPS), a large number of location-based applications, such as Foursquare and Flickr, have been developed. People are able to share their real-time events with friends anytime and anywhere as long as the Internet is available. For example, people can check into a specific location and can note their activities, and they can see their friends'shared real time information using the Foursquare application. These location-based applications induce that the amount of multi-attribute data, which at least consist of locations and time-stamps are dramatically increasing. In order to retrieve and manage this data effectively, different database management systems (DBMSs) have been developed. For traditional relational database management systems (RDBMSs), there are several index structures, such as k-dimensional (k-d) trees (Bentley 1975), quad trees (Finkel and Bentley 1974), and R-trees (Guttman 1984). However, RDBMSs are unable to deal with thousands of millions of queries efficiently

[a] National Chiao-Tung University, 1001 University Road, Hsinchu, Taiwan 300.
  Emails: wcpeng@cs.nctu.edu.tw; lywei.cs95g@nctu.edu.tw; shiyating@gmail.com;
  beats.1213@gmail.com
[b] The Pennsylvania State University, 1001 University Road, Hsinchu, Taiwan 300.
  Email: wlee@cse.psu.edu
* Corresponding author

when the amount of data is large (Nishimura et al. 2013). On the other hand, distributed relational database management systems (DRDBMSs) have been developed and are able to deal with multi-attribute accesses. However, DRDBMSs are unable to maintain and retrieve data among servers efficiently because they take a lot of time to make sure the data is consistent by appropriately locking and updating the data.

To deal with a huge amount of data efficiently and flexibly, cloud computing is now playing an important role, and new cloud data managements (CDMs), which are NoSQL databases (Stonebraker 2010), have been developed. The most prevalent NoSQL CDMs, such as HBase (Khetrapal and Ganesh 2008), Cassandra (Lakshman and Malik 2010) and Amazon Simple Storage (Varia 2008), are developed based on a BigTable (Chang et al. 2008) management system. Compared with DRDBMSs, these management systems have the characteristics of high scalability, high availability and fault-tolerance because they can effectively and efficiently handle a large number of data updates even if failure events occur. In addition, a BigTable management system stores data as <key, value> pairs, and thus these BigTable-like management systems can retrieve data efficiently by the following characteristics: 1) each<key, value> pair is stored on multiple servers; and 2) each key owns multiple versions of a value. In other words, the first characteristic, benefits the efficiency of retrieving data, and the second characteristic eliminates the waiting time of making data consistent. Due to the inherent restriction of a BigTable data structure, however, these management systems only support some basic operations, such as **Get**, **Set** and **Scan**. A **Get** operation retrieves values mapped by a key; a **Set** operation inserts/modifies values according to a corresponding key; and a **Scan** operation returns all values mapped by a range of keys. However, these basic operations do not directly support multi-attribute accesses.

In this chapter, to support efficient multi-attribute accesses of skewed data on CDMs, we propose a novel multi-dimensional index, called the KR$^+$-index, on CDMs by designing Key names for leaves of the R$^+$-tree. A challenging issue is to filter out data after querying the results from large differences in the volume of data between grids. In order to describe it conveniently in this chapter, the volume of the data in the grid is represented by the grid size. However, dividing a map more meticulously could reduce the differences in the grid sizes but could also reduce the efficiency of accessing data. For example, for a range query, we need to retrieve more grids for the same spatial range. According to the aforementioned observations, we expect that the differences in the grid sizes could be smaller and the time of the grid accesses could be less at the same time. Consequently, how to divide a map into grids to reach a balance between the two points plays an important role for CDMs. In this chapter, we first use an R$^+$-tree (Sellis et al. 1987) to divide the data, and the rectangles in the leaf nodes of the tree index are treated as dynamic grids. The reasons for using an R$^+$-tree are described as follows. First, we could get a balance between the grids sizes and the times of grid accesses by adjusting the two parameters, M and m, of the R$^+$-tree. Second, compared with other variants of the R-tree, the leaf nodes do not overlap each other, and thus it is a benefit as there is no redundant retrieval of the same data from different keys and it is easy to define different keys for each rectangle of a leaf node. Moreover, the second challenge is how to design the key names of these grids to support efficient queries on BigTable management systems. We observed the characteristics of CDMs as follows: a CDM has a fast key-value search and it is fast to Scan keys which are in a dictionary order. Based on these characteristics, we propose an approach to define the key name of a grid to support efficient queries. In the experiment, we implement the proposed index on two well-known CDM systems, HBase and Cassandra, and we compare the performance of the proposed index with the existing index methods. The experimental results demonstrate that our proposed index outperforms the existing index methods via skewed data. We summarize the contributions of this chapter as follows:

- We propose an efficient multi-dimensional index structure, the KR$^+$-index, on CDMs to support efficient multi-attribute accessing of skewed data.
- Based on the KR$^+$-index, we define new efficient spatial query algorithms, range query and *k*-NN query.
- The experimental results show that the proposed KR$^+$-index outperforms competitors.

The remainder of the chapter is organized as follows. First, we illustrate the background of multi-attribute access, multi-dimensional indexes and Hilbert curves in section "Background". We next propose the KR$^+$-index in section "The Multi-dimensional Index Structure". In section "Experiments", we evaluate the performance of the proposed KR$^+$-index for multi-attribute accessing of CDMs. In section "Related Work", we illustrate the existing CDMs, traditional index techniques, and some cloud indexes on CDMs. Finally, we conclude the chapter in section "Conclusions".

Background

Multi-attribute Access

For multi-dimensional data searching, multi-attribute access is used to restrict multiple attributes at the same time. For instance, Range Query and *k*-NN Query are common queries of multi-attribute access and are widely used in location-based services.

Range Query

Given a set of data points P and a spatial range R, a range query can be formulated as "searching the data points in P that are located in the spatial range R". Note that, in this chapter, each data point has location information, e.g., a longitude and latitude. Without loss of generality, in this chapter, a spatial range is represented by a rectangular range.

For instance, in Fig. 1a, given 15 restaurants, marked by gray points, and a red query range R, the range query is to search for which restaurants are located in the range R. As shown in Fig. 1a, the result of the range query is $\{p_1, p_2, p_3, p_4, p_5\}$.

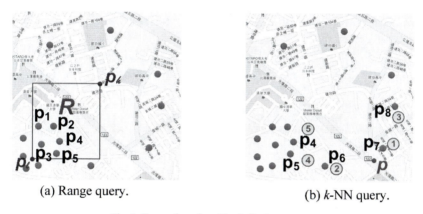

(a) Range query.   (b) *k*-NN query.

Fig. 1. Examples of multi-attribute access.

Color image of this figure appears in the color plate section at the end of the book.

k-NN Query

Given a set of data points P, a query location $p = (p_x, p_y)$ and a constant k, a *k*-NN query can be formulated as "searching the data points in P that are the k nearest data points of p".

For example, in Fig. 1b, given 15 restaurants, a user-specified location p marked in red and k = 5, a 5-NN query here is to search for the five nearest restaurants to p. Thus, the search result of this query is $\{p_4, p_5, p_6, p_7, p_8\}$ as shown in Fig. 1b.

Multi-dimensional Index Techniques

Tree Structures

R-trees, developed for indexing multi-dimensional data, are widely used in multi-attribute accesses. For instance, given fifteen checked-in restaurants in Fig. 2a, an R-tree index structure with two parameters M = 3 and m = 1 can be stored as shown in Fig. 2b and the top figure in Fig. 2d. Note that M and m are positive integers used to ensure that the number of elements of a node in a tree is in the range of [m, M]. Because an R-tree is a balanced search tree by dynamically splitting and merging nodes, and the number of elements in each node can be restricted to by controlling M and m, it is beneficial for searching skewed data. Moreover, to efficiently index different multi-dimensional data, different variations of R-trees have been developed, such as R^+-tree (Sellis et al. 1987), R^*-tree (Beckmann et al. 1990) and the Hilbert R-tree (Kamel and Faloutsos 1994). The R^+-tree developed a new rule of splitting and merging nodes to speed up multi-attribute accesses. For instance, given a set of data points in Fig. 2a, c shows an example of an R^+-tree with M = 3 and m = 1, and the corresponding index structure is illustrated in the bottom figure in Fig. 2d. As shown in Fig. 2c, the rectangles

(a) An example of skewed data.

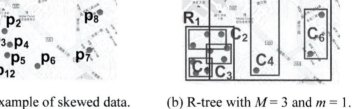

(b) R-tree with *M* = 3 and *m* = 1.

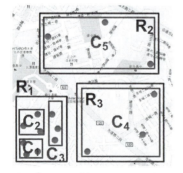

(c) R^+-tree with *M* = 3 and *m* = 1.

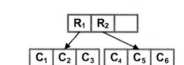

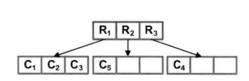

(d) The index structures for the R-tree in Fig. 2(b) and the R^+-tree in Fig. 2(c).

Fig. 2. Examples of R-trees and R^+-trees.

Color image of this figure appears in the color plate section at the end of the book.

do not overlap each other, which has the benefit of reducing searching time. The reason is that, compared with R-trees, we do not search duplicated results using R⁺-trees.

Quad-trees (Finkel and Bentley 1974) are another common tree structures for indexing multi-dimensional data. In quad-trees, each internal node has exactly four children. However, quad-trees are not balanced trees because a region is split into four sub-regions until the number of data points in the region is less than or equal to a given parameter M. For instance, given the data in Fig. 2a, Fig. 3 shows a quad-tree with M = 3.

Linearization

Linearization is a well-known technique for indexing multi-dimensional data by transforming it into one-dimensional data. One of the most popular methods of linearization is using space-filling curves, such as a Hilbert curve and Z-ordering. Given two-dimensional data, this method first divides the map into $2^n \times 2^n$ non-overlapping grids, where n is a parameter, and assigns a number for each grid according to the order of traversing all the grids. Note that the number of each grid is regarded as a key. Figure 4 illustrates an example of linearization using a Hilbert curve with $2^1 \times 2^1$ grids. In Fig. 4, the map is divided into four grids, and the keys of these grids are represented by 0, 1, 2 and 3 according to a Hilbert curve. However, using space-filling curves to index data may not be efficient. Take check-in records indexing as an example. If a value of n is set to be lower, it induces a larger grid size, and the grid would possibly cover more check-in records distributed in its area. Given a range query, all check-in records located in the grids overlapping

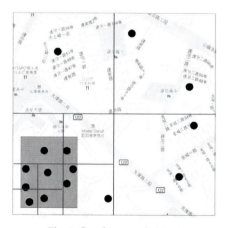

Fig. 3. Quad-tree with $M = 3$.

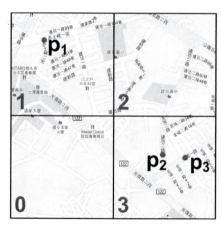

Fig. 4. A Hilbert curve on grids of a map.

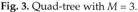

Color image of these figures appear in the color plate section at the end of the book.

the range would be retrieved, and then these records are further verified as they are indeed in the query range. Therefore, it would result in more unqualified check-in records, called false-positive, which should be pruned, thus taking more time to derive the query result. On the other hand, if n is set to be larger, it would increase the times to retrieve more grids. Thus, for this indexing technique, it is a trade-off to set a proper value of n for the sake of efficiency.

The Multi-dimensional Index Structure

CDMs provide key-value searching, which retrieves a value by a given key, based on the CDM data model. CDMs support basic operations to access data, but these operations do not directly support multi-attribute access. To deal with the problem of multi-attribute access, we have developed a multi-dimensional index structure for CDMs. Furthermore, in this chapter, we apply our developed index structure for range query and *k*-NN query on CDMs.

KR+-index

To design a multi-dimensional index, we observe three characteristics of CDMs: 1) the time of retrieving a key that has n data is far less than the time of retrieving n keys that each have one datum; 2) the time of retrieving n data by one key is large when n is large; 3) the operation Scan is more efficient than multiple **Get** which both retrieve the same keys. Considering the aforementioned characteristics, for a query, we should make the number of false-positives be small according to the characteristic 2, and let the number of sub-queries be small according to the characteristic 1. An R+-tree is a balanced tree that has M and m to control the size of each dynamic rectangle. We could therefore use the M and m to meet the trade-off between false-positive and sub-queries. Considering the characteristic 3, we use the Hilbert curve to let the queried key to be as continuous as possible and then the rate of **Scan** is increased.

Figure 5 is the framework of the KR+-index. First, the data is constructed by the R+-tree with the given M, m and the restaurant records for each rectangle $\{R_1, R_2, R_3\}$, are maintained. In order to retrieve the restaurant records efficiently, we propose a mapping method for retrieving the queried rectangle keys. Second, the map is divided into uniform $2^n \times 2^n$ non-overlapping grids $\{G_1, G_2, G_3, G_4\}$. Then, for each grid we maintain a list of rectangles that overlap it. For instance, the grid G_2 overlaps rectangles $\{R_1, R_2\}$ so the KeyTable stores a record $<G_2, \{R_1, R_2\}>$. Thus, a query could conveniently transform into which grids need to be queried and then through the KeyTable they could easily get the required rectangles.

For these key-value storages, it is crucial to define the key, because we use the key to access the corresponding data. We construct the R⁺-tree to discover non-overlapping minimum bounding rectangles. Considering characteristic 3, we use a Hilbert-curve to define the keys because this method manifests superior data clustering compared with other multi-dimensional linearization techniques. For each leaf rectangle, we use the Hilbert-value of the geographic coordinate of the centroid of the rectangle as the key. Note that different rectangles may have the same key, since their central points fall on the same grid. Then, we split the space into uniform non-overlapping $2^n \times 2^n$ grids each of which has a Hilbert-value which is transformed by a Hilbert-curve. Figure 6 shows an example of KR⁺-index.

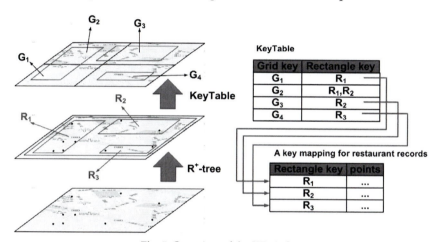

Fig. 5. Overview of the KR⁺-index.

Color image of this figure appears in the color plate section at the end of the book.

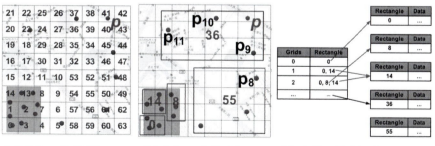

(a) A key definition for grids.

(b) A key definition for rectangles.

(c) A KeyTable for rectangles.

Fig. 6. An example of a KR⁺-index.

Color image of this figure appears in the color plate section at the end of the book.

First, in Fig. 6a, each grid is also given a Hilbert value and each solid point represents a data point. Note that the green area represents a query rectangle. For instance, the data point P is located in Grid 41 because the grid's Hilbert value is 41. In Fig. 6b, each rectangle is given a Hilbert value. For instance, Grid 1 in Fig. 6a overlaps with the rectangles {0, 14} so that <1, {0, 14}> is stored in the KeyTable as shown in Fig. 6c. We could get the rectangle information through the KeyTable and the multi-attribute access can retrieve the data efficiently.

The decision of (M, m) and the order of the Hilbert-curve will affect the efficiency of the range query. Thus, we decided to dynamically generate the values of (M, m) and the order. We knew that (M, m) influences the size of the rectangles and the order decides the grid size. We observed the relationship between the response times and the parameters (M, m), and o. As shown in Table 1, this is the average length and width of rectangles of ten million data size and the length of grids with different order; the average length and width of (M, m) = (250, 125) is closed to the grid length of order 7. According to the range query response times of the KR+-index, the range query with fixed (M, m) = (250, 125) had a better response time (i.e., 1.2 seconds) when order o = 7. Notice that, for the same range query with fixed (M, m) = (250, 125), the response time is 1.6 when o = 6. Similarly, for the same range query with fixed (M, m) = (250, 125), the response time is 1.9 when o = 8. We found that the closer the rectangle size and the grid size, the better the response time for that range query. Thus we proposed a new method that automatically determines the size of the parameters. It first decides a small value of (M, m), and evaluates the average size of the rectangles, then calculates the closest grid size generated by order. The objective function of o can be expressed as $\min_o(\,|\,len(o)-avgLen(M)\,|\,+\,|\,len(o)-avgWid(M)\,|\,)$. Thus, it determines the (M, m) and order automatically.

Table 1. The relationships between the parameters of the KR+-index.

(a) Average length and width of rectangles of KR+-index.

(b) The length and width of grids of different order.

M	m	Avg. len.	Avg. wid.
50	25	1128.4283	2656.2368
100	50	4928.2417	4671.1948
250	125	6941.6216	6280.1025

order	len.
6	15625
7	7812.5
8	3906.25

Range Query

Multi-dimensional range queries are commonly used in location based applications. Algorithm 1 is the pseudo code for a range query in HBase and Cassandra (p_l, p_h) is the range for the query, p_l is the lower bound and p_h is the upper bound. The Hilbert curve splits the space into grids, where each

---

Algorithm 1 Range Query

Input: p_l, p_h: the range for the query;
Output: points contained in the range;
 1: Coordinate $\leftarrow$ $ComputeCoordinateOfGrid(p_l, p_h)$;
 2: Keys $\leftarrow$ ϕ;
 3: RectKeys $\leftarrow$ ϕ;
 4: Result $\leftarrow$ ϕ;
 5: **for** each Coordinate c $\in$ Coordinate **do**
 6:     GridKeys $\leftarrow$ GridKeys $\cup ComputeContainGridKeys(c)$);
 7: **end for**
 8: RectKeys $\leftarrow$ $GetRectKeys$(GridKeys);
 9: **for** each Key k $\in$ RectKeys **do**
10:     Result $\leftarrow$ Result $\cup GetContainPoints(k)$);
11: **end for**
12: **return** Result;

Algorithm 2 $GetRectKeys$(GridKeys)

Input: GridKeys: the grid keys overlap with query range;
Output: the rectangle keys overlap with query range;
 1: RectKeys $\leftarrow$ ϕ;
 2: **for** each grid key gk $\in$ GridKeys **do**
 3:     RectKeys $\leftarrow$ RectKeys $\cup$ KeyTable(gk));
 4: **end for**
 5: **return** RectKeys;

---

grid has one grid key. The algorithm first computes the coordinate of grids overlapping the range query by the function Compute Coordinate Of Grid(). Grid Keys is the set of grid keys contained in the query range. For each coordinate of grid c, the function ComputeContainGridKeys() computes the corresponding grid keys via the Hilbert curve and adds it to the list, GridKeys. Then, according to the key table, we could find the rectangle keys in the query range. Lines 5–8 find the queried key and lines 9–10 fetch the points in the corresponding key. The function GetContainPoint() returns the queried data by first retrieving points from Cassandra and HBase with key k and then filtering out some points that are not in the query range.

As shown in Fig. 6, taking the green block as the range query, we will show an example of how range query works. Use the query range to get the geographic coordinates of the overlapping grids {(0, 0), (0, 1), (0, 2), (1, 0), (1, 1), (1, 2), (2, 0), (2, 1), (2, 2)}, then get the Hilbert values of each geographic coordinate {0, 3, 4, 1, 2, 7, 14, 13, 8}. Second, get the keys of the rectangles through the KeyTable such that grid 0 maps to rectangle {0}, grid 1 maps to rectangles {0, 14}, grid 2 maps to rectangles {0, 8, 14}, etc. Thus, we can get the queried rectangle keys {0, 8, 14}, by joining the rectangle sets obtained from the former steps. Finally, use the rectangle keys to retrieve data in the CDMs and then prune the unqualified data to get the query result.

k-NN Query

k-NN query is also commonly used in location based applications. Algorithm 3 shows the *k*-NN query algorithm in HBase and Cassandra, where K stores the result of *k* nearest neighbors, QueryRect stores the rectangles which could be scanned, dist is the range for the rectangle search, Rect$_{scanned}$ stores the rectangles that have been scanned, and the data structure of QueryRect is a queue. The *k*-NN query has two main parts: 1) set a range dist to search for rectangles which overlap a square range with centroid *p* and edge length 2·dist; 2) pick the nearest rectangle of *p* that is not scanned and add the nearest points in this rectangle into K. The algorithm keeps repeating steps 1 and 2 until the distance of the k-th nearest point and *p* is less than or equal to dist. Part 1) in Algorithm 3 is in lines 6–11, where RectInRegion() is used to find the rectangles in the square range and line 9 pushes the rectangles that have not been scanned into QueryRect; 2) is in lines 12–18, where line 12 pops the nearest rectangle, and line 14 will add the points of R into K. The function RectInRegion(*c*, dist) in Algorithm 4

Algorithm 3 *k*-NN Query

Input: *k*: *k* nearest neighbors; $p = (x, y)$: query point;
Output: *k* nearest neighbors of (x, y);

1: K← ϕ;
2: QueryRect ← ϕ;
3: dist ← 0;
4: Rect$_{scanned}$ ← ϕ;
5: **loop**
6:      **if** QueryRect== ϕ **then**
7:          Rect$_{next}$←$RectInRegion$(p,dist)−Rect$_{scanned}$;
8:          **for each** Rectangle R∈Rect$_{next}$ **do**
9:              Push(R, MinDist(p, R), QueryRect);
10:         **end for**
11:     **end if**
12:     R ←Pop(QueryRect);
13:     **for each** Point t ∈R **do**
14:         K ← K ∪ ¡t, Dist(p, t)¿ and sort K by dist;
15:     **end for**
16:     **if** dist(*k*-th point in K, p) ≤ dist **then**
17:         break;
18:     **end if**
19:     Rect$_{scanned}$ ←Rect$_{scanned}$∪R;
20:     dist←Max(dist, MaxDist(p, R));
21: **end loop**
22: return K;

Algorithm 4 *RectInRegion(p,dist)*

Input: $p = (x, y)$: query point; dist: means a square with edge length 2·dist and with p as its centroid $o = order$: the order of Hilbert;

Output: the keys of rectangles overlap with the input rectangle

1: RectKeys ← ϕ;
2: xl ←(x-dist) mod o;
3: xh ←(x+dist) mod o;
4: yl ←(y-dist) mod o;
5: yh ←(y+dist) mod o;
6: **for** i=xl → xh **do**
7:     **for** j=yl → yh **do**
8:         GridKeys ← GridKeys ∪ Hilbert(i, j);
9:     **end for**
10: **end for**
11: **return** RectKeys←KeyTable(GridKeys);

---

finds the rectangles which overlap with the input square. It is designed by our method for defining the key in the rectangles. Lines 6–8 find the grid keys which overlap the squares, and line 11 returns the rectangles which overlap grids through checking the KeyTable.

As shown in Fig. 6, take p as the query point, k = 3 and given an initial dist=0. First, we will get a rectangle 36 through the KeyTable with a square range of length 2·dist and then insert the location points $\{p_9, p_{10}, p_{11}\}$ of rectangle 36 into K. In that location points are ordered by the distance from p. Second, resize the dist to the minimum distance of k-th/ | K | -th location points in K from p, the dist=dist(3rd location point in K, p) in this example. The algorithm continues the first and second step; it will add the rectangle 55 into Rect$_{next}$ and add the location points in rectangle 55 into K. The algorithm is stopped by dist(3rd location point in K, p) ≤ dist, and we get the first three location points $\{p_{10}, p_9, p_8\}$ in K as the query result.

Experiments

In this section, we show the experiments about the time of range query and k-NN query in HBase and Cassandra with different methods: scan databases, Hilbert curve and our method, KR+, and compare the KR+HBase with MD-HBase (Nishimura et al. 2011).

We implement our experiments using HBase 0.20.6 with Hadoop 0.20.2 and Cassandra 0.8.2 as the underlying system. We have 8 machines, each consisting of 2 virtual machines, 2GB memory and 500GB HDD and 64bit Ubuntu 8.04.4. The arguments in our experiments with (M, m) = (1250, 625), (2500, 1250) and (5000, 2500) of KR+ are about 352, 176 and 88 rectangles respectively. The number of grids is divided into 24 × 24, 25 × 25 and 26 × 26. The ranges of the range query are 1 km × 1 km, 5 km × 5 km, 10 km × 10 km, 20 km × 20 km and 40 km × 40 km. The k-NN query with k = 10, 50, 100. The datasets have 440,912 GPS location points collected by our

Carweb, which is a data collection machine. The data collected by Carweb is extremely skewed. We also generate a uniform dataset that has 440,912 GPS location points. In the following we show the different results with different data distributions. Furthermore, we generate the different sizes of data points varying from 200,000 to 1,000,000 GPS location points to study the scalability of the proposed index method.

Before evaluating our method by comparing it with others, we have done some experiments on HBase and Cassandra. It is necessary to find the features of these CDMs, and we designed the index structure according to these features. We observe that it is more efficient to fetch a set of keys continuously than to fetch a single key repeatedly, and it has bad performance when one key stores too much data for these CDMs. Figure 7a shows the evaluation between scanning a set of data once and getting one key many times which indicates that scanning is quite outstanding. Figure 7b shows that the response time increases rapidly when the number of data n is increased from 25600 to 51200.

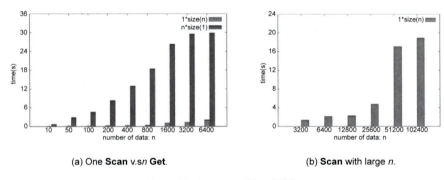

(a) One **Scan** v.s*n* **Get**.   (b) **Scan** with large *n*.

Fig. 7. The features of the CDMs.

Color image of this figure appears in the color plate section at the end of the book.

Tables 2 and 3 show the range query and the *k*-NN query on Cassandra respectively. We compare our KR⁺ with the Hilbert curves and no index. With there is no index, we scan the databases to find the location points in the query. The Hilbert curve with order 4, 5, 6 is uniformly dividing the map along the x-axis and y-axis into 24 × 24, 25 × 25 and 26 × 26. The method of Scan DB is obviously very slow, about 105s to 203s for the range query and 105s to 127s for the *k*-NN query. The Hilbert curve method for the range query is much faster than scan databases, the fastest for the range 1 km × 1 km is 4s and 40 km × 40 km is 9.4s with an order of 4. The time increases as the order of Hilbert curve increases since the number of sub-queries increases as the order increases. Our KR⁺ with order 4 grids is much faster than the Hilbert curve since it has the feature of balancing the number of

false-positives and the number of sub-queries. We observe that the KR⁺ with
M = 1250 and m = 625 is slower than with M = 2500 and m = 1250 since the
former has more sub-queries, and the KR⁺ with M = 5000 and m = 2500 is
slower than with M = 2500 and m = 1250 owing to the former having more
false-positives. Thus, the KR⁺ with arguments about M = 2500 and m = 1250
is closest to the target of the trade-off between the number of false-positives
and the number of sub-queries. The *k*-NN query has the same effect with
the range query in the Hilbert curve and KR⁺, but it is slower than the range
query because it may need to resize the distance of searching rectangles and
query again as the number of location points is less than k.

Table 4 and Table 5 show the range query and *k*-NN query on HBase.
We compare KR⁺ with the Hilbert curve and also with no index. As there

Table 2. Cassandra-Range Query (second).

km^2	1·1	5·5	10·10	20·20	40·40
KR⁺: M1250 m625	0.118	0.696	1.221	2.003	4.899
KR⁺: M2500 m1250	0.007	0.036	0.206	0.642	1.734
KR⁺: M5000 m2500	0.234	0.776	1.454	1.665	3.748
Hilbert: order4	4	7.8	7.2	7	9.4
Hilbert: order5	5	8.9	9.1	9.1	11.3
Hilbert: order6	7.6	13.7	13.5	13.7	19.8
Scan DB	105	110	127	155	203

Table 3. Cassandra-*k*-NN Query (second).

k	10	50	100
KR⁺: M1250 m625	0.675	1.387	1.668
KR⁺: M2500 m1250	0.039	0.088	0.657
KR⁺: M5000 m2500	0.433	0.944	1.329
Hilbert: order4	8	11.201	14.35
Hilbert: order5	10.348	15.455	19.349
Hilbert: order6	14.6	18.866	24.545
Scan DB	105	110	127

Table 4. HBase-Range Query (second).

km^2	1·1	5·5	10·10	20·20	40·40
KR⁺: M1250 m625	1.446	5.172	6.290	10.713	13.114
KR⁺: M2500 m1250	1.019	2.656	6.481	12.810	18.552
KR⁺: M5000 m2500	3.508	5.129	10.895	14.998	21.617
Hilbert: order4	44.65	47.93	50.04	52.16	56.47
Hilbert: order5	48.82	50.99	54.77	59.83	67.12
Hilbert: order6	54.10	57.31	62.50	66.32	79.51
Scan DB	135.06	134.99	135.08	134.84	135.01

Table 5. HBase-*k*-NN Query (second).

k	10	50	100
KR⁺: M1250 m625	2.13	3.558	7.729
KR⁺: M2500 m1250	1.679	2.896	8.354
KR⁺: M5000 m2500	6.732	80.24	14.2
Hilbert: order4	54.282	60.04	67.31
Hilbert: order5	57.354	64.13	73.52
Hilbert: order6	67.492	70.458	77.341
Scan DB	135.12	135.250	135.22

is no indexing, it needs to scan all of the data then filter out points which do not match the query. The Hilbert curve is also uniform dividing with order 4, 5, 6. As we have seen, the no index method has performs badly due to scanning all of the data. The Hilbert curve is much better than no index both for range query and *k*-NN query. When the order is 4, the time for the range query of 1 × 1 km is 44.6s, and the time for the range query of 40 km × 40 km km is 56.4s. When the order is 4 and k = 10, the time for the *k*-NN query is about 54.2s. With the same evaluation of Cassandra, KR⁺ is faster than the Hilbert curve and when M = 2500, m = 1250 the KR⁺ will get the best performance. The KR⁺ with M = 2500 and m = 1250 of range query for 1 km × 1 km is about 1.01s and *k*-NN query for k = 10 is about 1.67s.

Finally, we compare KR⁺ with MD-HBase for skewed data and uniform data. Figure 8a and b show range query and *k*-NN query for uniform data. KR⁺ slightly improves the efficiency compared with MD-HBase for uniform data. KR⁺ with M = 2500 and m = 1250 for 1 km × 1 km range query is about 1.2s and *k*-NN is about 2.0s when k = 10. In addition, with skewed data, Fig. 9a and b show that KR⁺ is much faster than MD-HBase. KR⁺ could balance the number of false-positives and the number of sub-queries so that it improves the efficiency of range query and *k*-NN query markedly. KR⁺ for range query 1 km × 1 km is about 1.0s and *k*-NN query when k =

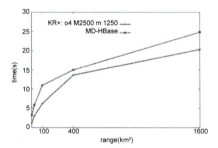
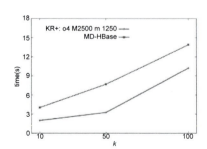

Fig. 8. Uniform data distribution.

Color image of this figure appears in the color plate section at the end of the book.

10 is about 1.7s. However, MD-HBase for range query 1 km × 1 km is about 6.2s and k-NN query when $k = 10$ is about 8.2s. The result of the evaluation shows that our KR⁺ has much better performance for range query and k-NN query. Besides, KR⁺ overcomes the trade-off between the number of points for getting one key and the number of keys for scanning so that it is more efficient than MD-HBase, especially for skewed data.

In addition, we study the scalability of the proposed index method by varying the data size (i.e., the number of points) in Fig. 10. In the experiments, we set the parameters as $M = 250$, $m = 125$, $o = 8$, and $k = 1000$. As shown in the experiments, the response time of the range query (or k-NN query) slightly increases as the data size is increased from 200,000 to 1,000,000. The response time increases as the data size increases because more points would need to be fetched.

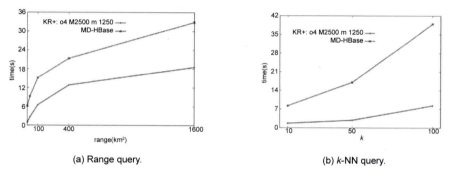

(a) Range query.   (b) k-NN query.

Fig. 9. Skewed data distribution.

Color image of this figure appears in the color plate section at the end of the book.

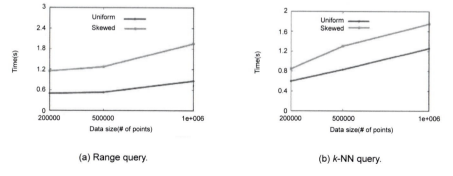

(a) Range query.   (b) k-NN query.

Fig. 10. Effect of data size.

Color image of this figure appears in the color plate section at the end of the book.

Related Work

We first introduce the data model and the basic operations of HBase and Cassandra. We then present the traditional multi-dimensional indexing techniques, linearization and index trees. In addition, we illustrate the existing multi-dimensional indexing techniques, RT-CAN and MD-HBase, developed for CDMs.

HBase and Cassandra

Data Model

HBase and Cassandra adopt the BigTable-like data model, which is a column-oriented data model. The data model of BigTable is constituted of columns, where each column is expressed as (name, value, timestamp), where a timestamp is an updated time of a column. For HBase, the data is structured in tables, row keys, column families, and columns. Specifically, each table comprises row keys and column families; and each column family contains one or more columns; each row consists of a key and columns mapped by the key. Given the data in Table 7, HBase stores the data as Table 6. For example, for key p_1, the corresponding data include three column families, and the second column family comprises two columns whose names are rest.name and rest.lng. In addition, for each key in HBase, the corresponding data can own multiple versions, identified by different timestamps. For instance, restaurant p_1 was checked by user u_1 and user u_2 on 8 May 2011 and on 8 August 2011, respectively. In addition, in HBase, the keys and columns' names are stored in lexicographical order.

Compared with HBase, the data stored in Cassandra are structured in two ways, column family or super column family. In Cassandra, a column family consists of keys and columns mapped by the keys, and a super column family consists of keys and the corresponding super columns, in which each super column is expressed by (super column name, columns). For instance, given Table 7, the data in Cassandra can be stored as Table 8a and b, which are column families. In addition, an example of a super column

Table 6. A table of HBase.

keys	timestamp	column family: rest.name	column family: rest.lan	column family: rest.lng	column family: uid
p_1	2011/05/08	Friday	24.805	120.995	u_1
	2011/08/08	Friday	24.805	120.995	u_2
p_2	2011/08/30	McDonald's	24.794	121.002	u_2
	2011/10/10	McDonald's	24.794	121.002	u_3
p_3	2011/11/07	KFC	24.794	121.005	u_4

Table 7. An example of check-in records.

cid	rid	rest.name	rest.lat	rest.lng	uid	Time
1	p_1	Friday	24.805	120.995	u_1	2011/05/08
2	p_1	Friday	24.805	120.995	u_2	2011/08/08
3	p_2	McDonald's	24.794	121.002	u_2	2011/08/30
4	p_2	McDonald's	24.794	121.002	u_3	2011/10/10
5	p_3	KFC	24.794	121.005	u_4	2011/11/07

Table 8. Data in Cassandra.

(a) A column family for the simplified check-in information.

keys	columns	
	name	value
1	rid	p_1
	uid	u_1
	time	2011/05/08
2	rid	p_1
	uid	u_2
	time	2011/08/08
3	rid	p_2
	uid	u_2
	time	2011/08/30
4	rid	p_2
	uid	u_3
	time	2011/10/10
5	rid	p_3
	uid	u_4
	time	2011/11/07

(b) A column family for restaurants' information.

keys	columns	
	name	value
p_1	name	Friday
	lat	24.805
	lng	120.995
p_2	name	McDonald's
	lat	24.794
	lng	121.002
p_3	name	KFC
	lat	24.794
	lng	121.005

family is illustrated in Table 9. Similarly, in Cassandra, the keys, columns' names, and super column family names are stored in lexicographical order. In addition, for HBase and Cassandra, the data of columns are distributed on servers. The difference between Cassandra and HBase is that Cassandra allows a column or a super column to be added arbitrarily, but a column family in HBase cannot be added arbitrarily after a table has been created.

The data model of RDBMSs are different to the data model of HBase/ Cassandra. The data stored in RDBMSs are structured in tables, fields and records. Specifically, each table consists of records and each record consists of one or more fields. Because RDBMSs guarantee the ACID properties, i.e., atomicity, consistency, isolation, and durability, RDBMSs are not scalable to support large data. For instance, if there are multiple records to be updated in a single transaction, multiple tables will be locked for modification. If

Table 9. A key mapping for restaurants.

keys	super columns		
	super columns' name	columns	
		name	value
1	p_1	name	Friday
		lat	24.805
		lng	120.995
3	p_2	name	McDonald's
		lat	24.794
		lng	121.002
	p_3	name	KFC
		lat	24.794
		lng	121.005

those tables are spread across multiple servers, it will take more time to lock the tables, update the data and release the locks. However, for CDMs, making data consistent is easier due to the data being stored on multiple servers. In addition, CDMs should ensure the CAP (Gray 1981) theorem, stating that a distributed system satisfies at least two of the three guarantees: Consistency, Availability and Partition tolerance, at the same time. HBase and Cassandra guarantee CP and AP, respectively. Compared with RDMBSs, CDMs can handle scalable data well.

Basic Operations

Based on the BigTable-like's data model, HBase and Cassandra develop new basic operations for reading, writing, updating and deleting data. Different from the language used in RDBMSs', i.e., structured query language (SQL), HBase and Cassandra provide key-value based queries that retrieve a record by specifying keys. The basic operations in HBase/Cassandra are described in detail as follows. Note that these operations are performed by given keys, column names or super column names.

In HBase, there are four primary basic operations as follows: **get**: returns attributes for a specified row; **set**: either adds new rows to a table (if the key is new) or updates existing rows (if the key already exists); **scan**: allows iterations over multiple rows for specified attributes; and delete: removes a row from a table.

In Cassandra, the basic operations are as follows: **get**: gets the column or super column at the given column's name or super column's name in a column family; **get_slice**: gets the group of columns contained by a column family name or a column family/super column name pair specified by the given columns' names or a range of columns' names; **multigetslice**:

retrieves a list of map <key, columns> or <key, super columns> for specific columns' names or super columns' names in a column family on each of the given keys; **get_range_slices**: returns a list of map <key, columns> or <key, super columns> within the range of keys in a column family; insert: inserts a column in a column family or a super column family; **batch_mutate**: inserts or removes the rows, the super columns or the columns from the row specified by keys; and **remove**: removes data from the row specified by a key in a column family or a super column family. The data could be an entire row, a super column or a column.

However, these operations could not support the operation of retrieving rows given two or more restrictions. For instance, in Table 6, we could not retrieve the result that satisfies "satisfyrest.name=Friday" and "uid=u_1" using only one operation from the basic operations provided by HBase or Cassandra. Although we could retrieve rows by scanning all rows and post-filter unqualified data to get the result, it is time-consuming.

Multi-dimensional Index

Due to the high scalability of cloud data managements, there have been more and more works for constructing indexes on cloud data managements recently. The B-tree is a commonly used index structure. The work in (Wu et al. 2010) presented a scalable B-tree based indexing scheme which builds a local B-tree for the dataset stored in each compute node and a Cloud Global index, called the CG-index, to index each compute node. However, the B-tree index can not support multi-dimensional queries effectively. Besides, much of the work on R-tree index structures for multi-dimensional data had been done, such as (Wang et al. 2010; Zhang et al. 2009; Liao et al. 2010). Wang et al. (Wang et al. 2010) present RT-CAN, a multi-dimensional indexing scheme. RT-CAN is built on top of local R-tree indexes and it dynamically selects a portion of local R-tree nodes to publish onto the global index. Although it uses R-tree indexing, it builds the R-tree on the authors' own distributed system epiC. The work (Zhang et al. 2009) combined R-tree and k-d tree to be the index structures and the work in (Liao et al. 2010) presented an approach to construct a block-based hierarchical R-tree index structures. These works all build an index structure on the Hadoop distributed file system or Google's file system to support multi-dimensional queries.

MD-HBase (Nishimura et al. 2011; Nishimura et al. 2013) is a data management system, based on HBase, using Quad trees and k-d trees coupled with Z-ordering to index multi-dimensional data for LBSs. The keys of MD-HBase are the Z-values of the dimensions being indexed. It uses the trie-based approach for splitting equal-sized space, and builds Quad tree and k-d tree index structures on the key-value data model. Moreover, MD-HBase proposed a novel naming scheme, called longest common prefix naming, for efficient index maintenance and query processing. Although the experiment of MD-HBase shows that the proposed indexing method is efficient for multi-dimensional data, it has some constraints. Before describing these constraints, we have discovered a characteristic of cloud managements for data access through experiments. A trade-off exists between the number of points for getting one key and the number of keys for scanning; a reduction in the number of points for getting one key results in an increase in the number of keys for scanning and vice versa. The way of splitting the space of the Quad tree and k-d tree is fixed, which may make some nodes store zero points. In addition, the Quad tree and the k-d tree cannot balance the number of stored points for each node, because they do not restrict the minimum number of points in one space. Therefore, if we regard one node as one key, it will make the keys store unbalanced data points, especially as the data is not uniform. Figure 11 is a Quad tree example of space splitting for MD-HBase. According to the data points in the map, the Quad tree will split the whole space into three. The red line shows the splitting results, and each black grid has its Z-ordering value. For instance, the Z-ordering value of (0,0) is 000000 and (1,0) is 000010. Then, the key of each region split by the red line is the prefix of the Z-ordering value of its sub-regions. Consequently, there are 10 keys, 000000, 000001, 000010, 000011, 0001*, 0010*, 0011*, 01*, 10* and 11*. However, there may be no data points in some regions. As mentioned above, the Quad tree and k-d tree cannot deal with multiform distribution data efficiently.

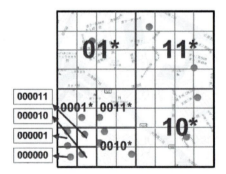

Fig. 11. A key formulation in MD-HBase.

Color image of this figure appears in the color plate section at the end of the book.

Conclusions

We proposed a scalable multi-dimensional index, KR+-index, based on the existing CDMs, such as HBase and Cassandra. It supports efficient multi-dimensional range queries and nearest neighbor queries. We use R+ to construct the index structure and have designed the key for efficient data access. In addition, we have redefined the spatial query algorithm, including range query and k-NN query for our KR$^+$. KR$^+$ took the characteristics of these CDMs into account so that it is shows much more efficient than other index methods in experimentation. Our evaluation using a cluster of 8 nodes handles the range queries and k-NN queries efficiently, and we also compared it with the state-of-the-art, MD-HBase, and the result showed that KR$^+$ has better performance than MD-HBase, especially for skewed data. In this chapter, we study the settings of the parameters (i.e., M, m and o) of the proposed index method in an empirical way. However, in the real world, there are different data distributions (such as skewed data) and different data sizes. To achieve a better performance, the parameters of the proposed index method may need to be changed when the data distribution or the data size varies. To apply the proposed index method to real-world data well, we will study how to set proper parameters of the proposed index method with respect to different data distributions and different data sizes in the future. In addition, ad-hoc queries in the real world would have various ranges, and different query rectangles would affect the performance. Thus, we will study the effect of different query rectangles on the performance and study proper parameter settings for ad-hoc queries.

References

Beckmann, N., H.P. Kriegel, R. Schneider and B. Seeger. 1990. The r*-tree: an efficient and robust access method for points and rectangles. Proc. of the 1990 ACM SIGMOD international conference on Management of data. 322–331.

Bentley, J.L. 1975. Multidimensional binary search trees used for associative searching. Communications of the ACM. 18(9): 509–517.

Chang, F., J. Dean, S. Ghemawat, W.C. Hsieh, D.A. Wallach, M. Burrows, T. Chandra, A. Fikes and R.E. Gruber. 2008. Bigtable: A distributed storage system for structured data. ACM Transactions on Computer Systems. 26(2): 1–26.

Finkel, R.A. and J.L. Bentley. 1974. Quad trees a data structure for retrieval on composite keys. Actainformatica. 4(1): 1–9.

Gray, J. 1981. The transaction concept: Virtues and limitations. Proc. of the Very Large Database Conference. 144–154.

Guttman, A. 1984. R-trees: a dynamic index structure for spatial searching. Proc. of the 1984 ACM SIGMOD International Conference on Management of Data. 47–57.

Kamel, I. and C. Faloutsos. 1994. Hilbert r-tree: An improved r-tree using fractals. Proc. of the 20th International Conference on Very Large Data Bases. 500–509.

Khetrapal, A. and V. Ganesh. 2008. Hbase and hypertable for large scale distributed storage systems. Dept. of Computer Science, Purdue University.

Lakshman, A. and P. Malik. 2010. Cassandra: a decentralized structured storage system. ACMSIGOPS Operating Systems Review. 44(2): 35–40.

Liao, H., J. Han and J. Fang. 2010. Multi-dimensional index on hadoop distributed file system. Proc. of the Fifth IEEE International Conference on Networking, Architecture and Storage. 240–249.

Nishimura, S., S. Das, D. Agrawal and A. El Abbadi. 2011. MD-HBase: A scalable multidimensional data infrastructure for location aware services. Proc. of the 12th IEEE International Conference on Mobile Data Management. 7–16.

Nishimura, S., S. Das, D. Agrawal and A. El Abbadi. 2013. MD-HBase: design and implementation of an elastic data infrastructure for cloud-scale location services. Distributed and Parallel Databases. 31(2): 289–319.

Sellis, T., N. Roussopoulos and C. Faloutsos. 1987. The r$^+$-tree: A dynamic index for multidimensional objects. Proc. of the 13th International Conference on Very Large Data Bases. 507–518.

Stonebraker, M. 2010. Sql databases v. nosql databases. Communications of the ACM. 53(4): 10–11.

Varia, J. 2008. Cloud architectures. Technical report, Amazon Webservices.

Wang, J., S. Wu, H. Gao, J. Li and B.C. Ooi. 2010. Indexing multi-dimensional data in a cloud system. Proc. of the 2010 ACM SIGMOD International Conference on Management of Data. 591–602.

Wu, S., D. Jiang, B.C. Ooi and K.L. Wu. 2010. Efficient b-tree based indexing for cloud data processing. Proceedings of the VLDB Endowment. 3(1–2): 1207–1218.

Zhang, X., J. Ai, Z. Wang, J. Lu and X. Meng. 2009. An efficient multi-dimensional index for cloud data management. Proc. of the First International Workshop on Cloud Data Management. 17–24.

CHAPTER 4

NoSQL Geographic Databases: An Overview

Cláudio de Souza Baptista,[a,][*] *Carlos Eduardo Santos Pires,*[a]
Daniel Farias Batista Leite,[a] *Maxwell Guimarães de Oliveira*[a]
and *Odilon Francisco de Lima Junior*[b]

Introduction

Web 2.0 refers to a new form of using the World Wide Web in which users are encouraged to add value to applications as they operate them (Amer-Yahia and Halevy 2007). Examples of such applications include blogs, wikis, and social networks. The amount of data has increased exponentially, as well as the need to store, retrieve and manipulate it effectively. As a result, databases of a size that have never existed before have emerged. These databases can no longer be contained in one physical system, but must run in a distributed system.

Traditional Relational Database Management Systems (RDBMSs) have reached their limitations as they face the problem of flexible scaling required for Web 2.0 applications (Cattell 2010). In fact, it is possible to scale a relational database; however, the costs are very high. For instance,

---

[a] Information Systems Laboratory (LSI), Department of Systems and Computing, Federal University of Campina Grande (UFCG), Av. AprígioVeloso 882, Bloco CN, BairroUniversitário–58.429-140, Campina Grande–PB–Brazil.
Emails: baptista@dsc.ufcg.edu.br, cesp@dsc.ufcg.edu.br, daniel.leite@ccc.ufcg.edu.br, maxwell@ufcg.edu.br
[b] Center for High Education of Seridó (CERES), Department of Mathematical and Applied Sciences, Federal University of Rio Grande do Norte (UFRN), RuaJoaquimGregório, S/N, Penedo–59.300-000, Caicó–RN–Brazil.
Email: odilonflj@gmail.com
[*] Corresponding author

it is necessary to acquire powerful servers, pay the salaries of database administrators, and invest in caching technologies such as Memcached.[1] As a result, new systems that can support such requirements have been proposed. An implicit requirement of these systems is to scale to a large number of nodes without degrading the quality of the services they deliver and with a low budget.

These systems were called NoSQL systems (or data stores). Currently, they are becoming increasingly popular among Web companies. The NoSQL systems represent an alternative solution for the needs of modern interactive software systems. They are designed to scale to thousands or millions of users performing simple data operations such as updates and reads. Proponents of NoSQL systems state that these systems provide simpler scalability and improved performance in contrast to traditional relational database systems. These are crucial features for Web 2.0 companies such as Facebook, Amazon, and Google. In addition, NoSQL systems excel at storing large amounts of unstructured data known as Big Data (Floratou et al. 2012). A recent report released by the Gartner team identifies NoSQL systems as one the top 10 technology trends that will have impact on information management (Gartner 2013). Some examples of NoSQL systems include MongoDB, CouchDB, BigTable, Cassandra, and Neo4j.

In principle, NoSQL systems were proposed to manage large amounts of non-spatial data. However, the spatial dimension has proven to be of particular interest in Web 2.0 applications (Schutzberg 2011). This can be observed with the emergence of location-based applications (e.g., Foursquare)[2] and others that allow users to geocode shared content (e.g., Twitter,[3] Facebook,[4] Flickr,[5] and Panoramio[6]). All of these applications use a NoSQL system in the background.

As a result, NoSQL systems began to allow the storage and retrieval of spatial data (Baas 2012). Some NoSQL systems include support for geospatial data either natively or with an extension. Others were not originally designed for geospatial applications but have been extended to support geospatial data. Examples of NoSQL systems that provide support for spatial data include: CouchDB, MongoDB, BigTable, and Neo4j. In this chapter, we have chosen to compare these four NoSQL systems, as they represent the main players that support spatial data. In the past, Cassandra coped with the spatial dimension, but that is not the case anymore.

---

[1] Memcached, http://memcached.org/
[2] Foursquare, http://foursquare.com/
[3] Twitter, http://twitter.com/
[4] Facebook, http://www.facebook.com/
[5] Flickr, http://www.flickr.com/
[6] Panoramio, http://www.panoramio.com/

The remainder of this chapter is organised as follows. The next section presents an overview of NoSQL systems. Then, the desirable characteristics of a NoSQL system with spatial data support are highlighted. The following sections focus on different NoSQL spatial systems: CouchDB, MongoDB, BigTable, and Neo4j. Finally, we conclude the chapter and compare the several systems addressed herein.

NoSQL Systems

NoSQL is a term in information technology that describes database management systems, which depart from classic Relational DBMS altogether, or in some parts. The term was first used in 1998 for lightweight open source databases that did not use SQL (Structured Query Language) as a database language (Strozzi 2007). The name "NoSQL" could indicate that these databases do not support SQL, but in this case it actually means "Not Only SQL" (NoSQL 2013). During the past years, the term has been widely used to designate a class of database systems that offer an alternative solution to the Relational Model (Codd 1970) in terms of availability, scalability, and performance for the management of large amounts of data. The goal of NoSQL systems is not to replace RDBMs as a whole, but to be used in cases in which more flexibility is required with respect to the database structure (Sadalage and Fowler 2012; Redmond and Wilson 2012).

A Brief History

Google's BigTable (Chang et al. 2008) and Amazon's Dynamo (DeCandia et al. 2007) seem to be the starting point for the NoSQL movement. Many of the design decisions and principles used in these systems can be found in later ones. Google's BigTable was launched in 2004 as a high-performance proprietary database offering scalability and availability. Its original proposal was to relax the structure used by the Relational Model. Amazon's Dynamo was launched in 2007 as a high-performance non-relational database system, to be used by Amazon Web Services. In 2008, Cassandra was presented as a highly scalable distributed database system, designed by Facebook developers to deal with large amounts of data (Lakshman and Malik 2009). Cassandra joined the distributed systems technologies from Dynamo and the data model from Google's BigTable. In the beginning of 2010, Twitter replaced MySQL with Cassandra (Lai 2010). At the same time, Apache CouchDB was launched, an open-source NoSQL system. It allowed applications to store data without having to adhere to the structure of a pre-defined schema (Anderson et al. 2009). In 2009, MongoDB (Chodorow and Dirolf 2010), a NoSQL system with characteristics similar to CouchDB, was released.

NoSQL Main Characteristics

NoSQL incorporates a wide range of different systems. In general, these systems have been created to solve a particular problem, which for various reasons RDBMSs have not been appropriate (Stonebraker 2010). A typical NoSQL system presents the following characteristics (Näsholm 2012):

- **Data partitioning**: NoSQL systems are often distributed systems where several nodes (or servers) cooperate to provide applications with data. Each piece of data is commonly replicated over several machines to allow redundancy and availability;
- **Horizontal scalability**: nodes can often be dynamically added to (or removed from) the system without any downtime, giving linear effects on storage and overall processing capacities. In general, there is no (realistic) upper bound on the number of nodes that can be added;
- **Built for large volumes**: most NoSQL systems were built to be able to store and process large amounts of data quickly;
- **Lack of schema definition**: the structure of data is usually not defined through explicit schemas. Instead, applications store data as they desire, without having to adhere to a predefined structure. This provides application flexibility, which ultimately delivers substantial business flexibility;
- **Simple API for data access**: NoSQL systems provide APIs to simplify data access, allowing applications to quickly manipulate data;
- **Eventual consistency**: data consistency is not always preserved among nodes in a cluster. This characteristic is based on the Brewer's CAP theorem (Brewer 2000) which states that a distributed system can have at most two of the three properties: Consistency, Availability and Partition tolerance. For a growing number of applications, having the last two properties is most important. Building a database with these properties while providing ACID properties—Atomicity, Consistency, Isolation and Durability— is difficult. That is why Consistency and Isolation are often forfeited, resulting in the well-known BASE approach (Pritchett 2008). BASE stands for Basically Available, Soft state, Eventual consistency, with which one means that the application that is available basically all the time, is not always consistent, but will eventually be in some known state (Vogels 2009);

- **Non-relational data models**: data models vary, but generally, they are not relational. Usually, they allow for more complex structures and are not as rigid as the relational model.

NoSQL Categories

Although NoSQL systems have several common characteristics, they also present some particularities (Leavitt 2010). As a result, there have been various approaches to classify and subsume NoSQL databases, each with different categories and subcategories (Hoff 2009; North 2009). Although there is no universally accepted taxonomy for NoSQL systems, often suggested taxonomies divide these systems, at least partly, based on their data models. For instance, the data model-based taxonomy proposed by (Cattell 2010) is composed of four families:

- Key-value NoSQL systems: store values indexed for retrieval by keys. They are highly scalable and offer a good performance, but are difficult to query and to implement real-world problems. Example: Amazon Dynamo;
- Document-oriented NoSQL systems: store and organize data as collections of documents, rather than as structured tables with uniform-sized fields for each record. Any number of fields of any length can be added to a document. The documents are indexed and a simple query mechanism is provided. Examples: CouchDB and MongoDB;
- Column-based NoSQL systems: store extensible records that can be partitioned vertically and horizontally across nodes. They are also known as "wide column systems". Examples: Cassandra and BigTable;
- Graph-oriented NoSQL systems: store objects and relationships in nodes and edges of a graph. For situations that fit this model, like hierarchical data, this solution could be faster than the other ones. Example: Neo4j.

NoSQL Geographic Data: Desirable Features

We highlight some desirable features of spatial database systems, aiming to enable a comparison between NoSQL geographic systems. We choose these features based on the ISO SQL/MM (Stolze 2003). The features analyzed are:

i. **Basic concepts**: we are interested whether the NoSQL system provides any support for using different spatial reference systems (SRID); whether reprojection can be performed. It is important to have different systems to choose from, according to application requirements. Another important issue concerns the multi-dimensionality of data

(1D, 2D, 3D). Finally, we would like to find out the internal data model used by the analyzed NoSQL system, whether it is based on graph, document, key-value, or column;

ii. **Indexing**: indexing techniques are known to speed up query processing. It has been argued in the literature that traditional indexing techniques, such as B-Tree, B+-Tree, and hashing, are not suitable for multidimensional spatial data. Therefore, special indexing techniques for spatial data have been proposed such as R-tree, R*-tree, quadtree, and gridfiles. It is important to know how spatial indexing is performed in such NoSQL spatial systems;

iii. **Vector Data types**: the Open Geospatial Consortium (OGC)[7] has proposed a standard for vector data types, known as OpenGIS Implementation Standard for Geographic Information—Simple Feature Access for 2D. This standard addresses the data types and the corresponding spatial functions.

In that OGC standard, the Geometry class has several functions as presented in Fig. 1, e.g., dimension, equals, and distance. In order to simplify our comparison, we have separated the spatial functions into three categories: topological, analysis, and set;

iv. **Topological functions:** topological functions are invariant concerning scale, rotation, and zooming. In our reference model we are interested in the following topological functions:

- equals(theGeom): returns TRUE if the given geometry (theGeom) represents the same geometry; and FALSE, otherwise;
- disjoint(theGeom): returns TRUE if there is no intersection between the geometries, i.e., the geometries do not share any common space;
- intersects(theGeom): returns TRUE if the geometries spatially intersect; and FALSE, otherwise;
- touches(theGeom): returns TRUE if the geometry boundaries intersect, but their interiors do not intersect; and FALSE, otherwise;
- crosses(theGeom): returns TRUE if the geometry interiors intersect but are not equal; and FALSE, otherwise;
- within(theGeom): returns TRUE if the geometry theGeom is completely inside the other geometry; and FALSE, otherwise;
- contains(theGeom): returns TRUE if no points of theGeom lie in the exterior of the other geometry, and at least one point of the interior of theGeom lies in the interior of the other geometry; and FALSE, otherwise;
- overlaps(theGeom): returns TRUE if the geometries share space, are of the same dimension, but are not completely contained by each other; and FALSE, otherwise.

---

[7] OGC, http://www.opengeosatial.org/

Fig. 1. OGC geometry class.

v. **Analysis and Metric Functions:** these functions enable to perform some spatial analysis on the data, including:

- length(): returns the length of the geometry (linestring or multilinestring);
- area(): returns the area of the geometry (polygon or multi-polygon);
- distance(theGeom): returns the distance between the geometries;
- buffer(distance): returns a geometry that represents all points whose distance from this geometry is less than or equal to distance;
- convexHull(): returns the minimum convex geometry that encloses all geometries within the set.

vi. **Set functions:** these functions are used to perform set oriented functions on spatial data, including:

- intersection(theGeom): returns a geometry that represents the intersection of the geometries;
- union(theGeom): returns a geometry that represents the union of the geometries;

- difference(theGeom): returns a geometry that represents the part of this geometry that does not intersect with theGeom;
- symDifference(theGeom): returns a geometry that represents the portions of the geometries that do not intersect.

vii. **Input/Output format:** it is important to know which data input and output formats the NoSQL system provides, including GML, WKT, WKB, SVG, JSON, and shapefile. GML (Geography Markup Language) is an Open Geospatial Consortium (OGC) standard format that uses XML to express geographical features. GML be also used as an open interchange format for geographic transactions on the Internet. WKT (Well-known Text) is also developed by OGC, represents geographical information as text. WKB (Well-Known Binary) is developed by OGC, WKB is similar to WKT, however it represents geographical information as binary. SVG (Scalable Vector Graphics) is an XML based vector format for two dimensional graphics proposed by the W3C. JSON (JavaScript Object Notation) is a lightweight data-interchange format based on two structures: a collection of name/value pair and an ordered list of values. Shapefile is a proprietary geospatial vector data format from the ESRI company. It is considered a *de facto* geospatial standard.

NoSQL Geographic Systems

Let us now focus on different implementations of NoSQL Geographic Systems. The following NoSQL Geographic Systems are described: CouchDB, MongoDB, Neo4j, and BigTable. In addition, the desirable features previously mentioned are addressed, so that a comparison among the analysed NoSQL systems can be made ahead.

CouchDB

CouchDB is a document-oriented NoSQL system based on schema-free documents, using a RESTful HTTP interface (Fielding 2000). CouchDB has been an Apache project since 2008, originally implemented in C++ but later ported to Erlang for concurrency reasons (Cattell 2010).

Main Features

In contrast to relational database systems, a database in CouchDB has no predefined schema; hence, its data structure can change dynamically. Basically, a CouchDB database is a collection of documents. A document is a JavaScript Object Notation (JSON) object formed by an unordered collection of name/value pairs. Each document is identified by a unique identifier.

CouchDB replicates and propagates data changes across network nodes. Among the options set by the CAP theorem (Brewer 2000), CouchDB prioritises high availability and partition tolerance.

CouchDB uses the MapReduce model (Lee et al. 2012), defined as JavaScript functions, to query (select and aggregate) documents. MapReduce enables the development of scalable parallel applications to process vast amounts of data on large clusters of commodity machines. The model relies on a Distributed File System (DFS) that is accessed by all nodes. Two functions are used, "map" and "reduce", which are applied to each document separately in such a manner that queries can be distributed over multiple nodes in parallel.

CouchDB takes advantage of the fact that these functions produce key/value pairs. These pairs are inserted in a B-Tree that is sorted by key. Thus, searching by either key or key range is extremely efficient.

Traditional relational database systems use locking, mainly the two-phase locking protocol (2PL), to ensure data consistency. In contrast, CouchDB uses Multi-Version Concurrency Control (MVCC) to manage concurrent access to data (Anderson et al. 2009). In MVCC, locks acquired for querying (reading) data do not conflict with locks acquired for writing data. As a result, reading operations never block writing operations and vice-versa. Documents in CouchDB are versioned. Every change in a document creates a new version of the entire document. After the modification, there are two versions of the same document: the old and the new version. Thus, every read operation will always see the latest version in the database.

CouchDB achieves *eventual consistency* between databases by incremental replication, in which document changes are periodically copied between servers. During the replication process, when CouchDB detects that a document has been modified in both databases being synchronized, the document is marked as being in conflict. CouchDB chooses a document to be considered the latest version, while the other document is labeled as the earlier version. Thus, both documents can be accessed after conflict resolution. The process of conflict resolution is done automatically in both databases. After synchronization, each node becomes independent and self-sufficient. CouchDB offers libraries for several programming languages such as Java, .Net, C, PHP, Python, Ruby, among others.

GeoCouch

GeoCouch[8] extends CouchDB to index and search geospatial data. GeoCouch was developed in Erlang and implements an R-Tree for indexing spatial

---

[8] GeoCouch, https://github.com/couchbase/geocouch/

data. Although R-Tree enables multi-dimensional indexing, GeoCouch implementation supports only two dimensions.

GeoCouch provides spatial views in CouchDB that enables CouchDB to perform geospatial queries, using bounding boxes.

CouchDB documents are stored in the JSON format (GeoJSON for documents with spatial data) and therefore allow representing Point, LineString, Polygon, MultiPoint, MultiLineString and MultiPolygon geometries.

GeoCouch has no internal utilities to import spatial data, but there are external tools like shp2geocouch[9] that creates GeoCouch databases from ESRI Shapefiles.

GeoCouch only allows bounding boxes search and do not provide different output formats beyond GeoJSON. However there are some CouchApps that offers extra features. The GeoCouch-Utils is a CouchApp that provides spatial functions and a set of scripts. For instance, the radius function (similar to buffer analysis functions) takes as input parameters a point and a radius in meters, and filters the bounding box result set by circular radius (only works on points). Furthermore, GeoCouch-Utils allows other output formats including: KML, CSV, and GeoJSON.

Other spatial features of GeoCouch are described below:

- SRID: the coordinate system is obtained from the input data (GeoJSON) and re-projection operations are not supported;
- Functions: GeoCouch does not support topological functions (equals, disjoint, intersect, touches, crosses, overlaps) or analysis functions (buffer, convexHull) or set functions (intersection, union, difference, symDifference).

MongoDB

MongoDB is an open source, document-oriented, NoSQL system written in C++. MongoDB was developed by the company 10gen.

Main Features

A database in MongoDB contains one or more collections of documents. Documents are stored in the BSON format (similar to JSON, but in a binary representation for efficiency reasons and to support additional data types).

---

[9] Shp2geocouh, http://github.com/maxogden/shp2geocouch

MongoDB documents may contain field and value pairs where the value can be another document, an array of documents or basic types such as Double, String, and Date.

Data in MongoDB has a schema-free model. Documents within a collection can be heterogeneous. MongoDB defines indexes on a collection level. Indexes can be created either on a single field or on multiple fields. All MongoDB indexes are implemented as a B-Tree data structure.

Database replication with MongoDB helps to ensure redundancy and high availability (failover). Replication occurs through groups of servers known as replica sets. Replica sets are groups of MongoDB nodes; one set is designated as the primary and the others are designated the secondary members. The primary node is responsible for writing operations, while the secondary members replicate data from the primary one asynchronously. If the primary node fails, one of the secondary members is elected automatically as a new primary node (automatic failover).

As such, MongoDB is strongly consistent. If clients change the setting to allow reading in secondary nodes, then MongoDB becomes eventually consistent where it is possible to read outdated results.

MongoDB supports horizontal scaling via automatic sharding. Sharding partitions a collection and stores the different parts on different machines. Sharding automatically implements load balancing across machines in the cluster.

Applications communicate with MongoDB through a client library or driver. There are MongoDB drivers available for the following programming languages: JavaScript, Python, Ruby, PHP, Perl, Java, Scala, C#, C, C++, Haskell, and Erlang.

MongoDB Geospatial Indexes

MongoDB provides the following geospatial index types to support geospatial queries: 2D and 2DSphere. Currently, MongoDB does not support 3D geospatial indexing.

The 2D indexes are used for data stored as points on a two-dimensional plane (Euclidean plane). A 2D index should not be used for data stored as GeoJSON objects. The 2D index supports calculations on a Euclidean plane and also distance-only calculations on a sphere, but for geometric calculations on a sphere, it is necessary to store data as GeoJSON objects and to use the 2DSphere index type.

The 2DSphere index supports all MongoDB geospatial queries: queries for inclusion (within), intersection, and proximity to a GeoJSON point. Thus, MongoDB only supports the following topological functions: within and intersects. There is no support for analysis functions (buffer, convexHull) or set functions (union, difference, symDifference).

The default coordinate reference system for an earth-like sphere (GeoJSON object) in MongoDB is the WGS84 datum.

MongoDB stores data in a binary JSON-like format called BSON. MongoDB supports the following GeoJSON objects: Point, LineString, Polygon, MultiPoint, MultiLineString, MultiPolygon, Feature, FeatureCollection, and GeometryCollection.

The following topics summarise the spatial characteristics of MongoDB:

- SRID: the coordinate system expected by MongoDB is WGS84. MongoDB does not provide re-projection operations;
- Functions: MongoDB provides support for topological functions (within and intersects). There is no support for analysis functions (buffer, convexHull) or set functions (intersection, union, difference, symDifference).

Neo4j

The Neo4j—Network-oriented database for Java—is an open source NoSQL graph database system developed using the Java Programming Language. Neo4j was officially launched in 2010 by Neo Technology.

Neo4j is called embeddable because it can be added to a Java application like any library. The data model adopted to express the stored information is based on the graph theory (Diestel 2005), comprised of nodes, properties and relationships eventually connecting some of the nodes.

The graph database is suitable for processing linked data. This type of database is usually found in applications related to areas such as telecommunications, logistics, social networks, bioinformatics, and detection of fraud (Hunger and Rathle 2012).

Main Features

Neo4j has a native storage manager completely optimised for storing graph structures to provide maximum performance and scalability. The system can handle graphs containing several billion nodes, relationships and properties stored on a single machine (Hoff 2009).

Neo4j provides a graph-oriented data representation rather than static tables with rows and columns. Therefore, it is possible to work with a network composed of flexible graph nodes, relationships, and properties. The edges of a graph represent relations and can be directed or not. The vertices are labeled and generally have properties represented by one or more key/value pairs, and can hold several edges simultaneously, each one representing a specific relationship with other vertices (Baas 2012).

The combination of nodes, relationships between them, and the properties of these nodes form a space node, a structured network representing a graph database.

Figure 2 illustrates an example of data stored in Neo4j. It is possible to see five identified nodes, their relationships, some properties like "name", "address.lat", "address.lon", and spatial attributes. In addition, the figure depicts properties in relationships like "knows" that connects the fourth node to the second node and has the properties "disclosure" and "time". Another relationship "knows", connects the second node to the third node with the property "disclosure". The link labeled "knows" represents a direct relationship that depicts which people know each other. From this example, it is possible to observe, for instance, that Bob (whose age is unknown) is the boyfriend of Kate (age 31) and resides in a location where the coordinates (lat, long) are known, and he also knows Peter (age 52).

The graph shown in Fig. 2 makes it clear that both nodes and relationships can have different properties. The properties are strings formed by key/value pairs and their values can be any Java primitive type or an array of primitive Java types such as int, int [], String, float, float [], and so on.

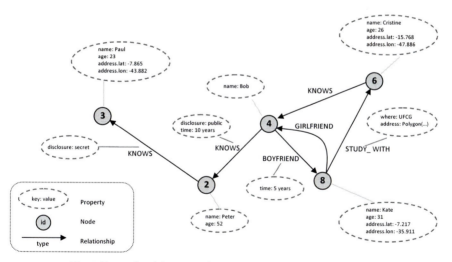

Fig. 2. Example of data stored in Neo4j (adapted from Bass 2012).

Neo4j does not support declarative queries. Instead, the search for information starts from either an initial node that may be provided or an arbitrary node, and follows the relationships to other nodes that satisfy the search criteria. This navigation is known as traversing. Neo4j provides an

API that allows the programmer to specify the traversal rules. The returned result can be empty, a single node, or a subgraph composed of nodes and their relationships interconnecting them.

Regarding the development platform, in addition to Java, Neo4j has links with the Python, Jython, Ruby and Clojure programming languages. For other platforms, Neo4j provides access through a REST interface (Hoff 2009).

Neo4j Spatial

Neo4j supports the storage and manipulation of spatial data through the library Neo4j Spatial,[10] developed in Java. The Neo4j Spatial enables spatial operations on data stored in Neo4j, for instance, operations to locate data within a specific region or an area close to a point of interest (PoI). Furthermore, it allows integration with the GeoTools library,[11] GeoServer,[12] and uDig.[13]

Neo4j Spatial may import spatial data in both ESRI Shapefile (.SHP) and Open Street Map (.OSM) formats. The first type that defines a geometry collection is a Layer type. Each layer has an index for queries and can be presented as an EditableLayer to edit its geometry. The DefaultLayer pattern uses geometric WKB functions for storing all types of geometry as properties of the type byte []. There is also a layer named OSMLayer that supports map files from Open Street Map and stores the OSM model as a single fully connected graph. The OSMLayer supports geometries of type Point, Linestring and Polygon, but has the restriction of supporting only one geometry by layer. Multi-geometries from OSM files are considered to be multi-layered. On the other hand, the layer containing data from a shapefile supports Point, Linestring, Polygon, Multipoint, Multilinestring, and Multipolygon geometries.

An interesting feature of the Neo4j Spatial library is the ability to divide a single layer into multiple sub-layers or views through the use of pre-configured filters. This ability can be interesting when working with large datasets.

The Neo4j Spatial library contains the Java Topology Suite, an API written in Java that implements geometries and 2D geometric operations. Thus, it is possible to use the full capacity of the JTS (Java Transaction Service) in geometric operations with instances obtained from Neo4j Spatial.

---

[10] Neo4j Spatial, https://github.com/Neo4j/spatial
[11] GeoTools, http://www.geotools.org/
[12] GeoServer, http://geoserver.org/
[13] uDig, http://udig.refractions.net/

Other spatial features of Neo4j Spatial include:

- SRID: each spatial data layer has its own configuration of the coordinate system. This configuration is obtained from input files (SHP or OSM); however, re-projection operations are not supported;
- Indexing: the current release of Neo4j Spatial uses RTree indexing, but if other types of indexes are needed, the extensible library offers support to add them. The indexing works well with 2D geometry and also supports 3D geometry, although 3D support is still incomplete and inaccurate;
- Topological functions: Neo4j Spatial provides the following topological query methods: Contain, Cover, Covered by, Cross, Equals, Disjoint, Intersect, Intersect Window, Overlap, Touch, Within, and Within Distance;
- Analysis functions: Neo4j Spatial provides the following analysis methods: Area, BBox, Boundary, Distance, Buffer, Centroid, ConvexHull, and Envelope;
- Set functions: Neo4j Spatial provides Difference, Intersection, Union and SymDifference methods.

Neo4j Spatial has native support for one output format: SLD-styled PNG. However, as the library offers integration with the GeoTools library, GML, GeoJSON, and KML can be used with minimum programming efforts.

BigTable

BigTable is a column-based database implemented by Google, Inc. BigTable, launched in 2004, is the database behind the Google App Engine (GAE), a system that allows anyone to develop and run applications using the Google infrastructure. Currently, many applications and services like Google Earth, YouTube, and Google Analytics use BigTable.

Main Features

BigTable is presented as a scalable, distributed and fault-tolerant NoSQL database. Data are stored in tables and each table consists of:

- row key: contains an arbitrary string;
- column key: grouped into column families, that contains the basic unit of access control;
- timestamp: to store multiple copies of each cell with distinct timestamp, and thus maintain a record of change over time.

BigTable scalability and fault-tolerance is achieved by dynamic partitioning of tables, in which each row range is called a "tablet". Multiple copies of the same information are stored through the Google File System(GFS).The GFS is a distributed file system designed to supply efficient and reliable access on large distributed commodity hardware (Ghemawat et al. 2003). This dynamic partition enables efficient and reliable access to data. With GFS, the MapReduce model is used to distribute the data processing.

BigTable has some query restrictions. For example, it is not possible to use an inequality operator (<, <=, >=, >,and !=) on more than one property (field) in a query. Concerning data types, only String, int32 (encode integers as 32-bits bit strings), int64 (encode integers as 64-bits bit strings), double (64-bits bit strings), and Boolean are supported in BigTable.[14] As a result, to provide geospatial data indexing and processing, BigTable requires GeoModel, which is detailed next.

Batch data insertion in BigTable is possible through the python command "appcfg.py". Only CSV and XML file formats are allowed.

GeoModel

GeoModel is an open source project aiming to provide support for storage and manipulation of geospatial data in Google App Engine. The GeoModel project offers an API for programming languages such as Java[15] and Python.[16]

GeoModel supports only the Point data type, the GeoPT. Furthermore, GeoModel only provides spatial queries out of the box: proximity search (i.e., retrieve all points within a distance from a target location) and bounding box queries (i.e., retrieve all points within a pre-defined rectangular bounding box). Other spatial features of the GeoModel Spatial are described below:

- Indexing: GeoModel uses GeoHash for Geospatial indexing. The coordinate system divides the world into rectangular cells (called geocells);
- Data Types and functions: GeoModel does not support the Linestring, Polygon and MultiGeometry data types (only Point). Additionally, topological functions (equals, disjoint, intersect, touches, crosses, overlaps); analysis functions (buffer, convexHull) and set functions (intersection, union, difference, symDifference) are not supported;
- Input/Output format: JSON and KML formats are supported.

---

[14] https://developers.google.com/appengine/articles/storage_breakdown
[15] https://code.google.com/p/javageomodel/
[16] https://pypi.python.org/pypi/geomodel

GeoModel does not allow the submission of queries using GQL (the SQL-like query language for Google App Engine) because GQL does not support table "joins" or "having" clause commands. These restrictions become impractical for supporting a query by proximity using GQL. To implement this type of search, a GeoModel class needs to be used that defines the proximity search or bounding box query. These methods are: proximity_fetch and bounding_box_fetch.

Experiments

In this section we describe the database used in the experiments as well as the code that implements the proposed queries in the different NoSQL systems.

Database Description

The database used in the experiments on the NoSQL systems comprises all types of spatial information. Figure 3 shows the database schema used in the experiments. The geometric type Point is represented by Points of Interest (PoIs) located in the American territory. These data were obtained by means of GPS data. The Points of Interest are stored in the POIS entity. Among the properties of this entity, we highlight the geometry property, which represents the geometry of the point (the same information can be obtained with the longitude and latitude attributes, but in another format), and the elevation and depth, indicating the height of the point with respect to sea level. The PoIs also have a type property, which labels schools, bars, hospitals, police stations, army bases, gas stations, warnings, and so on.

POIS	
PK	ID
	label
	notes
	type
	elevation
	depth
	proximity
	temperature
	waypoint
	comment
	longitude
	latitude
	URL
	geometry

MUNICIPALITIES	
PK	ID
	name
	area
	geometry

HIGHWAYS	
PK	ID
	name
	length
	geometry

Fig. 3. Database schema used for evaluating the NoSQL systems.

We use polygons to represent the geometry of American municipalities (Municipalities entity) and the line type to represent American highways (Highways entity).

In the following, we provide a list of queries used to test the spatial functionalities provided by the analysed NoSQL systems. The results are presented ahead.

Q1 - *Find all restaurants in New York City.*

Q2 - *Find all drugstores within a distance of one km from hospitals.*

Q3 - *How many gas stations are there along the Route 66?*

Q4 - *Which roads cross the city of Chicago?*

Most NoSQL systems analysed in this chapter only allow bounding-box queries. Queries Q1 and Q2 explore the "WITHIN" operation. Given a polygon, the "WITHIN" operation obtains all points belonging to it. In our example, the polygon represents the city of New York and the points represent restaurants. In Q2, the polygon is built by applying the "BUFFER" operation to a point. In our example, it is implemented by a buffer of 1km around each hospital and we look for drugstores within this area.

Queries Q3 and Q4 explore the relationships between lines and points as well as between lines and polygons, respectively. In Q3, we request the number of gas stations on the Route 66, while Q4 uses the "CROSS" operation to find which roads (lines) cross the city of Chicago.

Database Experiments

In the following, we describe how each query was implemented in the analysed NoSQL systems. Some queries could not be implemented due to the spatial limitations of the NoSQL systems.

CouchDB

In this section, we describe the HTTP requests submitted to CouchDB+GeoCouch to answer the formulated queries. Among the analysed NoSQL systems with spatial support, GeoCouch is the most limited one because it only offers bounding-box based queries. Due to this limitation, only query Q1 is detailed. We also provide an idea of how query Q2 can be performed.

The loading of spatial data (ESRI Shapefiles) into CouchDB can be performed using the shp2geocouch tool. Consider a file containing points of interest (PoIs) stored in a shape file (pois.shp). Figure 4 shows the upload of this file into CouchDB.

```
1    shp2geocouch ./pois/pois.shp
```

Fig. 4. Importing a shapefile into CouchDB.

As a result, GeoCouch builds a spatial index. The status of the indexer can be checked by visiting the URL: http://localhost:5984/_utils/status.html.

In CouchDB, operations to create, read, update and delete are performed using HTTP requests. The GET request is used to retrieve documents. PUT or POST requests are used to create new documents. Updates are done through a PUT request. The HTTP DELETE is used to delete documents. As CouchDB implements MVCC to prevent blocks, when a GET query is performed the last version of the document is returned. CouchDB also provides a way to refer to older versions of the document by adding the parameter "&rev=version_value" in the URL.

CouchDB (and GeoCouch) is not able to perform a multi-dimension spatial query. In other words, CouchDB cannot perform a bounding-box (bbox) query with the addition of filtering by a specific field. One way to work around this limitation is to execute the bbox query and then perform an additional search in the query result.

Therefore, in order to execute query Q1 (retrieve all the restaurants in New York) it is necessary to request all PoIs in New York City and then filter the results by restaurants. The URL of the HTTP request of the first part of Q1 (retrieve all PoIs in New York City) is presented in Fig. 5. Then the query result must be refined by the type of PoI (restaurants).

The URL contains the address (IP and port) of the machine where CouchDB is running, the database name (POIS) and the parameters needed to apply the bbox filter. The values of the bbox for New York City were obtained through a query on the Municipalities database by a temporary view with the map function (Fig. 6). The function is applied by CouchDB for each document of the database. The documents that fulfill the query condition are exhibited in the query result. The query result contains documents in GeoJSON increased by the bbox geometry presented in the document.

```
1    http://localhost:5984/pois/_design/geo/_spatial/basic?bbox=-
2    74.2557,-40.4957,-73.6895,-40.9176
```

Fig. 5. URL that performs Q1 in CouchDB.

```
1    function(doc) {
2    if (doc.name == "New York") {
3    emit(doc.id, doc)
4    }
5    }
```

Fig. 6. Map function that retrieves the city of New York stored in the CouchDB.

When executing a query using a bounding box (bbox), GeoCouch returns the smallest bounding or enclosing box (rectangle) that contains the geometries that satisfy the query criteria. In the example described in Fig. 6, the function returns the smallest rectangle that involves the PoIs in New York City. It is important to mention that because we are using bbox, some points inside the bbox but outside the NY geometry may return. These points would be false positives.

A major problem of CouchDB and the GeoCouch extension is the lack of a multi-dimensional index. It is not possible to add filters based on spatial and non-spatial criteria in the same query. Therefore, query Q2 must be broken in parts and some filters must be executed in memory using a programming language. Figure 7 illustrates an example of solution to perform query Q2 using a pseudo-code.

The implementation of the buffer function (), used in the sixth line of Fig. 7, would make a HTTP request as illustrated in Fig. 8.

```
1    hospitals = getAllHospitals();       //Retrieve all hospitals
2
3    pois_near_hospitals = []
4
5    for each hospital in hospitals:
6    pois_near_hospitals.append(buffer(hospital)) // one km
7    buffer
8    in each
9    hospital
10   drugstores_near_hospitals = []
11
12   for each poi in pois_near_hospitals:
13   if poi.type == "drugstore":
14   drugstores_near_hospital.append(poi)
```

Fig. 7. Pseudo-code to perform query Q2.

```
1    http://localhost:5984/pois/_design/geo/_spatiallist/radius/poi
2    nts? bbox=x1,y1,x2,y2&radius=1000
```

Fig. 8. URL that performs a buffer for query Q2 in CouchDB.

MongoDB

Since MongoDB only provides spatial operators to verify inclusion ($geoWithin), intersection ($geoIntersects), and proximity ($near), only the queries Q1, Q2 and Q3 were implemented.

MongoDB provides no internal tool to perform the loading of ESRI Shapefiles or other spatial data sources. However, the og2ogr GDAL application can be used to convert the shapefile into the GeoJSON format. Figure 9 presents an example of uploading a set of Points of Interest (PoIs) into the shapefile format.

To import data into MongoDB, the import tool (mongoimport) can be used as shown in Fig. 10. The -d option indicates the database in which data will be imported, in our example, gisdb. The -c option specifies the data collection and the last parameter is the file input (pois.json).

Then, the indexes can be created to allow spatial queries, as illustrated in Fig. 11.

The following algorithms were written using the JavaScript API for the MongoDB shell. The algorithms assume that the data were imported into the "gisdb" database, the corresponding items were created, and the connection to the database was established in MongoDB.

The algorithm that implements query Q1 (retrieve all restaurants located in New York City) for MongoDB is presented in Fig. 12.

The first line of Fig. 12 retrieves the municipality of New York from the Municipalities collection. The result is stored in the New York City variable as a document of the Municipalities entity in the BSON format. In the fragment between lines four and ten, all of the PoIs inside the polygon of New York City (lines four to eight) that are of the type "Restaurant" (ninth line) are retrieved.

The algorithm that performs query Q2 (list all drugstores located within a radius of one km from hospitals) is presented in Fig. 13.

The query Q2 is performed similarly to Q1. The main difference is that the "$near" method (second line of Fig. 13) is used which requires a set of

```
1    ogr2ogr  -f  "GeoJSON"  ./pois/pois.json  ./pois/pois.shp  pois
2    -lco ENCODING=UTF-8
```

Fig. 9. Converting a shapefile into MongoDB.

```
1    mongoimport -d gisdb -c pois pois.json
```

Fig. 10. Importing the gisdb database into MongoDB.

```
1    gisdb.pois.ensureIndex({"geometry.coordinates":"2dsphere"})
```

Fig. 11. Creating a spatial index in MongoDB.

```
1    var newYorkCity = gisdb.municipalities.find(
2                                    {"name":"New York"});
3
4    gisdb.pois.find({ "geometry.coordinates" :
5            { "$geoWithin" :
6                    { "$polygon" :
7                       newYorkCity
8            } },
9    "type" : "Restaurant"
10   });
```

Fig. 12. Algorithm that performs Q1 on MongoDB.

geometries and a maximum distance in meters. A list with the coordinates of all hospitals must also be provided in the fifth line.

The algorithm that performs query Q3 (list the number of gas stations that exist along Route 66) is presented in Fig. 14.

Similarly to the query Q1, two steps are needed: (1) filter the element of interest (first line), in this case the Route 66; (2) search for PoIs which have a spatial relationship (Intersects) with the element obtained in the previous step, filtered by type (between the third and eighth lines). The number of PoIs that meet the query criteria is counted in the eighth line of Fig. 14.

```
1     gisdb.pois.find( { "geometry.coordinates" :
2     { "$near" :
3     { $geometry :
4     { type : "Point" ,
5     Coordinates : [ <longitude> , <latitude> ]
6     /*In   above   line   should   be   informed   an   array   with
7         thecoordinates of all hospitals */
8     }
9     },
10                          $maxDistance : 1000
11    },
12                          "type" : "Drugstore"
13    } );
14
```

Fig. 13. Algorithm that performs Q2 on MongoDB.

```
1     var highway = gisdb.highways.find({ "name" : "Route 66" });
2
3     gisdb.pois.find( { "geometry.coordinates" :
4     { "$geoIntersects" :
5     { $geometry" :
6     highway
7     } },
8     "type" : "Gas Station" } ).count();
```

Fig. 14. Algorithm that performs Q3 on MongoDB.

Neo4j

The Neo4j Spatial offers a ShapefileImporter class which is responsible for converting shapefiles into a Neo4j database. Figure 15 shows some examples of conversion from shapefiles to Neo4j.

The importFile method of the ShapefileImporter class (between the ninth and thirteenth lines of Fig. 15) is responsible for converting shapefiles. This method has an optional attribute that enables an explicit specification of the charset from the input file.

An excerpt of the algorithm implemented to answer query Q1 (retrieve all restaurants in New York) is presented in Fig. 16.

```
1    ...
2    GraphDatabaseService database = new
3            EmbeddedGraphDatabase(storeDir);
4
5    try {
6        ShapefileImporter importer = new
7                ShapefileImporter(database);
8
9        importer.importFile("shp/br_municip.shp",
10                "MUNICIPALITIES",
11                Charset.forName("UTF-8"));
12        importer.importFile("shp/pois.shp", "POIS");
13        importer.importFile("shp/highways.shp", "HIGHWAYS");
14
15    } finally {
16        database.shutdown();
17    }
18   ...
```

Fig. 15. Converting shapefiles into Neo4j Spatial.

```
1    ...
2    try {
3        SpatialDatabaseService spatialService = new
4                SpatialDatabaseService(database);
5
6        Layer layer1 = spatialService.getLayer("MUNICIPALITIES");
7        SpatialIndexReader spatialIndex1 = layer1.getIndex();
8
9        Search searchQuery1 = new SearchCQL(layer1,
10               "name = 'New York'");
11       spatialIndex1.executeSearch(searchQuery1);
12       List<SpatialDatabaseRecord> newYork =
13               searchQuery1.getResults();
14
15       Envelope nyBBOX =
16               newYork.get(0).getGeometry().getEnvelopeInternal
17               ();
18
19       Layer layer2 = spatialService.getLayer("POIS");
20       SpatialIndexReader spatialIndex2 = layer2.getIndex();
21
22       Search searchQuery2 = new SearchCQL(layer2, "type =
23               'restaurant' AND
24               bbox(nyBBOX.getMinX(),nyBBOX.getMaxX(),
25               nyBBOX.getMinY(), nyBBOX.getMaxY())");
26
27       spatialIndex2.executeSearch(searchQuery2);
28       List<SpatialDatabaseRecord> rests =
29               searchQuery2.getResults();
30
31       for (SpatialDatabaseRecord rest : rests ) {
32           Search searchQuery3 = new
33                   SearchIntersect(rest.getGeometry(),newYork.ge
34                   t(0).getGeometry());
35               ...
36       }
37       ...
38   } finally {
39       database.shutdown();
40   }
41   ...
```

Fig. 16. An excerpt of the algorithm that answers Q1 in Neo4j Spatial.

The first step for the creation of a query is to construct a "try" block with the "finally" option containing the database.shutdown() command (line 39). The command ends the connection with the database as soon as the code inserted in the "try" block is processed. The "database" variable, used to maintain the connection with the Neo4j Spatial database, was instantiated at some previous moment and is indispensable for the correct execution of the presented code. This can be observed in the third and fourth lines of Fig. 16, where we instantiate a service for accessing the spatial database. Using this service, two variables are instantiated to store the spatial data layers: the first variable represents the layer containing the data of the MUNICIPALITIES entity, while the second one represents a layer containing the data of the POIS entity (the sixth and nineteenth lines). The indexes of the instantiated spatial layers are stored in objects of the type SpatialIndexReader (the eighth and twenty lines). This object type has the executeSearch() method, responsible for processing the queries on a layer and storing the results in an object of the type Search. In addition to the results, the object of the Search type stores the constraint conditions for the execution of a conventional query on a spatial data layer. An example of defining constraining conditions is illustrated in line 10 of Fig. 16. The Search object stores the constraint of returning only the geometry of the New York municipality from layer one, which represents the MUNICIPALITIES entity. The results of the conventional queries on the layers consist of lists of objects of the type SpatialDatabaseRecord.

The lists of records contain the results of the conventional queries on the layers. The first list (line 13) contains the geometry of the New York municipality and the second list (line 28) contains only the PoIs with type "restaurant" in a bounding box of the New York City geometry. We use the object of the type SearchIntersect, a specialisation of the generic object type Search, to perform the spatial intersection of these two geometry lists (lines 31 to 36). The result of this procedure is a list containing only the PoIs of the type "restaurant" that intersect the geometry of New York City.

An excerpt of the algorithm implemented to answer query Q2 (list all drugstores located within a radius of one km from hospitals) in Neo4j Spatial is presented in Fig. 17. Similarly to query Q1, the first steps needed to perform Q2 focus on the instantiation of the object that accesses the requested data layer (in this case, the "POIS" entity) and on the instantiation of the objects that prepare a conventional query on this layer. In the third line of Fig. 17, we can observe the insertion of the constraint in "POIS" such that the variable "searchQuery" returns only the PoIs of the "hospital" type, which will then be made available in the "hospitals" variable (sixth line).

Using the startNearestNeighborLatLonSearch (Layer to search, pointCentroid, maxDistanceInKm) method, as shown in line 12 of Fig. 17, we can obtain a list of the PoIs within the radius of one km from a hospital. Because this method demands a Layer object instead of a Search object, an additional effort is needed to filter only the PoIs of the "drugstore" type from the PoIs returned by the method.

The GeoPipeFlow object type (line 17 of Fig. 17) consists of methods for interacting with the results of queries through the GeoPipeLine object. This type of object provides methods such as getGeometry() (that returns the spatial geometry) and getProperties() (that returns the set of metadata).

It is worth pointing out that the start NearestNeighborLatLonSearch() method performs the search for a neighborhood in each centroid point at a time, i.e., individually for each hospital. Thus, to obtain the drugstores located within the radius of one km from all hospitals, it is necessary to use a loop to invoke this method for each hospital.

An excerpt of an algorithm to answer Q3 (list the number of gas stations along Route 66) in Neo4j Spatial is presented in Fig. 18. Similarly to the

```
1    ...
2    Layer layer = spatialService.getLayer("POIS");
3    Search searchQuery = new SearchCQL(layer,
4                                       "type = 'hospital'");
5    ...
6    List<SpatialDatabaseRecord> hospitals =
7            searchQuery.getResults();
8
9    for (int x=0; x<hospitals.size(); x++) {
10
11       GeoPipeline pipe =
12               GeoPipeline.startNearestNeighborLatLonSearch(
13               layer,hospitals.get(x).getGeometry(),1.0)
14                       .sort("OrthodromicDistance")
15                       .getMin("OrthodromicDistance");
16
17       List<GeoPipeFlow> poisClosests = pipe.toList();
18       ...
19       // the result list needs to be filtered to drugstores
20    }
21    ...
```

Fig. 17. An excerpt of the algorithm that performs Q2 in Neo4j Spatial.

```
1    ...
2    Layer layer1 = spatialService.getLayer("POIS");
3    Layer layer2 = spatialService.getLayer("HIGHWAYS");
4
5    Search searchQuery = new SearchCQL(layer2,
6                                       "name = 'Route66'");
7    ...
8    List<SpatialDatabaseRecord> route66 =
9            searchQuery.getResults();
10
11   GeoPipeline pipe = GeoPipeline.startContainSearch(layer1,
12           route66.get(0).getGeometry());
13
14   List<GeoPipeFlow> poisContained = pipe.toList();
15   ...
16   // the result list needs to be filtered to gas stations
17   ...
```

Fig. 18. An excerpt of the algorithm to answer Q3 in Neo4j Spatial.

first two queries, the initial steps consist of the instantiation of the Search type objects, which enable the application of constraints in the access to data layers. The creation of the variables "gasStations" and "route66" was suppressed because they are similar to the "hospital" variable, illustrated in Fig. 17. The startContainSearch() method (line 11 of Fig. 18) obtains a list of all PoIs contained in the geometry of the Route 66. Thus, to obtain only the gas stations, it is necessary to filter this list by the "type" attribute of the "POIS" entity. Finally, a count method is used to obtain the number of gas stations in the filtered list.

An excerpt of the algorithm implemented to answer query Q4 (list the roads that cross Chicago) in Neo4j Spatial is illustrated in Fig. 19. The main difference from query Q3 is that we use the startCrossSearch() method, as observed in the ninth line of Fig. 19. This method allows the retrieval of all of the roads that cross the city of Chicago.

```
1    ...
2    Search searchQuery1 = new SearchCQL(
3            spatialService.getLayer("MUNICIPALITIES"),
4            "name = 'Chicago'");
5    ...
6    // chicago is a var. of type List<SpatialDatabaseRecord>
7    ...
8
9    GeoPipeline pipe = GeoPipeline.startCrossSearch(
10           spatialService.getLayer("HIGHWAYS"),
11           chicago.getGeometry());
12
13   List<SpatialDatabaseRecord> highwayCrosses  =
14           pipe.toSpatialDatabaseRecordList();
15   ...
```

Fig. 19. An excerpt of the algorithm to answer Q4 in Neo4j Spatial.

BigTable

The data was uploaded into BigTable through the python statement "appcfg.py", as illustrated in Fig. 20. This statement has some mandatory attributes. For instance, the "application" attribute requires the id of the user application registered for Google. The "filename" attribute requires the path of the csv file and "<app-directory>" refers to the application directory path in the Google App Engine. It is worth mentioning that the file "*pois.csv*" used the in example illustrated in Fig. 20 contains the same information of "*pois.shp*". In other words, it is the same file but represented in another format.

Due to the limitations of the GeoModel, which only supports the Point data type and bounding-box queries, we will provide details only for the queries Q1 and Q2. These queries make use of the functions proximity_fetch (query, center, max_result=10, max_distance=0) and bounding_box_fetch (query, bbox, max_results=1000, cost_function=None). Consequently, they need an instance of the GeoModel object, as exemplified in Fig. 21. In the first line, we import the geo.geomodel library to allow the creation of a class that inherits the GeoModel class. In the fragment between second and fourth lines, the Points of Interest (POIs) class is created; and in the fragment between seventh and eighth lines, the Municipalities class is created.

Hence, to find all restaurants in New York as requested in query Q1, we must store New York's bounding box through the geotypes.Box Geomodel method, as shown in the first two lines of the algorithm illustrated in Fig. 22.

Since GeoModel cannot store geometries of the type polygon, it is not possible to process the exact New York's polygon. An alternative solution is to apply the New York's bounding box even though restaurants that are not located in New York City can come as a result of the query. Hence, false positives may occur in the result set. It is important to highlight that there is no processing bounding box. Internally, the method bounding_box_fetch() compares if the latitude and longitude of a point is located between the latitudes and longitudes of the bounding box explicitly informed.

Next, in the fourth and fifth lines of Fig. 22, we obtain all restaurants within New York through the bounding_box_fetch method of the POIS object. The parameters of this method are the query, the delimiting bounding box, the maximum number of results, and the cost of executing the method. The bounding box of New York is passed as a parameter to the method after being computed in the first and second lines of the algorithm. The other

```
1    appcfg.py upload_data -application=<app_id> -kind=Permission--
2    filename=pois.csv <app-directory>
```

Fig. 20. Uploading csv data to BigTable.

```
1    from geo.geomodel import GeoModel
2    class POIS(GeoModel):
3    label = db.StringProperty()
4    type = db.StringProperty()
5
6
7    class Municipalities (GeoModel):
8    name = db.StringProperty()
9    ...
```

Fig. 21. Creation of the entities derived from GeoModel.

```
1    new_york_bb = geotypes.Box(-74.2557, -40.4957, -73.6895,
2        -40.9176)
3
4    restaurants = POIS.bounding_box_fetch
5        (POIS.all.filter('type =','restaurants'), new_york_bb)
```

Fig. 22. Algorithm to answer query Q1 in BigTable.

two parameters of the bounding_box_fetch method are omitted, and the default values are used (1,000 for the maximum number of results and no cost delimiter for the function).

The Fig. 23 presents the algorithm for query Q2 (retrieve all drugstores within a maximum radius of 1km from hospitals). First, it is necessary to filter the points of interest by the types "drugstore" and "hospital", as exemplified in the first and third lines of the algorithm, respectively. Next, we iterate each retrieved hospital and, based on its location, we use the proximity_fetch method with the following parameters: the query q, defined in the first line, containing all of the drugstores; the location of the current hospital; the maximum number of results, which was fixed at 10; and the maximum distance (this parameter must be in radians, therefore we used the constant 0.00157 that represents one km in radians). The drugstores found for each hospital are added to an external list which stores all drugstores within a maximum distance of one km from any hospital.

```
1    q = db.GqlQuery("SELECT * FROM POIS WHERE type = 'drugstore'")
2
3    hospitals = POIS.all.filter('type =', 'hospitals'))
4
5    allDrugstores = []
6
7    for h in hospitals:
8    hospital = db.GeoPt(h.latitude, h.longitude)
9
10   drugstores = proximity_fetch(q, hospital, maxResult=10,
11                                 max_distance=0.00157)
12
13       allDrugstores.extent(drugstores)
```

Fig. 23. Algorithm to answer Q2 in BigTable.

Conclusions

NoSQL systems have emerged to efficiently handle large volumes of data. In addition to being built to handle large volumes, a typical NoSQL system has mechanisms for data partitioning, horizontal scalability, lack of schema definition, non-relational data models, simple APIs for data access, and eventual consistency.

Location-based applications have become popular, generating large volumes of highly distributed spatial data daily and demanding high processing for the spatial queries. Traditional GIS technologies do not cope with this highly distributed and scalable demand. Therefore, several spatial extensions were included in many NoSQL systems, like the pioneers in the provision of spatial data support, MongoDB and CouchDB. Such extensions enable NoSQL systems to address the complexity and particularity of highly distributed spatial data.

In this context, we enumerated in this chapter some desirable features of a spatial database system, based on the ISO SQL/MM standard, aiming to assess how the NoSQL systems provide support for geographic data. Seven items were selected for analysis to measure the depth of the geographic support of each NoSQL system: basic concepts, indexing, vector data types, topological functions, analysis and metric functions, set functions, and input/output format.

The geographic NoSQL systems CouchDB (with the GeoCouch extension), MongoDB, Neo4j (with the Neo4j Spatial extension), and BigTable (with GeoModel and the GeoDataStore extension) were described, analysed according to the seven selected items, and then compared. Table 1 presents a comparison between the analysed systems.

As observed in Table 1, Neo4j (and Neo4j Spatial) is currently the most complete NoSQL spatial system.

GeoCouch (on CouchDB) has proven to be an unattractive alternative for dealing with geographic data. It is an extension developed by a third party with few spatial resources. Additionally, it has not undergone any evolution over the past two years.

MongoDB is very limited because it deals only with the point vector type, and its input/output is limited to GeoJSON.

Table 1. Comparison of the analysed geographic NoSQL systems.

	CouchDB	**MongoDB**	**Neo4j**	**BigTable**
Basic Concepts	Document-oriented	Document-oriented	Network-oriented	Column-oriented
Indexing	R-Tree Only2D	2D 2Dsphere	R-Tree 2D and partially 3D	B-Tree Only2D
Vector Data Types	Fully	Fully	Fully Basic Types and Limited MultiGeometry Types	Fully Basic Types
Topological functions	Only Within() Contains()	Only Within(Point) Contains(Point)	Almost fully	Not supported
Analysis and metric functions	Only Distance()	Only Distance (Point)	Fully	Only Distance()
Set functions	Not supported	Only Intersection (Point)	Fully	Not supported
Input/output Format	Input: .SHP Output: .KML .CSV .GeoJSON	Input: GeoJSON Output: GeoJSON	Input: .SHP .OSM Output: SLD styled PNG	Input .JSON .KML Output .JSON .KML

When dealing with spatial data, BigTable is as limited as MongoDB. Moreover, BigTable does not provide support for Multi-Geometries vector types, as it only implements the distance() function. Therefore, BigTable still needs considerable improvement to become a good candidate for dealing with spatial data.

The general conclusion from this comparison is that NoSQL geographic systems still need many improvements, especially concerning different data types and functions. As these systems evolve, GIS applications will be able to take advantage of their main characteristics, mainly high performance and scalability. Currently, Neo4j seems to be the best option for GIS applications with NoSQL storage, although this system still requires several improvements.

References

Amer-Yahia, S. and A. Halevy. 2007. What does Web 2.0 have to do with databases? Proc. 33rd International Conference on Very Large Databases. Vienna, Austria. 1443–1443.

Anderson, J.C., N. Slater and J. Lehnardt. 2009. CouchDB: The Definitive Guide. 1st Edition, O'Reilly Media.

Baas, B. 2012. NoSQL spatial: Neo4j versus PostGIS. M.S. Thesis. Geographic Information Management and Applications (GIMA), Delft University of Technology, Delft, Netherlands.

Brewer, E.A. 2000. Towards Robust Distributed Systems. Proc. 19th Annual ACM Symposium on Principles of Distributed Computing. Portland, USA. 7–7.

Cattell, R. 2010. Scalable SQL and NoSQL Data Stores. ACM SIGMOD Rec. 39(4): 12–27.

Chang, F., J. Dean, S. Ghemawat, W.C. Hsieh, D.A. Wallach, M. Burrows, T. Chandra, A. Fikes and R.E. Gruber. 2008. Bigtable: A Distributed Storage System for Structured Data. ACM Trans. Comp. Syst. 26(2).

Chodorow, K. and M. Dirolf. 2010. MongoDB: The Definitive Guide, O'Reilly.

Codd, E.F. 1970. A Relational Model of Data for Large Shared Data Banks. Commun. ACM. 13(6): 377–387.

DeCandia, G., D. Hastorun, M. Jampani, G. Kakulapati, A. Lakshman, A. Pilchin, S. Sivasubramanian, P. Vosshall and W. Vogels. 2007. Dynamo: Amazon's Highly Available key-value Store. ACM SIGOPS Oper. Syst. Rev. 41(6): 205–220.

Diestel, R. 2005. Graph Theory. 3rd Edition, Springer-Verlag, Heidelberg.

Fielding, R. 2000. Architectural Styles and the Design of Network-based Software Architectures. Ph.D. Thesis. University of California, Irvine, California.

Floratou, A., N. Teletia, D.J. DeWitt, J.M. Patel and D. Zhang. 2012. Can the elephants handle the NoSQL onslaught? Proc. The VLDB Endowment. 5(12): 1712–1723.

Gartner Group. 2013. Top 10 Technology Trends Impacting Information Infrastructure. http://www.gartner.com/DisplayDocument?ref=clientFriendlyUrl&id=2340315 (published in 19 February 2013; last access in 18 April 2013).

Ghemawat, S., H. Gobioff and S.-T. Leung. 2003. The Google File System. Proc. The 19th ACM Symposium on Operating Systems Principles. New York, USA. 29–43.

Hoff, T. 2009. A Yes for a NoSQL Taxonomy. http://highscalability.com/blog/2009/11/5/a-yes-for-a-NoSQL-taxonomy.html (published in 5 November 2009; last access in 18 April 2013).

Hunger, M. and P. Rathle. 2012. NoSQL, Big Data and Graphs: Technology choices for today's mission-critical applications. http://www.databaserevolution.com/2012/11/nosql-big-data-and-graphs/ (published in 23 November 2012; last access in 04 July 2013).

Lai, E. 2010. Twitter growth prompts switch from MySQL to NoSQL database. http://www. computerworld.com/s/article/9161078/Twitter_growth_prompts_switch_from_ MySQL_to_NoSQL_database (published in 23 February 2010; last access in 18 April 2013).

Lakshman, A. and P. Malik. 2009. Cassandra: structured storage system on a P2P network. Proc. The 28th ACM symposium on Principles of Distributed Computing. Calgary, Canada. 5–5.

Leavitt, N. 2010. Will NoSQL Databases Live Up to Their Promise? Computer. 43(2): 12–14.

Lee, K., Y. Lee, H. Choi, Y. Chung and B. Moon. 2012. Parallel Data Processing with MapReduce: a Survey. ACM SIGMOD Rec. 40(4): 11–20.

Näsholm, P. 2012. Extracting Data from NoSQL Databases—A Step towards Interactive Visual Analysis of NoSQL Data. M.S. Thesis. University of Gothenburg, Gothenburg, Sweden.

NoSQL. 2013. NoSQL Your Ultimate Guide to the Non-Relational Universe. http://NoSQL-database.org/ (last access in 18 April 2013).

North, K. 2009. Databases in the cloud. http://www.drdobbs.com/database/218900502 (published in August 03, 2009; last access in April 18, 2013).

Pritchett, D. 2008. BASE: An ACID Alternative. ACM Queue. 6(3): 48–55.

Redmond, E. and J.R. Wilson. 2012. Seven Databases in Seven Weeks: A Guide to Modern Databases and the NoSQL Movement. Pragmatic Bookshelf.

Sadalage, P.J. and M. Fowler. 2012. NoSQL Distilled: A Brief Guide to the Emerging World of Polyglot Persistence. 1st Edition, Addison-Wesley Professional.

Schutzberg, A. 2011. NoSQL Databases: What Geospatial Users Need to Know. http:// www.directionsmag.com/articles/NoSQL-databases-what-geospatial-users-need-to-know/164635 (published in 16 February 2011; last access in 18 April 2013).

Stolze, K. 2003. SQL/MM Spatial—The Standard to Manage Spatial Data in a Relational Database System. Proc. TheDatenbanksystemefür Business, Technologie und Web (BTW'03). Leipzig, Germany. 247–264.

Stonebraker, M. 2010. SQL Databases vs. NoSQL Databases. Commun. ACM. 53(4): 10–11.

Strozzi, C. 2007. NoSQL—A Relational Database Management System. http://www.strozzi. it/cgi-bin/CSA/tw7/I/en_US/NoSQL/Home%20Page (published in 10 October 2007; last access in 18 April 2013).

Vogels, W. 2009. Eventually Consistent. Commun. ACM. 52(1): 40–44.

CHAPTER 5

Web Services Composition and Geographic Information

Pasquale Di Giovanni,[a,b], Michela Bertolotto,[a] Giuliana Vitiello[b] and Monica Sebillo[b]*

Introduction

In recent years, the continuous development of the Web and network infrastructures has had a profound impact both on the way users' access information and utilize traditional computer applications, and on the manner these applications are designed and developed. The network development, in fact, has been a key factor for the transition from stand alone and centralized software architectures to distributed ones, enabling software components located in different geographic locations to communicate through the network in order to provide for coordinated services and joint results.

One of the main limitations of the first distributed systems was represented by the great difficulty or by the complete inability to communicate with other applications running on different technological platforms. The technical complexity met during the development of such distributed solutions has fostered the emergence of different middleware technologies, such as Microsoft Distributed Component Object Model

[a] School of Computer Science and Informatics, University College Dublin, Belfield, Dublin 4, Ireland.
  Emails: pasquale.di-giovanni@ucdconnect.ie; michela.bertolotto@ucd.ie
[b] Department of Management and Information Technology (DISTRA), University of Salerno, Via Giovanni Paolo II, 132-84084 Fisciano (SA), Italy.
  Emails: gvitiello@unisa.it; msebillo@unisa.it
* Corresponding author

(DCOM), Java Remote Method Invocation (RMI) and Common Object Request Broker Architecture (CORBA) which, although based on different methodologies, had the common goal of simplifying the development of distributed applications. However, although interesting, all these solutions had and have, on the whole, several limitations, such as the tight coupling between the various developed components and the presence of mature development tools, just realized for a specific platform. Furthermore, developing non-trivial applications by using these technologies is still complex, because they lack a reference implementation, a standard in the deployment of software solutions and a support for deploying applications in environments with strict security needs (Henning 2006). The above-mentioned difficulties and the growing trend in the research and development of loosely coupled solutions, meant to ensure both, a platform independence and a greater sharing and reusing of functionalities, have found in the Service Oriented Computing (SOC) paradigm and in the Service Oriented Architecture (SOA) a new way for designing and implementing distributed applications.

The general definition of *service* can be met by any software system, however the most widely adopted description of Web service comes from the World Wide Web Consortium (W3C) which represents the organization in charge of 'leading the World Wide Web to its full potential by developing protocols and guidelines that ensure the long-term growth of the Web' (http://www.w3.org/). According to W3C a Web service 'is a software system designed to support interoperable machine-to-machine interaction over a network.' It is also relevant to guarantee the interoperability property in environments handling large amounts of complex information, such as geographic information. Indeed, improving the effectiveness and the performances of any overall solution through the usage of distributed geographic information services represents a primary goal for the geographic community that has devoted many efforts to this achievement, through the initiatives carried out by the Open Geospatial Consortium (OGC) which represents the reference organization for 'the development of international standards for geospatial interoperability' (http://www.opengeospatial.org/). In particular, the opportunity to use the enormous amount of geographic information accessible via OGC services within W3C services is one of main reasons stimulating the recent efforts to seamlessly combine these two different worlds. In fact, the geographic community has recognized that a more complete integration among the underlying protocols would allow for employing the well-established standards specified for the W3C platform. To this aim, OGC has set a special working group in order to provide general recommendations and guidelines for making differences between OGC and Web services transparent when the service composition is applied to them. However, despite the simplicity behind the general idea of services composition and some basic assumptions, such as the subdivision of the involved services into categories, and the use of XML for messages exchanges, services composition is still a complex task and several facets must be wisely considered.

The goal of this chapter is to investigate and discuss the issues arising during the composition of W3C and OGC services, that is how two or more services can work together in order to obtain a higher level service.

The remainder of this chapter is organized as follows. In the "Service-oriented Computing" section an overview of SOC and SOA is given. In "W3C Web services" and "The Open Geospatial Consortium Web Services" sections the main characteristics and differences of the service platforms proposed by the W3C and the OGC are discussed. After a brief introduction to the concept of service composition, the "Services Composition" section discusses issues related to the Web services and OGC services orchestration, while the "Web and OGC Services Integration" section reviews some solutions useful for their integration. "The Representational State Transfer paradigm" section covers the main characteristics of the Representational State Transfer (REST) paradigm, as a possible alternative for the development of distributed systems. Some conclusions are drawn in the last Section.

Service-oriented Computing

A service can be seen as an independent software module that performs certain (more or less complex) operations, such as an invoicing and a booking of a plane ticket. One of the key features of a service consists in the client capability to invoke it through the service public interface, disregarding details about the service internal implementation. To ensure such a fundamental characteristic, a service has to meet some basic requirements (Papazoglou 2003):

- being loosely coupled, namely a low level of interconnection among services;
- being technology neutral, which implies that its public interface can be described by using standard technologies available to anyone in any operating environment;
- supporting location transparency, namely the possibility for a client to invoke it irrespective of its real location.

A service that meets these characteristics can be then invoked by clients belonging to different organizations and can be seen as an independent basic building block to create new types of distributed applications.

The definition of service along with the aforementioned features represent the core of the SOC paradigm whose benefits have become clear in many application domains and in particular in Geographic Information

Systems (GIS), where this new computing model would represent a possible solution to some of the classic problems due to the complexity of geographic information, such as retrieving of geographic data, accessing newer versions of the same information and, above all, supporting geographic interoperability (Zhao et al. 2007). In fact, before the massive spread of SOC, a user wishing to access geographic data could either download an entire dataset or require the data provider to send such a dataset on a digital or magnetic medium. Typically, the end user was also provided with a catalogue to look for desired data (Zhao et al. 2007). However, a similar approach presented some important problems. First of all, the geographic information was provided in a proprietary format whose granularity often did not coincide with the real user's needs and did not allow for data sharing. Moreover, an additional drawback was related to the geographic information validity, which might be out-dated or not updated to the latest version (Zhao et al. 2007).

As known, being based on open standards, accessible by anyone, geographic services are at the base of the revolution of how spatial data are obtained and processed. However, as autonomous building blocks, services have to be properly managed when combined to realize complex systems. In the following subsection, a way to integrate different services is presented.

Service Oriented Architectures

SOA 'represents a model in which automation logic is decomposed into smaller, distinct units of logic' (Erl 2005). In its simplest implementation, SOA defines an interaction among software agents as a simple exchange of messages among different participants. In particular, a basic implementation of a SOA is a relationship among three kinds of participants, namely the service provider, the service registry and the service requestor (the client). A desirable characteristic is the ability of the messaging framework to preserve the low coupling property. In fact, as soon as a service sends a message it loses control of what happens to the message itself. Moreover, to facilitate services interaction a type of inter-service communication infrastructure is also required. The Internet is the most commonly used communication infrastructure. Services that make up a SOA exhibit the following main characteristics (Papazoglou and van den Heuvel 2007):

- all functionalities in a SOA are defined as services;
- all services are autonomous;
- the service location is irrelevant.

W3C Web services

According to the definition of the W3C working group (Booth et al. 2004), a Web service is

> 'a software system designed to support interoperable machine-to-machine interaction over a network. Other systems interact with the Web service in a manner prescribed by its description using messages, typically conveyed using HTTP with an XML serialization in conjunction with other Web-related standards.'

Standards are one of the main reasons behind the massive adoption and deployment of Web services as a means to create a framework for application-to-application interaction by the Web publication of the business functionalities and a universal Web access to them. In particular, three main roles can be associated with the interacting applications, namely the service provider that owns a Web service and makes it accessible through the network, the service client which represents a potential user of a Web service, and finally, the service registry corresponding to the entity that allows service clients to locate the service and obtain information on how to invoke its functionalities (Yu et al. 2008). Figure 1 illustrates the interaction among the above-mentioned roles.

Among the advantages offered by the Web services technology the following can be mentioned (Tsalgatidou and Pilioura 2002):

- easy and fast deployment compared to the efforts required to realize a traditional enterprise application;
- interoperability through the adoption of common XML based standards;

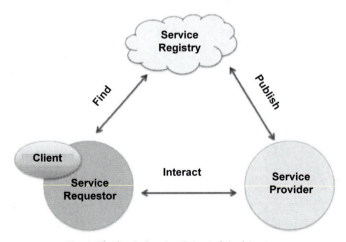

Fig. 1. The basic Service Oriented Architecture.

- just in time integration thanks to the intrinsic loosely coupled nature of services;
- reduced complexity through a common interface that hides the implementation differences behind each Web service.

Although the service oriented paradigm and the SOA concept preceded the arrival of Web services, these latter have become, over time, the most common implementation of SOA principles so that some concepts and technologies behind them have influenced and contributed to a number of new SOA characteristics. However, when embedded into a SOA environment, a Web service needs a supporting infrastructure to offer its functionalities, which has to guarantee the following basic activities (Tsalgatidou and Pilioura 2002):

- Web service creation,
- Web service description,
- Web service publishing to intranet or the Internet repositories for potential users to locate it,
- Web service discovery by potential users,
- Web service invocation, binding,
- Web service un-publishing in case it is no longer available or needed.

Finally, as for the protocols that allow for the effective exchange of information among applications, they can be divided into the following three categories:

- Communication protocol,
- Service description, and
- Service discovery.

For each of them, XML based standards have been defined: the SOAP protocol for the communication, the Web Services Description Language (WSDL) for the description, and the Universal Description Discovery and Integration (UDDI) standard for the discovery.

In the following subsections, the SOAP protocol and the WSDL standard are recalled. In order to be effective, readers should be confident with both Extensible Markup Language (XML) (Fawcett et al. 2012; Zisman 2000) and the Hyper Text Transfer Protocol (HTTP) (Fielding et al. 1999).

SOAP

SOAP is

'a lightweight protocol for exchange of information in a decentralized, distributed environment. It is an XML based protocol that consists of three parts: an envelope that defines a framework for describing what is in a message and how to process it, a set of encoding rules for expressing instances of application-defined data types, and a convention for representing remote procedure calls and responses.'

(Gudgin et al. 2001)

SOAP main design goals are simplicity and extensibility and, although it can be potentially used in conjunction with any type of transport protocol, currently the only widely used binding is the HTTP protocol. In general, SOAP enables different platforms to communicate each other thus allowing, for example, Web services implemented with Java platform and Web services implemented with Microsoft .NET platform to seamlessly communicate. Being completely XML based, it heavily relies on its standards, like XML Schema and XML namespaces for data type definitions and messages syntax.

SOAP messages are fundamentally one-way transmissions from a sender to a receiver, which can be also combined to implement more complex communication patterns, either synchronous (e.g., Request / Response) or asynchronous (e.g., Fire and Forget). Figure 2 illustrates the general structure of a SOAP message, which comprehensively refers to these three basic components: an Envelope, a Header and a Body.

The Envelope is the top-level element and can be seen as the container of the message itself. An Envelope can contain an optional Header field but has to contain exactly one Body field. A Header, if present, must necessarily appear before the Body.

Fig. 2. The general structure of a SOAP message (Source: www.w3.org).

The Body contains the actual information of the SOAP message intended for the ultimate recipient of the communication. The Body subject can be any well-formed XML content, provided that such a content has a namespace and does not contain any processing instruction or Document Type Definition (DTD) reference (Erl 2005). A SOAP fault is a special type of message used to communicate information about errors that may occur during the message processing.

As for data encoding of the Body content, SOAP supports two main styles: SOAP Remote Procedure Call (RPC) style and SOAP Document Style. RPC is a style that offers the greatest simplicity and the idea behind it is quite simple: a function available on a remote machine is invoked as if it was a local function and the parameters it needs are sent through the network. The returned value of the function is usually the expected result of the computation. SOAP RPC exactly replicates this behaviour: the functionality of a Web service is invoked by a SOAP message containing in its body the parameters needed by the remote method. Differently, in a SOAP Document Style the body content sent to a remote machine contains an entire XML document without even requiring a returned value. When compared to the RPC style, the Document Style has several advantages. In the RPC messaging any change in the signature of the remote function involves changes in all clients that use such a function; in contrast Document Style rules are less restrictive. Moreover, the RPC style can also require higher parsing times because of the necessity of having, every time, to perform the marshalling of parameters (McCarthy 2002). The Document Style also promotes loose coupling between producer and consumer messages, which represents the commonly accepted solution.

As for the third component of a SOAP message, the Header field, it contains important information about how the message should be processed and provides a generic mechanism for adding features (such as security and transaction support) in a decentralized manner. It also provides options to indicate whether the information contained in the Header is optional or mandatory. Each element in the Header field is called Header block and the namespace of these blocks may be also different from the SOAP namespaces, thus making the general structure of a SOAP Header quite flexible. Such flexibility and extensibility are the basis of many important features present in contemporary (Web services based) SOA implementations. In fact, a key feature of the SOAP framework is its emphasis on creating messages that are as self sufficient as possible, a characteristic of the utmost importance in a completely loosely coupled environment such as Web services (Erl 2005). This independence of messages is achieved by adding other information directly in the Header blocks thus avoiding services to store message specific logic. Moreover, since the information is fully contained in the message itself, the parts that communicate with each other do not need a prior agreement

(Erl 2005). SOAP also provides precise Header blocks for the management of intermediaries that have been taken into account since the early stages of the protocol design. Intermediaries are software systems capable either to receive or to forward a SOAP message, and are usually used to provide value added services or to support the scalability of a distributed environment. The set of a SOAP sender, the zero or more SOAP Intermediaries and the ultimate SOAP receiver is known as SOAP message path.

The Web Services Description Language

The Web Services Description Language (WSDL) is

> 'an XML format for describing network services as a set of endpoints operating on messages containing either document-oriented or procedure-oriented information. The operations and messages are described abstractly, and then bound to a concrete network protocol and message format to define an endpoint.'
>
> (Christensen et al. 2001)

In other words, WSDL is an XML based language for Web service description and accessing. It uses XML Schema for both the definition of the type system and the definition of SOAP messages (although their use is not compulsory) and, furthermore, it separates the abstract aspects of a service description from the concrete aspects, such as the binding with a certain network protocol. Essentially, a WSDL description contains the three fundamental properties of a Web service (What it does, how it is accessed, where it is located) through the following elements: Types, Message, Operation, Port Type, Binding, Port, and Service.

The former four elements refer to the Abstract Description of a Web service, which specifies the interface characteristics with no references to any specific technology platform; the latter three refer to its Concrete Description which allows to connect the abstract interface of a Web service to a real technology and to transport protocol. One of the main advantages of this distinction is that the "public interface" of the Web service can be preserved by changes in the underlying technology. Figure 3 depicts a WSDL Document structure as made up of an Abstract Description and a Concrete Description.

The Open Geospatial Consortium Web Services

The Open Geospatial Consortium (OGC) is 'an international industry consortium of more than 479 companies, government agencies and universities participating in a consensus process to develop publicly available interface standards.' The main goals of this consortium are to

Fig. 3. The structure of a WSDL Document.

promote the benefits of integrating location resources into commercial and institutional processes and to facilitate the collaboration of developers and users of spatial data products.

In the recent years, the broad consensus observed around the OGC proposals has made services realized in OGC compliance the de facto standard for the exchange of geospatial data in distributed environments. In particular, the OGC services represent 'an evolutionary, standards-based framework that enable seamless integration of a variety of online geoprocessing and location services' (Doyle and Reed 2001). Moreover, by using established technologies such as XML or HTTP, the OGC services can provide a 'vendor-neutral, interoperable framework for web-based discovery, access, integration, analysis, exploitation and visualization of multiple online geodata sources, sensor-derived information, and geoprocessing capabilities' (Doyle and Reed 2001).

Many factors have influenced the design of the OGC services; these include, for example, the need to interconnect services often provided by different organizations, the need to ensure requirements, such as access control and security, and the necessity for a client to know what service can be used with a specific type of data. However, although OGC services use HTTP as the transport protocol and XML as the lingua franca for the exchange of data, they are often incompatible with the platform proposed by W3C, mainly due to the fact that these two standards were developed in parallel by different organizations.

Basically, the first important difference between the two types of services is represented by the strong standardization imposed by OGC regarding the public interface of a geographic service. As a matter of fact, unlike W3C services, each OGC service represents a separate standard designed to handle a specific kind of data (Ioup et al. 2008). Each of these standards describes how a service should perform its tasks following a neutral approach for technology implementation, describes the service public interface and specifies additional parameters and data structures needed in all request and response operations. Furthermore, each OGC service that implements a particular standard presents a fixed interface whose functionality (and the type of data returned) is defined a priori. On the contrary, each W3C service can expose its own interface in a WSDL document, so that two services offering the same functionality could have two totally different public interfaces.

As for the OGC specifications, the consortium has proposed over time a quite wide and complete set of them, where the most widespread and commonly used are: Web Map Service (WMS), Web Feature Service (WFS) and Web Coverage Service (WCS). In particular, to facilitate the development process of these three types of services, the OGC has developed a Common Standard that defines all aspects that should be common to these three implementations in addition to their single operational requirements (Whiteside and Greenwood 2010). These common structures relate to some of the parameters and data types used in the various request and response operations. Moreover, every OGC service must provide a standard way to describe its capabilities to its clients and a client may obtain such a description invoking the standardized GetCapabilities operation. The implementation of such an operation is mandatory and, for a client, there is no other way to know what the capabilities offered by a specific server are.

The Common Standard requires that a request message for the GetCapabilities operation is encoded in an XML document or using the Key-Value Pair (KVP) encoding. A KVP is a set of two data items linked each other: a unique identifier (the key) for some data item and the value associated to it. This technical feature represents another important difference between the two types of services under discussion. W3C services only use XML, while OGC services are not limited to it. The response to a GetCapabilites request is represented by a document containing metadata about the operations supported by the server that implements a particular OGC specification. Such documents are usually encoded in XML, and use XML Schemas to specify the correct document content and structure. Moreover, the OGC Common Standard describes how the encoding of the request and response operations should be implemented when HTTP is used as transport protocol. The supported HTTP methods are GET and POST and, for each operation, the server must support at least one of these methods. In particular, for the request operations, the standard states that the encoding of parameters and operations within the POST method can be done by using either KVP encoding or an XML document. When using XML the operation request can be transferred "as is" over HTTP or contained in a SOAP envelope. On the other hand, with the GET method, only a KVP encoding can be used since a valid HTTP GET request can be seen just as a URL with a fixed prefix where additional parameters are added. Upon receiving a correct request, the service must send a response or an exception report if unable to correctly serve the query. In the following subsection, a quick overview of the Geographic Markup Language (GML), the cornerstone of the whole OGC architecture is given. Then, a brief description of the basic characteristics of the three main types of services is provided.

The Geographic Markup Language

Geography Markup Language (GML) is 'an XML grammar written in XML Schema for the description of application schemas as well as the transport and storage of geographic information' (Portele 2007). In other words, GML is an XML language to manage geographic information, where the GML schema describes the document and the instance document contains the actual data.

The main goal of GML is to provide a means to exchange and manipulate geographic information in a standard way that is guaranteeing a programming language and source format independency. In general terms, the role played by GML in the context of geographic information can be split into three main categories (Lake 2005):

- an encoding standard for the transport of geographic information from one system to another,
- a storage format for geographic information, and
- a modelling language for describing geographic information types.

As for the first two aspects, GML can be considered as the OGC answer to the need of representing geographic information in a standard manner in order to facilitate communication and data exchange among a wide variety of autonomous and distributed data sources thus contributing to the reduction of costs related to the management of spatial information. Like XML, GML represents information in a textual form, focusing on the content description and relying on other mechanisms for data visualization. Another advantage is that being XML-based, people can immediately use the plethora of available XML tools to perform all common XML operations. Additionally, the OGC standard lets users decide whether to store geographic information directly in GML or use some other storage format and convert it to GML only for data transportation purposes.

As for the third aspect, the only key point is the definition of the feature concept. According to the ISO19101 Reference Model, a feature 'is an abstraction of real world phenomena'; if such an abstraction is associated with an Earth location, the term *geographic feature* is used. Features are fundamental objects in GML (Burggraf 2006) and are described through a list of geometric and non-geometric properties. The former represent the basic geometry primitives and their aggregates, such as points, lines, curves, and polygons, while the latter describe the objects semantics and are expressed in the standard XML form. In GML a feature is represented by an XML element whose individual children elements describe a property. Furthermore, a feature can be defined as the result of the composition of other features. Therefore, the use of GML in conjunction with OGC services allows the implementation of infrastructures for sharing geospatial information in a globally accessible manner, independently of the different proprietary formats used (Burggraf 2006).

Web Map Services

A Web Map Service (WMS) provides an HTTP based interface for requesting georeferenced map images from one or more distributed geospatial databases (de la Beaujardiere 2006). Basically, the requester specifies elements of interest, such as an area, and the response consists of a map image referred to that area, returned in a picture format. Appropriate MIME types (e.g., "image/png") usually accompany the response objects.

According to the OGC standard, a generic WMS can be classified as a Basic or a Queryable WMS. A Basic WMS supports the GetCapabilities and the GetMap operations, while a Queryable WMS also supports the GetFeatureInfo operation, whose goal is to provide more information about the features contained in the pictures of a map. Textual output (usually XML documents) is also available and can be used to provide for an error description or response to information requests about the features shown on a map. A WMS has to support the HTTP GET method and may support the HTTP POST method.

Finally, the WMS specification represents another difference between an OGC service and a W3C service, namely the former can return binary documents (as well as XML documents), the latter always relies on pure XML documents (although, as described later, there are several ways to encode binary documents in a SOAP message, such as images and PDF documents).

Web Feature Services

While a WMS gives users the possibility to retrieve maps from multiple sources, a Web Feature Service (WFS) provides interfaces to access and manipulate the previously mentioned geographic features. The guidelines for implementing an OGC compliant WFS can be found in (Vretanos 2005). According to the standard, besides the mandatory GetCapabilities operation, a WFS has also to support operations allowing to Insert, Update, Delete or Discovery geographic features expressed in GML. To accomplish these tasks, five operations are defined:

- DescribeFeatureType: an operation used to describe the structure of any feature that a WFS can service. The only mandatory output is a GML application Schema (presently GML version 3). Such a Schema describes the features encoding (either for input or output operations) expected by the WFS.
- GetFeature: a function used to request and retrieve feature instances. A client has to be able to specify the desired properties for each feature it wants to fetch. WFS may respond to a GetFeature request in two ways. It can return either a complete document or simply a counter corresponding to the number of features that the GetFeature request would return. The optional resultType attribute in the request message is used to specify how a WFS should respond to a GetFeature request.
- GetGmlObject: a client specifies the identifier (ID) of a GML object and the WFS returns it.
- Transaction: a transaction request is made up of operations (create, update, delete) that may alter the state of one or more features.
- LockFeature: when supported, a client can ask WFS to lock one or more instances of a feature type during a transaction. This operation is crucial to ensure consistency in scenarios where a client modifies a feature and sends it back to the server (using the above-mentioned Transaction operation).

Depending on the supported functions, WFS can belong to two major categories: Standard (or read only) and Transactional. A Standard WFS has to support the DescribeFeatureType and the GetFeature operations, a Transactional WFS has to implement the Transaction operation and optionally the GetGmlObject and the LockFeature operations. The encoding of the requests can be done by using either KVP values or XML, while the state of geographic features should be encoded using GML.

In a WFS the use of three normative namespaces is also defined, namely:

- http://www.opengeospatial.net/wfs for the WFS interface vocabulary.
- http://www.opengeospatial.net/gml for the GML vocabulary.
- http://www.opengeospatial.net/ogc for the OGC Filter vocabulary.

Finally, as for the underlying transport protocol, at least one between the HTTP GET and the HTTP POST methods has to be supported and response messages should be accompanied by the appropriate MIME type and by other appropriate HTTP entity headers. Moreover, with the HTTP POST method, the usage of SOAP is also possible. In fact, a client may send requests by using a SOAP message and the WFS may respond with another SOAP message. Nothing is mentioned about the structure of the SOAP Header.

Web Coverage Services

The Web Coverage Service (WCS) standard defines an interface for the exchange of geospatial information representing phenomena known as coverages that can vary in space and time, a specialized class of features (Baumann 2010). Like WMS and WFS, in WCS a client has the possibility to specify the desired criteria for its queries. The mandatory operations that a WCS service has to support are: GetCapabilities, DescribeCoverage (a client submits a list of coverage identifiers and the service returns, for each identifier, the description of such coverage) and GetCoverage (a client requests the processing of a particular coverage from a WCS service). The use of the HTTP GET with KVP encoding or HTTP POST with XML encoding are both supported in WCS. In addition, the "OGC WCS XML/SOAP Protocol Binding Extension document" specifies how WCS clients and servers can communicate using the SOAP protocol.

Services Composition

One of the primary properties that should be guaranteed when realizing services consists in their seamless integration. Indeed, in addition to being platform or operating system independent, one of the hallmarks of service philosophy is the possibility to compose two or more services, developed by different entities, in order to obtain a higher-level service that can meet needs of a company or an institution. However, despite the simplicity behind the general idea of services composition and some basic assumptions, such as the subdivision of the involved services into the three general categories, and the use of XML for messages exchanges, services composition is a complex task and several facets must be wisely considered. As clearly explained in

Dustdar and Schreiner (2005), the business logic does not lie entirely in a single monolithic entity, that is the operations and the algorithms that handle data and information exchange may also result from the composition of different services. This fundamental characteristic leads to the need of considering some key issues, such as the necessity to coordinate the sequence of operations to ensure the accuracy of computation and avoid inconsistencies, and the necessity to use a transaction protocol (e.g., the WS-Transaction (OASIS 2009) protocol used in Web services stack) suitable either for short or long running tasks. Besides these general problems, another important challenge described in Dustdar and Schreiner (2005) is related to the choice of the composition strategy that could be automated or manual, static or dynamic, that is whether the composition takes place during the design phase of the whole architecture or at run time. In addition, any composition mechanism, besides providing a dynamic and flexible composition model, has to satisfy several non-functional requirements such as scalability, security and dependability (Milanovic and Malek 2004).

Among the various approaches proposed in literature, two types of service composition are widely used, namely choreography and orchestration.

As for service choreography, each service involved in the composition knows its role in the whole interaction. Choreography implementation is fairly simple, but discovering the source of a malfunction can be a difficult task. The Web Service Choreography Interface (WSCI) is one of the most widespread standards for the choreography of services (Barros et al. 2005). Service orchestration, instead, describes how Web services can interact by exchanging messages, including the business logic and the execution order of the interactions, thus benefiting from loosely coupled services. For simple orchestrations involving few components, a possible and quite basic approach consists of the usage of a traditional programming language to link the various components. However, since programming languages are mainly focused on the definition of classes, methods or structures rather than on the overall execution process, for complex orchestrations it has been necessary to develop service composition standards accepted and used by all the involved entities. Among the current standards and proposals for Web services orchestration, the Web Services Business Process Execution Language (WS-BPEL) (Alves et al. 2007) represents a robust and widely adopted solution. In the following subsections, a brief overview of the WS-BPEL process is given along with the description of its main features. Then, in order to build the basis for an integrated solution, requirements and properties for the OGC service composition are detailed and several approaches are presented. Finally, an initial analysis of issues related to the Web services and OGC services orchestration is performed also with relation to the OGC program for the interoperability.

Elements of a WS-BPEL Web Service composition

WS-BPEL defines

> 'a model and a grammar for describing the behaviour of a business process based on interactions between the process and its partners. The interaction with each partner occurs through Web service interfaces, and the structure of the relationship at the interface level is encapsulated in what is called a partnerLink.'
>
> (Alves et al. 2007)

From a general point of view, WS-BPEL can be seen as a scripting language to create applications by composing existing Web services. Like other Web services standards, WS-BPEL is expressed by using XML, through XML Schema metadata, and depends on several W3C specifications, the most important of which is WSDL. In particular, the data model used in WS-BPEL processes is specified using either XML Schema or the message element of WSDL documents. There can be two types of WS-BPEL composition (named WS-BPEL process), namely Abstract (useful, for example, to describe a process "template") and Executable (fully specified and executable).

Figure 4 shows the structure of a typical basic WS-BPEL definition. The <PROCESS> element represents the root element of the definition. Each

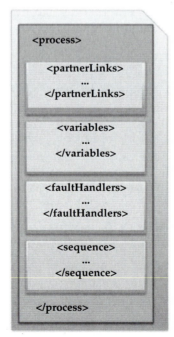

Fig. 4. The definition of a typical WS-BPEL Process.

<PARTNERLINKS> element contains the <PARTNERLINK> children, a concrete reference to WSDL services (named Partner Services) that will take part in the execution of the business process. In particular, for each <PARTNERLINK> element, the partnerLinkType attribute identifies the portType of the referring service. The <VARIABLES> element defines the data variables used during the process workflow and their definition is provided in terms of WSDL Message Types, XML Schema Types or XML Schema Elements. The <FAULTHANDLERS> element is used to define what to perform in case of fault (for example, during the invocation of a service). Finally, the <SEQUENCE> element lets process designers organize a series of building blocks (named activities) executed in a sequential and predefined order. Besides the sequential order, the whole process logic can be structured in several other ways, such as conditional branching (<if> element), iterations, repetitions (<while> or <repeatUntil> element) and parallel processing (<flow> element).

Composing OGC services

Due to its strong reliance on WSDL, it is quite difficult to use WS-BPEL for the composition of OGC services. As a matter of fact, there is no need for OGC services to be equipped also with a WSDL document. Moreover, WS-BPEL lacks support for the direct management of non-XML data, e.g., the binary data, such as the image files returned by a WMS. For these reasons researchers have investigated other modalities to manage orchestration of geographic services.

A valid example of an OGC services orchestration is provided in Stollberg and Zipf (2007). After describing difficulties of using WS-BPEL for the orchestration of OGC services, in this paper the authors propose the use of the OGC Web Processing Service (WPS) standard introduced in Schut (2007) as a feasible solution.

The WPS standard provides for a standardized interface that aims at supporting publication and discovery of geospatial processes. According to the standard (Schut 2007), a geospatial process includes 'any algorithm, calculation (e.g., polygon intersection) or model that operates on spatially referenced data'. In particular, the purpose of the WPS interface is to standardize the way processes and their input/output are described, how a client can invoke the execution of a process and how the output should be treated. Data that can be used during the elaboration can vary from image data formats to data exchange standards such as GML.

To achieve this goal, three mandatory operations are specified:

- GetCapabilities,
- DescribeProcess that allows a client to request and receive detailed information about processes that can be run on the service instance,
- Execute that allows a client to run a specified process implemented by the WPS.

In the proposed example, based on the search for adequate evacuation shelters in a bomb threat scenario, the authors first show how the use of WPS can supply all the traditional well-known GIS functionalities. Then, they also point out the need to combine all the developed services to represent them as a single application, which is the main idea behind the service composition (Stollberg and Zipf 2007). To implement such a composition, they observe that, although the specification for the WPS mainly focuses on the implementation of geoprocessing methods, there are no restrictions on what can actually be implemented as a "WPS process", that is a WPS can also be used as a service that coordinates an orchestration of geographic Web services (Stollberg and Zipf 2007). Moreover, they observe that by using WPS for composition purposes, three different approaches are possible. The first two belong to general categories known as Centralized Service Chaining and Cascading Chaining. In the former, a single service controls the entire workflow invoking all other services in order to achieve a goal; in the latter, services communicating each other can directly exchange data. It is clear that these concepts are quite similar to the general definitions of orchestration and choreography. The third option available with WPS is to combine all functionalities into a single WPS implementation.

The ability to compose and orchestrate OGC services is of paramount importance for the development of Spatial Information Infrastructures (SII), distributed systems based on SOA principles that allow the processing of spatial data in order to provide useful information to the final user (Rautenbach et al. 2013).

Some guidelines that describe the general structure of a service-oriented SII can be found in an OGC best practice paper (Whiteside 2005) which 'summarizes the most significant aspects of the Open Geospatial Consortium (OGC) web services architecture'. The proposed architecture is based on the following properties (Whiteside 2005):

a) A multiple tiers subdivision of the various service components.
b) Support for Transparent, Translucent or Opaque service chaining.
c) Use of open standards for the definition of the service interface.
d) Use of open Internet standards for the service communication.
e) Independence from specific hardware or software platforms.

While the last four points have been previously mentioned and discussed, the technical choices carried out for the first point deserve some further considerations. From a high level perspective, geographic services are loosely arranged in four tiers as shown in Fig. 5. These tiers are geographic data independent although each of them includes specific services that deal with geographic data. Each tier can offer its functionality both to other tiers and also directly to final clients. In fact, a complete separation in four tiers is not always required and clients can bypass the tiers that are not needed. Such a characteristic is useful especially when a complete separation could be inefficient for the purposes of the final system.

The first tier, the Information Management Services includes services that provide access to geographic datasets and are usually used to retrieve a subset of such data useful for the client. WMS and WFS are typical examples of services that belong to this layer. The Processing Services tier contains services designed to process data. Services in this tier can either use other services of the same tier or services from the Information Management Services layer. The Application Services tier encompasses those services useful to provide support functions to final clients of the system, in particular web browser clients.

The development of a SII is a non-trivial task exhibiting different challenges that, currently, can be overtaken only by adding some form of customization to the various services functionalities, thus proving the lack of flexibility and generality of current standards.

A complete example that demonstrates how the use of the various OGC standards can constitute a possible basis to perform distributed geoprocessing and that investigates the difficulties arising during the

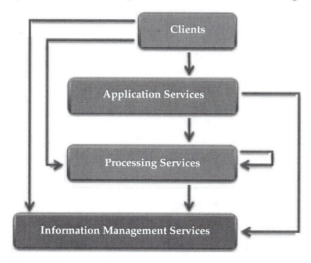

Fig. 5. OGC Proposed Architecture–Service tiers subdivision.

development of such a type of systems can be found in Friis-Christensen et al. (2007). In this paper the authors describe a use case for the development of an application that computes forest fire statistics. By it they present their architecture for the distributed computation and analyze and discuss some challenges related to the use of a SOA to perform geographic computation. The proposed architecture is a traditional multilayer system where the data access services (WMS, WFS) provide access to data stored in the various distributed geodata repositories. An additional discovery service layer is used to provide users with a catalog service to allow them to discover what data could be useful for their needs. The execution of the diverse steps for the actual calculation of the various statistics is performed in the geoprocessing service layer. Finally, the client of the application is a traditional web application.

The prototype and the underlying architecture are used by the authors to make some interesting considerations that are worth mentioning here. First of all, the various steps needed for the effective calculation of the statistics are performed in a single geoprocessing service. This is a simple choice and guarantees good performance but limits the general flexibility since the different operations required cannot be reused in another application. A good solution could be the development of a single service for each functionality, which then have to be chained together in order to get the desired result. The three types of service chaining identified by the authors correspond to those supported by the proposed OGC architecture (transparent, translucent, opaque). In the first type the workflow is managed by a human user, in the second type a service that controls the chain is invoked and the human is aware of the various steps, in third type an aggregated service is invoked and the user does not know the single steps. The first two types of service chaining are analyzed.

The first approach provides a high flexibility but the client application must continuously interact with the service and the continuous transmission of input data can increase the total computation time. The translucent approach could solve some of the transparent approach issues, since the entire workflow is sent to the service instance in a unique step. However, the WPS specification would require some adaptation in order to support such an approach. The second important observation concerns the amount of time required to perform the processing of voluminous spatial data. This issue does not only affect the final user experience but might also have an impact on the services involved in the process. According to the authors the problem is caused by the synchronous communication mechanism on which the proposed architecture is based. A possible solution is the use of asynchronous messaging (supported by the WPS specification) for time-consuming operations. In such a modality the service response is provided at a later time in a different communication session (Friis-Christensen et al. 2007). The retrieval of information about the status of the process can be performed in two different ways: the pull and the push mechanisms. In the former the client carries out a periodic check, in the latter the service provider sends a notification about the status of the process. According to the authors the push mechanism is more convenient to alert humans (e.g., by email or SMS) while the pull mechanism is more suitable for the machine-to-machine communication. The pull mechanism is directly supported by the WPS standard. In such a mechanism the response of a WPS to the Execute request is an XML file containing a link pointing to a constantly updated Execute response document. During the processing this document contains the status of the operation, when the processing is finished the provided URL contains the link to the final result. However, the asynchronous approach raises other research questions, such as what happens when an asynchronous call is performed in a service chain or the need to define a policy to deal with the data referenced by the URL contained in the XML response file.

Another type of difficulty that can arise during the development of an SII is represented by the lack of supporting infrastructure for geodata processing. An analysis of available frameworks for the orchestration of Geospatial services can be found in Rautenbach et al. (2012) where two different platforms are discussed and analyzed: the 52° North framework and the Zoo project. The characteristics of these two solutions are tested by using the production of thematic maps as case study. Both the frameworks are operating system independent, available as open source and compatible with the current WPS standard. The 52° North framework is Java based and offers a web admin tool that helps the uploading of WPS processes, an orchestration API and a graphical modelling tool for the organization of the geoprocessing workflow. On the other hand, the Zoo project is made up of three main components: the Zoo kernel, the Zoo services and the Zoo API. The Zoo kernel is the module that allows the WPS creation and management. The Zoo services communicate with the Zoo kernel and are composed of two parts: a configuration file that describes the service, and the code that the final user wants to turn into a Web service. Finally, the Zoo API is a Javascript library that can be used for process creation and chaining. The two frameworks are compared against a wide set of characteristics, from the available documentation to the ease of integration with other GIS applications. Each of them has strengths and weaknesses (e.g., the lack of semantic information and support to the BPEL standard) but the results show that it is possible to use OGC standards such as WMS, WFS and WPS in order to orchestrate a thematic map service.

A more detailed example of OGC services orchestration to produce thematic maps can be found in Rautenbach et al. (2013). In addition to the traditional OGC services, for the development of thematic Web services the authors discuss also the use of the Common Query Language (CQL) and some non-standard extensions to the Styled Layer Descriptor (SLD). The former is a formal language to express queries to information retrieval systems, the latter is an OGC XML Schema useful for describing the appearance of a map. The described extensions can be used in conjunction with GeoServer, a Java based server that supports the editing and sharing of geospatial data on the Web. The previously mentioned 52° North framework has been used for the orchestration of OGC services. The issues raised by the authors include the impossibility to run asynchronous operations, the need to perform some programming tasks for wrapping, in a WPS, all the statistical processing needed for the generation of a thematic map and the use of some GeoServer extensions to SLD. As such customizations are not part of an OGC standard, the portability of the proposed solution is limited.

The OGC Program for the Interoperability

When compared to W3C services, OGC services represent a totally different standard. However, the growing popularity of SOAP, WSDL and WS-BPEL and the awareness of the great advantages that could result from the possibility of seamlessly combining these two worlds, led the OGC to set a special working group in order to provide general recommendations and guidelines for adding WSDL/SOAP support to existing and future OGC services.

The first result is the awareness that for OGC services there is the need to define an Interface Definition Language that is a language to describe the interface of a software component, usually in a language independent way (Schäffer 2008). A possible choice is, of course, the use of the WSDL. Currently, in OGC services the role to describe the available operations is carried out by the GetCapabilities function although a complete intersection with the WSDL specification is not possible. The main difference is that WSDL focuses mainly on the description of the explicit interface providing for both the list of available operations and the types of input and output messages, while the GetCapabilities provides only for the list of all operations along with meta-information. The proposed solution is that the GetCapabilities should list a path to a WSDL file that describes the OGC service. A complementary approach is the possibility for an OGC service to be discovered by a WSDL document and then, additional metadata could be fetched by using the traditional GetCapabilities operation. Other important differences outlined in (Schäffer 2008) are the binding type and the binding time of operations. In W3C services the message payload is completely defined at design time while in the OGC services the type of a response message can dynamically vary based on the client requests (e.g., in the WFS GetFeature operation).

The last problem discussed here is how the SOAP protocol can be used in conjunction with traditional binary data, such as the images returned by a WMS. Binary data, usually called "opaque data" (Powell 2004) often constitute a problem for (Web) services based solutions. First of all, the serialization (i.e., the translation of an object into an XML stream) of such data into XML documents is not always an easy or feasible solution. For example, documents with digital signatures could lose their integrity (Powell 2004). To deal with this problem two predominant techniques can be followed (Bosworth et al. 2003), namely either embedding, in some way, the opaque data in an XML element or referencing it as an external entity. The former is the currently used solution because the latter is inapplicable in a W3C compliant service environment, being based on XML features prohibited by the SOAP standard (such as Document Types Declarations). In XML, the support to binary data is usually achieved by using the Base64 or the hexadecimal text encoding. The result for both is a sequence of octets (Bosworth et al. 2003). However, although these two encoding solutions are very simple to implement, a well-known problem concerns the size increase of the binary data as well as the overhead caused by the processing time needed to perform the encoding and decoding operations. To overcome these performance issues other proposals have been suggested, such as SOAP with Attachment (SwA) (Barton et al. 2000). SwA relies on the fundamental concept of MIME multipart messages, which simply means that a message is split into two or more parts and hence can include multiple attachments. The MIME standard specifies how these parts should be combined to form a single message. In SwA, the traditional SOAP message constitutes the root part of the MIME multipart message and the SOAP Body element contains explicit references to other parts of the MIME multipart message, which may contain arbitrary data (Schäffer 2008). A problem with this approach is that it does not work with other fundamental components of the W3C service stack, such as the WS-Security protocol (Powell 2004; Nadalin et al. 2006). Therefore the solution currently used (recommended also by the previously mentioned OGC working group) is represented by the Message Transmission Optimization Mechanism (MTOM). MTOM is a W3C recommendation that provides an efficient mechanism for exchanging large binary data by using SOAP messages and is based on another W3C recommendation, namely the XML binary Optimized Packaging (XOP) that, among other advantages, does not require the time consuming task of the Base64 encoding (Schäffer 2008). Opaque data in MTOM are treated in a manner similar to what happens in SwA but, in this case, the SOAP message consists of the whole MIME multipart message (Schäffer 2008).

This makes it compatible with all other high level protocols of the W3C service stack (Powell 2004).

Unfortunately, due to the large number of existing geographic services not supporting SOAP or WSDL, the integration of W3C and OGC services is still challenging. However, as discussed in the next section, the growing necessity to integrate the two worlds has led researchers to investigate possible solutions that would allow the current W3C and OGC services to communicate.

Web and OGC Services Integration

The opportunity to use the enormous amount of geographic information accessible via OGC services within W3C services is one of main reasons stimulating the efforts to seamlessly combine these two different worlds. In fact, geographic community recognizes that a more complete integration with the SOAP and WSDL protocols would allow for employing all standards specified for the W3C platform, such as those relating to security and rights management.

To reach this aim, some issues have to be faced which cannot be solved by a mere mechanical process meant to make a translation from one service standard to another by simply transforming an OGC service interface into a WSDL document. Indeed, other aspects should be considered, e.g., the management of the metadata returned with Capabilities documents. In addition, it is relevant to avoid, as much as possible, both moving the computational complexity of the whole operation to the service client, and making any substantial changes to already existing services.

The focus of the following subsection is on a review of current solutions presented in literature. They share, as an underlying idea, the design of a service wrapper or a proxy meant to provide for geographic information in a W3C compliant way, keeping the structure of existing W3C or OGC services unchanged. Then, the INSPIRE directive is presented, which aims at creating a common spatial data infrastructure for facilitating the sharing of environmental information. Finally, a brief discussion about SOAP performances is done.

An overview of current solutions

A first proposal can be found in a discussion paper from OGC (Gartmann and Schäffer 2008). In this work, the authors propose a generic approach to equip OGC services with a SOAP binding that allows for the transformation of any HTTP GET or POST request into a SOAP request. The proposed solution might be used as a basis for the construction of a wrapper to be applied to all OGC services, thus making the SOAP transformation completely transparent to the client application. The proposed architecture consists of a server-side proxy and a client-side proxy. The client-side proxy receives the HTTP GET and POST requests and transforms them into a SOAP message, while the server-side proxy receives this SOAP message and restores the original HTTP GET or POST request. For the response messages, the contents must be properly XML encoded in order to incorporate them in a SOAP document and, in particular, the previously discussed MTOM standard could be used for binary data and for those services that can also return plain or html text, such as the WMS.

A more complete analysis of problems arising during the integration of W3C and OGC services can be found in (Ioup et al. 2008). In this paper, some techniques for 'dividing OGC services and mapping capabilities into multiple SOAP services' are presented and discussed. The core idea is to split a single OGC service into multiple atomic W3C services each representing a single geospatial dataset. As for the effective implementation, also these solutions are based on the creation of a wrapper aimed at solving a series of integration problems that the authors categorize into Data Handling, Functionalities Mapping and Metadata Management issues. The authors' work arises from a critique to the solution proposed in Gartmann and Schäffer (2008). Their evaluation is based on the observation that OGC services are based on a two steps process: a generic client, first queries a server to know its capabilities and then uses the various functionalities of the OGC service to get the real data. The authors argue that by adding a simple SOAP transformation a W3C service client should perform three steps: get the WSDL, get the server capabilities and then get the real data. Therefore, on the basis of such a possibility, they discuss issues that have to be faced to realize this split. The first issue to deal with concerns the Data Handling since, as previously discussed, OGC services are not limited to XML as a data exchange format. Moreover, different OGC services may return different data types. A viable solution could be the possibility for a W3C service to return only string data types. Although simple, this option is not admissible when it is necessary to return data in binary format. A different solution for binary encoded data could be to simply return a URL pointing to actual data. Through it, a client could directly contact the OGC service and retrieve the binary information. By this simple approach, a Web service is not in charge of managing data in binary format and might not act as a proxy for OGC services specific data. However, this approach transfers the computational complexity to the clients, a not always performing solution. Finally, a relevant part of the communication occurs outside the W3C service stack, thus blocking the use of the aforementioned standards, such as WS-Security. These motivations make this simple to implement proposal unfeasible. The only way to overcome these issues is the creation of a full wrap around the OGC service, thus forcing clients to communicate exclusively with the W3C compliant wrapper. Based on such a solution, a W3C service could be able to handle even binary data. For this task the authors propose to use the standard MTOM or, in case the Web service environment does not support this standard, the use of Base64 encoding (Ioup et al. 2008).

The second problem the authors discuss concerns the mapping of functionalities. In this case, some issues arise due to the fixed set of functionalities that OGC services have and that are described in the Capabilities document. To solve this problem, in Ioup et al. (2008) two main methods are discussed. In the first method the WSDL document simply lists all the available operations of an OGC service, thus the same WSDL specification can be used for all the OGC services of the same type. However, a drawback exists, namely a direct mapping would lose information since the exact dataset of each OGC service can be retrieved only by parsing the document returned by the GetCapabilities function. For this reason the authors propose an alternative solution which represents the second method that they discuss in the paper. It consists in a direct mapping between the OGC service dataset and the W3C service functionalities, i.e., the mapping does not occur at individual functionality level. In particular, a wrapper can expose directly the OGC service data layer. This task can be performed in two different ways: 1) all data layers available in a single OGC function are mapped into a single and atomic W3C service; 2) each data layer is mapped into a different W3C service; in this case function names will be the same for each Web service, but every service will return a single and different data layer. However, from a Metadata Management point of view, in the W3C services realm there is no standard for the management of spatial metadata while almost every OGC service requires them, thus removing them would mean to lose a lot of their usefulness. Then, it needs to provide a way to offer such metadata in a WSDL document and in Ioup et al. (2008) the authors suggest to consider that, although WSDL documents do not contain metadata (except for those contained in functions provided by the service) such documents are, to some extent, extensible. Therefore, the proposed method includes metadata in the extensible part of a WSDL document. In particular, they suggest to include them in the <SERVICE> element of the document and to use XML Schema to encode the limits of the input parameters. Some constraints have nevertheless to be satisfied, namely the whole set of available metadata of an OGC service should be supported and they must not interfere with the proper use of the WSDL document. A further benefit of this solution concerns the possibility of preserving the two steps process of getting the capabilities and then executing an operation.

The difficulties to integrate efficiently Web services and OGC services are analyzed also in Amirian et al. (2010). By using as a case study the development of a software system for the management of the Urban Services

Data (USD) of a large city, the paper investigates the communication issues that arise in environments where the underlying data must be accessed by different types of users, in a reliable and up to date manner. In this case, further difficulties are represented by the needs of each system user and their way to access such data, namely different computing platforms and communication technologies, while performance and reliability aspects must also be taken into account. As for the integration issues, a classification into Data Types handling, Functionality Mapping and Metadata Delivery issues is adopted as in Ioup et al. (2008). To deal with the first task (Data Types handling), four different approaches are discussed and evaluated, namely the translator Web service, the physical Web service, the wrapper Web service and the common back end.

In the first approach, a Web service receives SOAP messages from a client and translates them in a format suitable for the target OGC service. This suitable format is sent back, in a SOAP message, to the client, which uses it to retrieve data directly from the OGC service. In the physical Web service approach, the Web service translates the SOAP request and sends it directly to the OGC service; the OGC service generates the binary file and stores it in a permanent location on the server. The link to this location is sent back to the W3C service, which forwards it to the client in a SOAP message. Finally, the client uses this physical address to retrieve the binary data. In the wrapper Web service approach, the wrapper service catches the response message from the OGC service and sends it back to the client by using only W3C encoding. The whole communication between client and service is totally Web service based. As for the binary data returned by an OGC service, the preferred choice is, again, the use of the MTOM standard. In the last approach, the common back end, a W3C service and an OGC service provide for 'two direct gateways to the same server engine' (Amirian et al. 2010). A client can either send requests directly to the OGC service or can query the W3C service, which will provide responses in a W3C compliant way. Moreover in the latter case, according to the need of the client, the W3C service can return the possible binary data using their physical address, the Base64 encoding or the MTOM standard. Flexibility and performance are the main advantages of this approach. As for the mapping of functionalities these two approaches and those discussed in Ioup et al. (2008) are quite similar. In the former approach, a Web service replicates the same methods of an OGC service in a SOAP compatible way. In this case a client must first invoke the GetCapabilities method, parse the Capabilities document and finally invoke the desired OGC functionality replicated by the W3C service. The main disadvantage of this method is that a service consumer is able to exploit a service by exclusively using the published service contract, namely its interface. In the latter approach, the one to many mapping, each data layer of an OGC service is mapped by a specific Web service resource. In particular, a single Web service can expose a single data layer (*multiple services method*), alternatively, a Web service can expose a function for each single OGC service data layer (*facade service*) (Amirian et al. 2010). Finally, for the metadata delivery, three possible solutions are analysed, namely GetCapabilities function, WSDL extension, and Metadata exchange. In the first approach, each Web service provides a GetCapabilities function and a client has to parse the returned values. In the second approach, all relevant metadata of an OGC service are put in the extensible part of a WSDL document. The third approach, instead, uses the W3C WS-Metadata-Exchange specification (Ballinger et al. 2008), that describes a standard format to encapsulate metadata. The main advantage of the third solution is that the usage of a standard and documented way for delivering service metadata drastically reduces the need of developing customized solutions for metadata retrieval.

In Sancho-Jiménez et al. (2008), the integration of OGC and W3C services is addressed from another point of view. A method to automatically retrieve the SOAP interfaces and the WSDL metadata starting from the mandatory operations (GetCapabilities, Describe Process, and Execute) of any WPS is proposed. In order to provide these interfaces, the underlying idea is the creation of an intermediary proxy which is made up of two sub-modules. The former generates the WSDL metadata used to describe the WPS interface, the latter adapts a SOAP message in a request suitable for the WPS interface. As the proposed solution is an automatic derivation method, in order to generate the WSDL document, a request to the proxy must contain the URL of the WPS. Through this URL, the proxy uses the WPS public interface to retrieve all the information needed for the generation of the WSDL document. Moreover, the proxy offers the possibility to get both one document containing all the operations offered by the WPS, and a single document for each of them. Each generated WSDL document will include both the GetCapabilities and the DescribeProcess specification along with an Execute method for each operation offered by the particular WPS. In addition, in order to accurately define the parameters of the Execute operation, a parsing of the DescribeProcess response is performed. Finally, on receiving a SOAP request, the proxy first parses the message, gets the WPS URL and then generates the request and invokes the WPS public interface.

The INSPIRE Directive

Another important example of difficulties met when combining the two standards in a heterogeneous environment handling large amounts of data can be found in the extensive documentation provided by the Infrastructure for Spatial Information in Europe (INSPIRE) (Villa et al. 2008a,b; Villa et al. 2009).

The INSPIRE project is an effort of the European Community to create a common spatial data infrastructure aiming at facilitating both the sharing of environmental spatial information among public sector organizations and the public access to spatial information across Europe. The fulfilment of such a common spatial data infrastructure is a non-trivial task that requires overcoming some fundamental issues, such as the fragmentation of datasets and sources, the lack of harmonization between datasets and the duplication of information. Moreover, the open standards compliance, the definition of a common contract for all interfaces, and the data description and representation constitute other key aspects addressed by INSPIRE. In particular, for the pressing need to use wide adopted standards, the INSPIRE Network Services SOAP Framework (Villa et al. 2008b) describes core ideas behind the proposal of a SOAP framework for the INSPIRE infrastructure as well as issues and solutions related to different geospatial domains. In Villa et al. (2008b) the authors observe that although based on open standards the existing OGC services support a mix of protocols and technology bindings. Unfortunately, such a technology mix could slow the integration and the implementation process, thus it should be avoided in order to get the maximum benefit from the offered services. Then, on the basis of all possible solutions, risks and requirements, SOAP has been proposed as the default communication protocol and binding technology for the INSPIRE services. The authors (Villa et al. 2008b), also present some criticisms to the OGC about the Consortium decisions concerning the main aspects of a hypothetical SOAP framework, namely there is not a common choice for data encoding, data transport and representation and for a profitable use of SOAP Headers. The proposed INSPIRE framework focuses on the following topics (Villa et al. 2008b):

- Standard compliances
- Underlying protocols binding
- Use of the SOAP Header
- Exception report in SOAP encoding
- Data encoding style and use
- Binary data transport and representation.

As for the standard compliance, beside the choice of SOAP as communication protocol, significant role is given to the Web Services Interoperability Organization (WS-I) Basic Profile recommendations (Ballinger et al. 2004) in order to provide the highest level of interoperability. The WS-I Basic Profile consists 'of a set of non-proprietary Web services specifications, along with clarifications, refinements, interpretations and amplifications of those specifications which promote interoperability'.

In contrast, the OGC proposals are not WS-I compliant. As for the use of the SOAP Headers, Villa et al. (2008b) present some concrete examples of how such headers can be effectively used. In fact, although their use in the OGC integration proposals is not compulsory, the authors propose to use them to manage, in a modular way, some common aspects of the INSPIRE Web services, including for example the use of SOAP Headers for security purposes. In fact, a header block could be used for carrying security-related information to a specific recipient. Another way to take advantage of SOAP Headers consists of using them for checksums and signature purposes: a message producer could want to provide recipients with a message along with a means to determine whether a message was altered during its path. Moreover, a Header block could carry information useful to a receiver to check the integrity of the binary data attached to the SOAP message.

A SOAP Header could also be used to provide for human readable information without interfering with the content of the Body of the SOAP message. Finally, a SOAP Header could be used to transport all metadata or information not strictly connected with the message encoded in the Body element. INSPIRE suggests this solution also to solve the multilingualism issue.

SOAP performances

As previously discussed, the advent of the SOC paradigm has revolutionized the way distributed applications are designed and implemented. Among the various standards upon which such solutions are based, SOAP represents one of the cornerstones for the achievement of interoperability among heterogeneous systems. However, the massive adoption of SOAP has stimulated a debate around a fundamental aspect, namely its performance during the implementation of complex and large scale distributed applications.

A quite complete description of the SOAP performance issues and a review of the research efforts aimed at SOAP performance enhancement can be found in Tekli et al. (2012). For the purposes of this section, the main performance issues are mentioned that could occur when SOAP is used to transfer large amounts of data, such as in distributed geographic environments. According to Tekli et al. (2012), the performance metrics of service-oriented environments can be grouped into three main categories: the response time, the throughput and the network traffic. The response time (latency time) can be seen as the time perceived by a client to get a response; the throughput can be seen as the number of requests fulfilled in unit of time, and the network traffic is the total size of messages exchanged during the whole communication. By taking these metrics into account, the use of SOAP exhibits several problems mainly due to the XML message encoding and decoding, the verbosity of XML and its redundant textual characteristics. First of all, the latency and the network traffic produced by SOAP are considerably higher compared to technologies like CORBA or Java RMI. These issues can have an even greater performance impact when, for example, the client is a mobile device where the available bandwidth can be low and the latency very high. Moreover, the process of converting a memory object into an XML object and vice versa is a quite expensive computational task, able to consume over 90 percent of the entire end to end SOAP processing time. The performance problems are amplified also by difficulties encountered by traditional hardware architectures to simultaneously evaluate multiple conditions, which represents a central issue in XML string and character processing. Finally, the addition of security policies to SOAP messages adds another source of overhead. In fact, the use of WS-Security protocol has a big impact on both the processing and response time and on the SOAP message size. In this case, the problem is due to the need of providing for a message level security instead of the traditional channel-level security, where SSL/TLS on the HTTP protocol is used.

An empirical proof of the performance problems occurring when using W3C service technologies can be found in Zhang et al. (2007). In this paper, a prototype system built to evaluate the performance of composition and invocation of geospatial functionalities by using Web services is discussed. The experimental results agree with the above discussion and show that the trade-off between convenience and overheads are acceptable for small volumes of geospatial data, although the large response time for ample data volumes represents a problem that cannot be underestimated.

The Representational State Transfer paradigm

A totally different architectural style for the development of distributed applications is represented by the Representational State Transfer (REST) paradigm (Webber et al. 2010). Many researchers and practitioners judge this paradigm as a feasible way to realize distributed applications eliminating the intrinsic complexity of the Web services standards (Pautasso et al. 2008).

Two fundamental notions in this context are the concept of *resource* (i.e., any meaningful information that can be addressed) and the representation of it (i.e., a document that captures its current state).

The REST architectural style is based on four principles (Pautasso et al. 2008):

a) Resource identification through URI: each resource is identified by a URI, which basically represents the endpoint where the desired resource is located.
b) Uniform Interface: the manipulation of resources is obtained through a fixed set of operations (create, read, update and delete) through the usage of the standard HTTP PUT, GET, POST and DELETE methods.
c) Self-descriptive messages: since resources are decoupled from their actual representation, they can be accessed in more than one format (e.g., PDF, JPEG, etc.).
d) Every interaction with a resource is stateless.

One of the advantages of the REST paradigm is the ease of development (the services description, for example, does not rely on standards, such as WSDL, it is more informal and human oriented) and the fact that the required standards (such as HTTP) and related infrastructure are already pervasive. Moreover, the possibility to choose lightweight data-interchange formats, such as the JavaScript Object Notation (JSON) makes it easier to optimize the performance of the service (Pautasso et al. 2008). Support for caching or load balancing is also available.

The OGC growing interest towards the use of this paradigm to perform distributed geoprocessing has led to the proposal of a new standard, the GeoServices REST API briefly summarized here. The proposed REST API

'provides a standard way for web clients to communicate with geographic information system servers based on the Representational State Transfer (REST) principles. Clients issue requests to the resources on the server identified by structured URLs. The server responds with map images, text-based geographic information, or other representations of resources that satisfy the request.'

(Portele and Sankaran 2012)

This OGC candidate standard traces its roots to the ESRI GeoServices REST Specification Version 1.0 (ESRI 2010) originally developed by ESRI and released in 2010 as a non-proprietary open specification (Portele and Sankaran 2012). An interesting aspect concerns the fact that this proposal is based both on REST principles and on pragmatic considerations about the effective support of the HTTP protocol by current web browsers or environments. For example, the use of the HTTP POST method is needed when the size of a URL is greater than 2048 characters, a quite common event in the context of geographic information.

The GeoServices REST API is stateless and each request contains all the information necessary for a successful processing. The standard representation for resources is JSON (although, developers may support resource representations in other different data formats) but in order to be compatible with the majority of existing platforms, the API supports only the GET and POST methods. A hierarchy of endpoints (URL) provides access to all resources and operations exposed by the GeoService. Some resources, in particular, are "controller" resources and can be used to perform operations, such as querying and editing information stored on the server.

Conclusions

The SOC paradigm represents a radical change in the way distributed applications are designed and implemented. By overcoming difficulties exhibited by previous middleware technologies, such as CORBA, this approach has promoted the development of loosely coupled solutions capable to guarantee the sharing and reusing of functionality. The cornerstone behind the new computing platform is the idea of service, an independent software module that performs certain operations. Among the various software proposals matching the general definition of service, in this chapter the proposals from W3C and OGC are discussed, whose success has been determined by both the simplicity of access and the ubiquity of the components, and the common goal of standardization and interoperability. However, as shown throughout the chapter, despite sharing the same objectives, these services are based on different standards, namely W3C services are based mainly on the SOAP and WSDL standards, while the OGC services specifications are based on standards developed independently from SOAP or WSDL. In particular, each OGC service represents a separate standard designed to handle a specific kind of data.

These differences must be carefully kept into account when dealing with service composition, an important feature of a services platform. Services composition is without doubt one of the hallmarks of this new computing paradigm. In fact, service-based applications can satisfy client requests by dynamically binding two or more services in a manner totally transparent to the final user. Among the proposed approaches for the services composition, services orchestration is one of the most commonly used. However, because of their different standards, a seamless orchestration of W3C and OGC services is still a complex and challenging task. As described in this chapter, the current solutions that try to realize such integration classify the problems that need to be addressed into three main categories: Data Handling, Functionalities Mapping and Metadata Management. Such a subdivision is, in our opinion, a valid starting point for the identification of future research directions. In particular, for the Metadata Management, the solution discussed in (Amirian et al. 2010) tries to solve this issue by using the WS-Metadata Exchange protocol, whose specifications can be used to develop a standardized framework for sending and receiving metadata information. The limitation of this approach concerns the W3C services stack, namely there is no a standardized way for the specific management of spatial metadata.

Another research topic concerns the benefits deriving from a greater use of the SOAP Headers for geospatial processing. In fact, although their usage is not compulsory for providing OGC services with a SOAP interface, they represent a fundamental part of the SOAP standard and their flexibility is the keystone behind the development of a broad range of other Web services protocols.

Finally, the SOAP performance issue is a fundamental aspect to take into account. When compared with middleware technologies, such as Java RMI, the SOAP performances usually constitute a bottleneck during the implementation of large scale distributed and complex applications. Therefore, overcoming this issue is of utmost importance for the development of geospatial data oriented effective solutions.

The chapter ends presenting a totally different way to develop distributed systems, represented by the REST paradigm. Such an architectural style represents a possible solution to the complexity of the W3C services stack, and OGC is considering the adoption of new standards for Web services by using the REST protocol, originally proposed by ESRI.

Acknowledgments

The research presented in this paper was funded by a Strategic Research Cluster grant (07/SRC/I1168) by Science Foundation Ireland under the National Development Plan. The authors gratefully acknowledge this support.

References

Alves, A., A. Arkin, S. Askary, C. Barreto, B. Bloch, F. Curbera, M. Ford, Y. Goland, A. Guízar, N. Kartha, C.K. Liu, R. Khalaf, D. König, M. Marin, V. Mehta, S. Thatte, D. van der Rijn, P. Yendluri and A. Yiu. 2007. Web Services Business Process Execution Language Version 2.0. OASIS. http://docs.oasis-open.org/wsbpel/2.0/wsbpel-v2.0.pdf (Accessed on March 20, 2013).

Amirian, P., A.A. Alesheikh and A. Bassiri. 2010. Standards-based, interoperable services for accessing urban services data for the city of Tehran. Computers, Environment and Urban Systems. 34: 309–321.

Ballinger, K., B. Bissett, D. Box, F. Curbera, D. Ferguson, S. Graham, C.K. Liu, F. Leymann, B. Lovering, R. McCollum, A. Nadalin, S. Parastatidis, C. von Riegen, J. Schlimmer, J. Shewchuk, B. Smith, G. Truty, A. Vedamuthu, S. Weerawarana, K. Wilson and P. Yendluri. 2008. Web Services Metadata Exchange 1.1 (WS-MetadataExchange). World Wide Web

Consortium. http://www.w3.org/Submission/WS-MetadataExchange/ (Accessed on March 20, 2013).

Ballinger, K., D. Ehnebuske, M. Gudgin, M. Nottingham and P. Yendluri. 2004. Basic Profile Version 1.0. Web Services Interoperability Organization. http://www.ws-i.org/profiles/BasicProfile-1.0-2004-04-16.html (Accessed on March 20, 2013).

Barros, A., M. Dumas and P. Oaks. 2005. A Critical Overview of the Web Services Choreography Description Language (WS-CDL). BPTrends Newsletter, Volume 3, Number 3.

Barton, J.J., S. Thatte and H.F. Nielsen. 2000. SOAP Messages with Attachments. World Wide Web Consortium. http://www.w3.org/TR/SOAP-attachments (Accessed on June 24, 2013).

Baumann, P. 2010. OGC WCS 2.0 Interface Standard-Core. Open Geospatial Consortium Inc. http://www.opengeospatial.org/standards/wcs (Accessed on March 20, 2013).

Booth, D., H. Haas, F. McCabe, E. Newcomer, M. Champion, C. Ferris and D. Orchard. 2004. Web Services Architecture. World Wide Web Consortium. http://www.w3.org/TR/ws-arch/ (Accessed on March 20, 2013).

Bosworth, A., D. Box, M. Gudgin, M. Nottingham, D. Orchard and J. Schlimmer. 2003. XML, SOAP and Binary Data. http://www.xml.com/pub/a/2003/02/26/binaryxml.html (Accessed on March 20, 2013).

Burggraf, D.S. 2006. Geography Markup Language. Data Science J. 5: 178–204.

Christensen, E., F. Curbera, G. Meredith and S. Weerawarana. 2001. Web Services Description Language (WSDL) 1.1. World Wide Web Consortium. http://www.w3.org/TR/wsdl (Accessed on March 20, 2013).

de la Beaujardiere, J. 2006. OpenGIS Web Map Server Implementation Specification. Open Geospatial Consortium Inc. http://www.opengeospatial.org/standards/wms (Accessed on March 20, 2013).

Doyle, A. and C. Reed. 2001. Introduction to OGC Web Services. http://www.opengeospatial.org/pressroom/papers (Accessed on March 20, 2013).

Dustdar, S. and W. Schreiner. 2005. A survey on web services composition. Int. J. Web Grid Serv. 1(1): 1–30.

Erl, T. 2005. Service-oriented architecture: concepts, technology, and design. Prentice Hall PTR, Upper Saddle River, New Jersey.

ESRI. 2010. GeoServices REST Specification Version 1.0. http://www.esri.com/library/whitepapers/pdfs/geoservices-rest-spec.pdf (Accessed on June 24, 2013).

Fawcett, J., L.R.E. Quin and D. Ayers. 2012. Beginning XML. John Wiley & Sons, Indianapolis, Indiana.

Fielding, R., J. Gettys, J. Mogul, H. Frystyk, L. Masinter, P. Leach and T. Berners-Lee. 1999. Hypertext Transfer Protocol—HTTP/1.1. World Wide Web Consortium. http://www.w3.org/Protocols/rfc2616/rfc2616.html (Accessed on March 20, 2013).

Friis-Christensen, A., M. Lutz, N. Ostländer and L. Bernard. 2007. Designing Service Architectures for Distributed Geoprocessing: Challenges and Future Directions. Transactions in GIS. 11(6): 799–818.

Gartmann, R. and B. Schäffer. 2008. OpenGIS Wrapping OGC HTTP-GET and -POST Services with SOAP. Open Geospatial Consortium Inc. http://www.opengeospatial.org/standards/dp (Accessed on March 20, 2013).

Gudgin, M., M. Hadley, J.J. Moreau and H.F. Nielsen. 2001. SOAP Version 1.2. World Wide Web Consortium. http://www.w3.org/TR/2001/WD-soap12-20010709/ (Accessed on March 20, 2013).

Henning, M. 2006. The rise and fall of CORBA. ACM Queue. 4(5): 28–34.

Ioup, E., B. Lin, J. Sample, K. Shaw, A. Rabemanantsoa and J. Reimbold. 2008. Geospatial Web Services: Bridging the Gap between OGC and Web Services. pp. 73–93. In: J.T. Sample, K. Shaw, S. Tu and M. Abdelguerfi (eds.). Geospatial Services and Applications for the Internet. Springer, New York.

Lake, R. 2005. The application of geography markup language (GML) to the geological sciences. Computers & Geosciences. 31(9): 1081–1094.

McCarthy, J. 2002. Reap the benefits of document style web services. http://www.ibm.com/developerworks/webservices/library/ws-docstyle/index.html (Accessed on March 20, 2013).

Milanovic, N. and M. Malek. 2004. Current Solutions for Web Service Composition. IEEE Internet Computing. 8(6): 51–59.

Nadalin, A., C. Kaler, R. Monzillo and P. Hallam-Baker. 2006. Web Services Security: SOAP Message Security 1.1 (WS-Security 2004). https://www.oasis-open.org/committees/tc_home.php?wg_abbrev=wss#technical (Accessed on March 20, 2013).

OASIS. 2009. Web Services Transaction (WS-TX) TC. https://www.oasis-open.org/committees/tc_home.php?wg_abbrev=ws-tx (Accessed on March 20, 2013).

Papazoglou, M.P. 2003. Service oriented computing: concepts characteristics and directions. Proc. Fourth Int. Conf. Web Information Systems Engineering. pp. 3–12.

Papazoglou, M.P. and W.J. van den Heuvel. 2007. Service oriented architectures: approaches, technologies and research issues. The VLDB J. 16(3): 389–415.

Pautasso, C., O. Zimmermann and F. Leymann. 2008. RESTful Web Services vs. "Big" Web Services: Making the Right Architectural Decision. Proc. of WWW2008.

Portele, C. 2007. OpenGIS Geography Markup Language (GML) Encoding Standard. Open Geospatial Consortium Inc. http://www.opengeospatial.org/standards/gml (Accessed on March 20, 2013).

Portele, C. and S. Sankaran. 2012. GeoServices REST API — Part 1: Core. Open Geospatial Consortium Inc. http://www.opengeospatial.org/standards/requests/89 (Accessed on June 24, 2013).

Powell, M. 2004. Web Services, Opaque Data, and the Attachments Problem. Microsoft Corporation. http://msdn.microsoft.com/en-us/library/ms996462.aspx (Accessed on March 20, 2013).

Rautenbach, V., S. Coetzee, M. Strzelecki and A. Iwaniak. 2012. Results of an evaluation of the orchestration capabilities of the ZOO project and the 52° north framework for an intelligent geoportal. ISPRS Annals of the Photogrammetry, Remote Sensing and Spatial Information Sciences. XXII ISPRS Congress.

Rautenbach, V., S. Coetzee and A. Iwaniak. 2013. Orchestrating OGC web services to produce thematic maps in a spatial information infrastructure. Computers, Environment and Urban Systems. 37: 107–120.

Sancho-Jiménez, G., R. Béjar, M.A. Latre and P.R. Muro-Medrano. 2008. A Method to Derivate SOAP Interfaces and WSDL Metadata from the OGC Web Processing Service Mandatory Interfaces. Proc. ER 2008 Workshops (CMLSA, ECDM, FP-UML, M2AS, RIGiM, SeCoGIS, WISM) on Advances in Conceptual Modeling: Challenges and Opportunities. pp. 275–384.

Schäffer, B. 2008. OWS 5 SOAP/WSDL Common Engineering Report. Open Geospatial Consortium Inc. http://www.opengeospatial.org/standards/dp (Accessed on March 20, 2013).

Schut, P. 2007. OpenGIS Web Processing Service. Open Geospatial Consortium Inc. http://www.opengeospatial.org/standards/wps (Accessed on March 20, 2013).

Stollberg, B. and A. Zipf. 2007. OGC web processing service interface for web service orchestration: aggregating geo-processing services in a bomb threat scenario. Proc. 7th int. conference on Web and wireless geographical information systems. pp. 239–251.

Tekli, J.M., E. Damiani, R. Chbeir and G. Gianini. 2012. SOAP Processing Performance and Enhancement. IEEE Trans. Services Computing. 5(3): 387–403.

Tsalgatidou, A. and T. Pilioura. 2002. An overview of standards and related technology in web services. Distrib. Parallel Databases. 12(2–3): 135–162.

Villa, M., G. Di Matteo, R. Lucchi, M. Millot and I. Kanellopoulos. 2009. SOAP primer for INSPIRE Discovery and View Services. European Commission. http://inspire.jrc.ec.europa.eu/index.cfm/pageid/241/documentid/487 (Accessed on March 20, 2013).

Villa, M., R. Lucchi, M. Millot and I. Kanellopoulos. 2008a. SOAP HTTP binding status. European Commission. http://inspire.jrc.ec.europa.eu/index.cfm/pageid/241/documentid/502 (Accessed on March 20, 2013).

Villa, M., G. Di Matteo, R. Lucchi, M. Millot and I. Kanellopoulos. 2008b. INSPIRE Network Services SOAP framework. European Commission. http://inspire.jrc.ec.europa.eu/index.cfm/pageid/241/documentid/494 (Accessed on March 20, 2013).

Vretanos, P.A. 2005. Web Feature Service Implementation Specification. Open Geospatial Consortium Inc. http://www.opengeospatial.org/standards/wfs (Accessed on March 20, 2013).

Webber, J., S. Parastatidis and I. Robinson. 2010. REST in Practice. O'Reilly Media, Sebastopol, CA 95472.

Whiteside, A. 2005. OpenGIS web services architecture description. Open Geospatial Consortium Inc.

Whiteside, A. and J. Greenwood. 2010. OGC Web Services Common Standard. Open Geospatial Consortium Inc. http://www.opengeospatial.org/standards/common (Accessed on March 20, 2013).

Yu, Q., X. Liu, A. Bouguettaya and B. Medjahed. 2008. Deploying and managing Web services: issues, solutions, and directions. The VLDB J. 17(3): 537–572.

Zhang, J., D.D. Pennington and W.K. Michener. 2007. Performance Evaluations of Geospatial Web Services Composition and Invocation. Proc. IEEE Int. Conf. on Web Services. pp. 1128–1135.

Zhao, P., G. Yu and L. Di. 2007. Geospatial Web Services. pp. 1–35. In: B.N. Hilton (ed.). Emerging Spatial Information Systems and Applications. Idea Group Publishing, Hershey.

Zisman, A. 2000. An overview of XML. Computing & Control Engineering J. 11(4): 165–167.

CHAPTER 6

Database Server Models for WMS and WFS Geographic Web Services

Nissrine Souissi[a,]* and *Michel Mainguenaud*[b]

Introduction

Geographical Information Systems (GIS) have improved their services in the past years, mainly due to two factors: the overwhelming spread of Internet (e.g., speed, security, data availability) and the use of web technologies. Due to these evolutions, new GIS services have appeared. They provide the diffusion (e.g., interactive maps), the analysis or updates of geo-localized data.

These systems produce and manipulate huge amounts of geographical data (e.g., Google Earth) which is stored all over the world in various places using various formats. These formats may be closed (i.e., proprietary dependent) or opened, normalized or not. Consequently, analysis and/ or decision-making may be difficult or time-consuming due to this heterogeneity of formats. To avoid these drawbacks and in order to facilitate accesses and usage of these data, the Open Geospatial Consortium (OGC) defined some propositions designed to ease the interoperability of cartographical servers thanks to interfaces defined in technical specifications

---

[a] Ecole Nationale de l'Industrie Minérale (ENIM)–Dept. Informatique, Avenue Hadj Ahmed Cherkaoui, B.P. 753, Agdal–10000 Rabat–MAROC.
   Email: nissrine.souissi@enim.ac.ma
[b] Institut National des Sciences Appliquées de Rouen–LITIS EA4108, 685 Avenue de l'Université, B.P. 08, F76801 Saint-Etienne du Rouvray–FRANCE.
   Email: michel.mainguenaud@insa-rouen.fr
* Corresponding author

(i.e., the implementations can be provided by several companies). One of these propositions is named "Web services" (OGC 2005). These propositions allow accesses to geographical data without taking into account physical aspects of communications (i.e., with an URL). They also propose a set of parameters, methods and communication rules to simplify accesses and manipulations as soon as clients and servers respect them.

The Web Map Service (WMS) (OGC 2006) and the Web Feature Service (WFS) (OGC 2010b), used in this chapter, are OGC standards for web services. WMS deals with the dynamic production of maps as images, built with geo-referenced data. WFS are designed to provide an access and a manipulation tool of geographical data within a map. End-users' interactions are reduced to a selection of an object on a map. This selection raises the GetFeatureInfo operation of WMS or the GetFeature operation of WFS and as a result these operations provide data associated with a geographical map or object. Information may be alphanumerical such as for example a town name or graphical such as for example the segments defining a border of a town.

Designing a database schema requires analyzing the information system and enumerating data which should be handled by the Data Base Management System (DBMS). The implementation of a database schema is divided into three main levels according to the ANSI/X3/SPARC recommendations (ANSI-SPARC) and (Brodie and Schmidt 1982). The most important level is the intermediate one. This level models all available information to be managed by a DBMS. The lower level defines the physical storage system depending on a set of parameters (e.g., operating system, update rate, and number of users). The upper level defines the specific views of a database schema, each of which is adapted to a specific application. Some data available in the DBMS are presented, some are hidden and others are re-organized depending on users' needs. As an example a tourist application may not be concerned by information about the number of television sets in a specific area but a marketing application is concerned by this information. In the first application, this information will be hidden and application users are not aware that this data is available in the information system. In the second case this data will be presented (may be as aggregated figures). Depending on the physical storage some data may be easy to obtain, some may require the resolution of a complex query and this process may take time.

In our case, we are in the context of distributed applications all over the world. Database administrators could not know in advance the different kinds of applications (i.e., their needs) that will query their databases. Therefore, due to the number of potential applications and users, providing the definition of external schemas as defined in the ANSI/SPARC (ANSI-SPARC) and (Brodie and Schmidt 1982) decomposition is not realistic at all.

As a consequence, end-users must be aware of the distant database schema and its semantics (with conventional problems such as aggregative/metrics units, clear definition of the scope, and semantics of the attributes). Some data available in the database schema are relevant for their needs, some are not.

GetFeatureInfo and GetFeature operations of the WMS/WFS services will query the database schema. They provide the set of data associated with a map or a geographical object. They present some limitations since they cannot query an external schema adapted to a specific user's need and no information about the time required to obtain data is available. They require the analysis of distant database schemas. Furthermore, people involved in this task are not the distant database designers since they are distant application designers. The results of WMS/WFS operations may therefore not be adapted to decision making from a specific application point of view (e.g., tourist, social, economical, transports).

In order to adapt available attributes to end-users' needs, we propose a solution that is independent of GetFeatureInfo and GetFeature operations of WMS and WFS. We still respect the normalized interface of these operations within these services. We propose an extension of the database schema in distant databases in order to precise answers that these operations may provide. The answers may be alphanumerical attributes or results of methods (i.e., functions) viewed as alphanumerical or graphical attributes. The proposed solution is to enhance the data model of distant databases with typed links on a conceptual lattice and information about the operational complexity required to obtain data. These types are defined by distant database administrators (i.e., people who have the best knowledge of available information and the required time to get these data). This solution provides a dynamic data model as a result of querying and is a semi-automatic process to define relevant data for these operations.

In the second section, we present the state of the art in the spatial enhancement of data models. In the third section, we present data model extensions suggested by us. In the fourth section, we present pre-defined rules to choose relevant attributes for these operations. In the fifth section, we present a conclusion and some perspectives.

State of the Art

We present in this section, research works devoted to enhance the semantic in geographical Web services. Two main approaches are possible: the extension of the norm or to define a service that is closer to end-users' needs.

In the first part, we present solutions promoted by the Open Geospatial Consortium (OGC 2005). In the second part, we present a model-oriented proposition for a Web service. In the third part, we present a user-oriented database schema transformation. We end this section with a synthesis of such propositions.

OGC propositions

Among the propositions defined by the OGC, we focus on specific operations associated with Web Services: the Web Map Service (WMS) and the Web Feature Service (WFS). We particularly focus on the GetFeatureInfo and GetFeature operations available in these services.

WMS

Information provided in this part is based on the standard documentation for WMS version 1.3.0 (OGC 2006). The WMS standard induces three main operations: GetCapabilities, GetMap, and GetFeatureInfo. Figure 1 shows these operations. First of all, we present relevant operations for our topic: GetMap and GetFeatureInfo. Then we present the limits of the GetFeatureInfo operation.

The most important facility for a cartographical server is to provide a map from its available layers. Parameters such as desired layers, representation styles, size, relevant geographical areas, the projection system or the output format must be provided to the GetMap operation. The obtained result is a map (in general an image that can be displayed by conventional web browsers). This operation corresponds to a visualization process.

The GetFeatureInfo operation corresponds to a query process. This is a mandatory facility for a Webmapping application. Webmapping is a word used to represent cartography on the web. Within this generic word different applications are grouped covering a wide range spectrum from

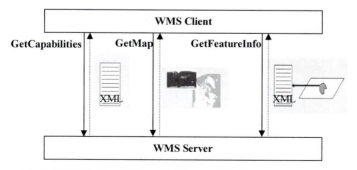

Fig. 1. WMS GetCapabilities, GetMap and GetFeature operations.

a simple visualization with a web browser to a dynamic and interactive cartographic tool. The dynamic component is provided for example by pan-zoom, modulations of displayed layers or some widgets such as tooltips: e.g., Google Maps or Bing Maps. To provide such facilities, the GetFeatureInfo operation requires a set of parameters. The server can therefore build a structured result based on the database schema. The query can be reformulated by: "What is there here?" Here is defined by a couple of coordinates. The output format is an array in HTML or a XML tree.

Current results of a WMS service are generally images. The intrinsic structure of a result reduces possible manipulations. An end-user does not manipulate structured data but an image. To be able to access data, the end-user must select an object on the image. The GetFeatureInfo operation provides this selection. Nevertheless, some limits appear in the sense that the provided schema is the set (or a sub-set) of available data for this/ these object(s). No link is performed with the environment of this object. Furthermore, an end-user must be aware of the database schema in order to select only a sub-set of available data.

WMS operations are graphic-result-oriented. WFS specifications want to deal with data access. In the following, we present operations based on the WFS specifications and in particular the GetFeature operation.

WFS

Information provided in this part is based on the standard documentation for WFS version 2.0.0 (OGC 2010b). WFS is an OGC specification for data access. It describes responses of a Web server to geographical data manipulation operations. These operations are based on the CRUD manipulations of geographic data based on alphanumeric/spatial constraints: creation (C), read (R), update (U) and deletion (D).

The formalism used to model data exchanges for the WFS specification is GML (Geography Markup Language). GML (OGC 2007a) is a XML dialect designed to encode, to manipulate and to exchange geographical data. Specified by the OGC, its main goal is to guarantee interoperability in the location-based data field.

The WFS specification defines five operations to send queries to a geographical data server and to get answers from it: GetCapabilities, DescribeFeatureType, GetFeature, Transaction and LockFeature. Figure 2 presents the basic operations of the WFS specification.

The most important operation, in our context, for a WFS server, is the GetFeature operation. This operation delivers data instances typed by features, identifies properties that should be delivered and provides the results of spatial and non-spatial queries.

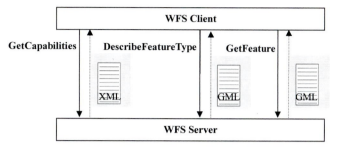

Fig. 2. Basic operations of the WFS Specification.

Selection criteria are defined using a filter. To get relevant data, a client specifies an object identifier in the filter. The WFS server receives the GetFeature query, determines the correct database, creates and sends a SQL statement to the database and formats the results. The filter is used to manage the "Where" clause of a SQL statement. Spatial data are handled with the spatial schema defined in the ISO 19107 norm. Before sending results to a client, the server transforms objects using a GML3 format.

The WFS specification emphasizes the exchange of geographical objects. Since data are distributed worldwide for a tremendous set of applications, there is no reason to define a unique and standard database schema. Therefore it is of prime importance that the data model should be provided with the data set. Unfortunately, that means that the cognitive charge of understanding the data model is directed towards end-users. In this context, we present some works devoted to the improvement of interoperability and the database schema definition.

Data-model-oriented web services

The proposition, named mdWFS, defined in Donaubauer et al. (2007a,b), defines a data-model-oriented web service. This proposition tries to take advantage of both: data interoperability as defined in OWS specifications (OGC Web Services) and the expressive power of conceptual data models. The main idea, used from conceptual data models, is the definition of a conceptual language to describe spatial data. Since the definition is provided at the conceptual level, it is independent of a specific implementation system or transfer format such as XML or GML. A wrapper (i.e., a compiler) is the ideal tool to move from a generic representation to a specific one. The semantic interoperability is provided by a representation of conceptual models. The representation of a conceptual model is translated from a schema A into a schema B. The model driven approach consists in four steps illustrated in Fig. 3 (Donaubauer et al. 2007a): specification of an application domain, specification of a conceptual schema associated with its UML meta-

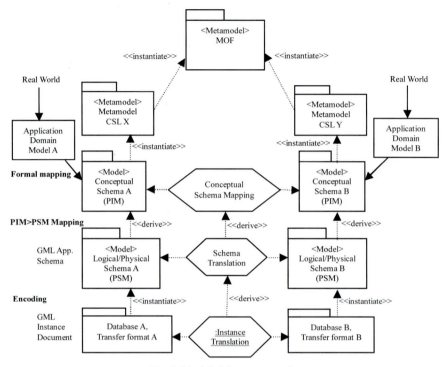

Fig. 3. Model-**driven** approach.

model, description of an application domain with a conceptual language and then derivation of the schema into various dialects.

The mdWFS proposition handles the storage/delivery of conceptual schemas and the semantic transformations depending on translation models. The first step is a semantic transformation. The second step is the configuration of a standard WFS service (OGC 2010b) to provide a data exchange service. The standard WFS service is parameterized with the destination data model but delivers data from the initial database. The WFS service must be able to store data and provide several conceptual schemas. OGC specifications should be extended.

A conceptual language, UMLT, is introduced for two main purposes: the data representation and the semantic transformations. This cartographic conceptual language is defined upon an independent extension of the UML2 meta-model. Elements of the language are specified with a UML2 model as a heritage of Activities. The textual notation of the language is defined by a set of grammatical rules (EBNF—Extended Backus Naur Form). Table 1 presents the extensions of the WFS specifications. Table 2 presents the elements of the UMLT language.

Table 1. mdWFS specifications.

Extensions	Description
Service (parameter)	In order to define a protocol, a new parameter in a query is defined: service = mdWFS.
GetCapabilities operation	The GetCapabilities operation is extended to obtain a SchemaList. This list contains all the available schemas in this service.
DescribeFeatureType operation	The DescribeFeatureType operation is extended to provide the XMI format (i.e., metadata exchange format in UML using XML) to transfer information on models.
DoTransform operation	A new operation. Its aim is to transfer the conceptual schema representation towards mdWFS and to set up the semantic transformations.

Table 2. Elements of the UMLT language.

Elements	Description
SelectionCriteria	Allows a data selection upon a logical expression.
VirtualAssociation	Allows the management of objects defined as being not linked with an association object. Imported objects may have some attributes or foreign keys (evaluated in real time) in order to provide calculated relationships.
TransformationAction	Defined as a heritage of an Opaque Action UML. It provides an element of activity that cannot be anymore structured. This is an elementary action of a transformation.
AssignmentDefinitions	Identifies expressions and primitive types as specifications.
MappingRule	Built as a composition of rule definitions, it provides cartographic objects.
AssociationBinding	Selecting associated objects, it is possible to define the way these associations are evaluated.
JoinType	An enumerate type to specify the type of join within an association link.

This new service provides a better semantic interoperability in the sense that it describes a methodology. The semantic transformations are formalized at the conceptual level. During a semantic transformation, virtual associations may be associated with, using a virtual link if necessary. The virtual association is defined to provide an explicit link for the join operation (in opposition to a conventional derived association). The joinCriteria property specifies the join conditions. The new service, mdWFS, allows a higher semantic interoperability. The relevant level is a transformation at the semantic level using a conceptual level of abstraction. The results of the data schema transformation process can be queried using a GetFeature operation. For the time, this specification is not an OGC normalized service. Foerster et al. (2010) and Wiemann et al. (2012) present different kinds of transformation based on the same principles.

User oriented schema transformation

An alternative of mdWFS service is to let the user guide transformations. Bucher and Balley (2007) define a methodology to get a transformation schema to adapt data to a specific application. The idea is, first, to transform the initial data model to adapt it to new requirements, then to get the geographic data instances. Modifications of the initial data model are performed step by step (e.g., selection, modification of the data structure) to obtain the final structure. These modifications define an algorithm to be applied to get instances from the database. This is equivalent to a workflow application illustrated in Fig. 4 (Bucher and Balley 2007), with an imperative programming language, step-by-step modifications are specified and then

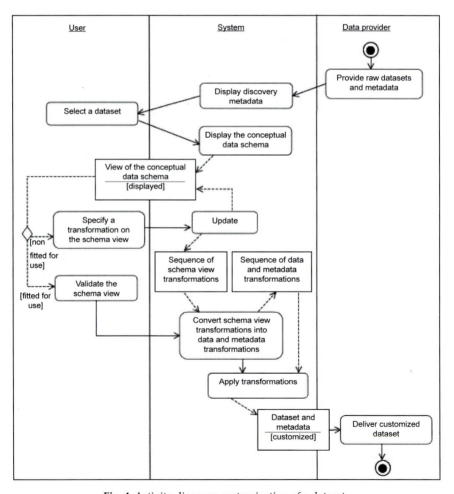

Fig. 4. Activity diagram: customization of a dataset.

applied. During a transformation, the system is in charge of the coherency of the dataset.

Proposed operations to manage the conceptual schema are applied on classes (in the sense of object oriented modeling), attributes, objects and relationships. The language allows manipulations upon classes with the definition of aggregates, renaming, deletion, specialization and generalization; upon attributes, with renaming and deletion; upon objects with aggregation and filter; upon relationships, with creation and deletion.

Some perspectives of this work are the semi-automation of the final schema, the use of web services to transform database schemas (direct or indirect—i.e., issued of a web service) or a standard manipulation to provide a predefined structure as results of the transformation.

One of the main differences between the approaches developed in (Staub 2007; Staub et al. 2008) and (Bucher and Balley 2007) is based on the transformation specification. Staub's approach relies on a mapping specification from the initial schema to the destination schema. The mapping is defined with selection and transformation operators. The user specifies transformations based on schema manipulations instead of defining an imperative approach of restructuring tasks. This approach has a better relevance whenever the destination schema is known and the transformation specification is based on a method rather than operations. The mix of these two approaches leads to promising perspectives.

Synthesis

Described propositions show weaknesses on the topic of data semantics in web services. OWS such as WMS and WFS are database schema dependent. OWS offer a syntactical interoperability (e.g., data exchange) but nothing about semantic representations (e.g., data model relationships). Conceptual data models are hidden, and semantic transformations are not available. One of the reasons may be similar to the introduction of external schemas in relational databases (i.e., view mechanism). The lack of semantic tools may be another one. We propose to reduce the second one by providing a way to introduce semantic aspects while exchanging data.

WMS services mainly provide graphic results since WFS deals with data access. The visual aspect, rather than complex functions, should be kept in mind while evaluating consultation tools based on Internet solutions. The choice is based on requirements and constraints on services: simple consultation (data and interactions with images) or real integration of data sources in complex applications (i.e., interactions with data based on the CRUD paradigm).

It is of prime importance to integrate the use of WFS operations. In the current state of OWS, a requirement based on the visualization of a huge volume of data leads to use a WMS service. This service is better fitted and widely sufficient. Furthermore, if the requirement is to access to specific data, to modify on the fly the representation of specific data or to update them then WFS operations are mandatory. Nevertheless, this service must be used with precautions since the involved volume of data may be important and therefore the response time may be significant.

The goal of our proposal is to improve OWS such as WMS and WFS, mainly the GetFeatureInfo and GetFeature operations. The aims are to provide a better relevance of alphanumeric data to improve spatial analysis and to provide a data model schema to handle such information. The second proposition we presented in this section deals with rules to determine a conceptual schema but does not consider instantiations of such models. This is the matter we want to tackle.

Data Conceptual Schema

A selection of a (geo)graphical object (e.g., city, restaurant) represented on a map, is translated into a query that will retrieve information about this object. This knowledge (schema and data) that can be considered as static, is provided by default to all users. It does not correspond in any way to users' point of view. We, therefore, suggest adapting the database schema associated with a query result, by withdrawals of irrelevant attributes and/or additions of relevant attributes. We propose a mechanism of refinement for a query. It is a way to improve the search performance of relevant attributes. It can be based, for example, on a process that adds to users' queries new attributes related to those initially present in these queries.

Provided information will be richer and more "dynamic" than with a simple use of a WFS call. Indeed, some additional attributes are the result of methods (here functions) applied, among other things, on the attributes associated with a spatial object. These methods are completely transparent for an end-user. Provided information will always be, by the end, presented as a set of attributes.

Obviously, this new database schema cannot be obtained from current distant database schemas. We are therefore interested in determining a more sophisticated database schema that integrates semantic links, attributes and methods. We stand here at the conceptual level and introduce a semantic extension that will allow the definition in a database schema, of relevant attributes and methods. This requirement is related to the impossibility of defining external schemas considering the heterogeneity of potential clients. Figure 5 illustrates our approach that consists in enriching the data schema.

Starting from an extended schema with our new concepts, we proceed in two phases. The first phase removes attributes that are considered as irrelevant with respect to a specific user's point of view. The second phase introduces new relevant attributes.

To illustrate these new extensions, we use an example of a conceptual schema formalized in UML and presented in Fig. 6, as a class diagram. It describes self-explaining geographical entities: cities, countries, hotels, restaurants, trees and green spaces. Each class is formalized by a set of attributes and a set of methods. SR stands for Spatial Representation defines on the OGC Geometry domain. Each method is defined as a function (non-function methods are not visible in the context of this chapter).

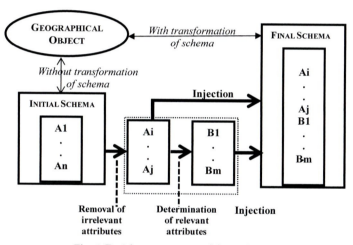

Fig. 5. Enrichment process of data schema.

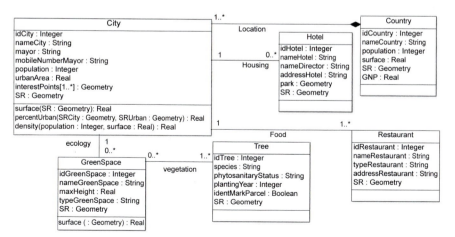

Fig. 6. Sample of a conceptual schema.

Notations: The following notations and conventions are used throughout the chapter. {} denotes the set constructor and () denotes the list constructor. Let $C = \{c_1,..., c_p\}$ be a set of classes. Let $A = \{a_1,..., a_m\}$ be a set of attributes. Let $M = \{m_1,..., m_n\}$ be a set of methods. For each class $c_i \in C$, we denote $c_i = (A_i, M_i)$ a class with $A_i \subseteq A$ and $M_i \subseteq M$. Let K_i be a set containing the attribute(s) defining a key for a given class c_i with $K_i \subseteq A_i$. The definition of a class c_i becomes $c_i = (A_i, M_i, K_i)$. The list $(p_1, .., p_q)$ denotes the parameters of a method m_k as $m_k \in M$. OP denotes the available set of spatial operators (e.g., orthometric distance noted $\sqrt{}$, inclusion noted $\subset$). FD denotes a Functional Dependency. RL denotes a Relevant Link.

This section is organized into five parts. In the first part, we present the concept of functional dependency and its weaknesses. In the following three sections, we present the three types of proposed Relevant Links. Finally, we conclude this section by giving in a fifth part a summary of our proposals.

Functional Dependencies

Functional Dependencies (FDs) are one of the key concepts of relational database schema definition.

Definition

The first semantic link introduced in the design of a relational database schema is the functional dependency. It expresses a dependency between two (sets of) attributes at the functional level. The formal definition in the relational database context is the following: X functionally determines Y if, whatever relation r is the current value for R (a relation), it is not possible that r (an instance of R) has two tuples that agree in the components for all attributes in the set X yet disagree in one or more components for attributes in the set Y (Ullman 1995). Table 3 presents a representation of Functional Dependencies in a class context.

Table 3. Formal definition of a functional dependency.

Let $c_i = (A_i, M_i, K_i)$ be a class where $c_i \in C$. Let $A_s = \{a_{s1},..., a_{sn}\}$ be a set of attributes. Let $A_c = \{a_{c1},..., a_{cm}\}$ be a set of attributes
<u>Pre-conditions</u>: $A_s \subset A_i$ and $A_c \subset A_i$ and $A_s \cap A_c = \varnothing$
There is a Functional Dependency between A_s and A_c noted $A_s \rightarrow A_c$ then:
$\forall\ I_1, I_2$ two instances of c_i,
$(I_1.a_{s1}, ..., I_1.a_{sn}) = (I_2.a_{s1}, .., I_2.a_{sn}) ==> (I_1.a_{c1}, .., I_1.a_{cm}) = (I_2.a_{c1}, .., I_2.a_{cm})$

As an example, there is a FD between the attribute mobileNumberMayor and the attribute mayor of the City class. This FD is written with an arrow: "mobileNumberMayor → mayor". The semantic interpretation of such an FD denotes that a mobile phone number relates only to a single mayor (but that does not mean anything about the fact that a mayor may have several mobile phone numbers).

Weaknesses

Functional Dependencies do not express all data semantics. They do not in fact represent any link that may exist between data from the application point of view. For example, regarding the Tree class it is not possible to link the attributes "species" and "phytosanitaryStatus" with a FD. Indeed, a value of the attribute "species" does not determine a unique value of the attribute "phytosanitaryStatus" (several trees can be of the same species with different phytosanitary status). With the "species" attribute of a tree, it would be useful to obtain, from an agronomic point of view, its phytosanitary status but the identifying mark of the parcel is not relevant (for which there is no FD).

A FD does not allow establishing a link between the attribute "population" and the attribute "density". The value of density is calculated from the attribute "area" and the attribute "population" whose value is obtained using the "surface" function. Once the population is available, it would be useful, from a social point of view, to provide the density.

The independence of attributes (e.g., "Park" of Hotel and "typeGS" of GreenSpace) does not allow, for the application of a spatial operator (e.g., "inclusion" on spatial representations), obtaining information through a functional dependency. With a park, it would be useful to obtain from a tourist point of view, types of green space associated with.

The Functional Dependencies are not sufficient to recover, at the request of a user, the relevant attributes (single or calculated). It is important to establish a link that allows, from a database schema, obtaining a set of relevant attributes with respect to the user's point of view. This requirement is related to the impossibility of defining external schemas considering the heterogeneity of potential users.

We present in that sense, three types of links called relevant Links (RL). The first type is called RLintA-A and concerns relevant links defined between attributes of the same class (the meaning of the acronym is R(elevant)L(ink) int(ernal)A(ttributeTo)A(ttribute)). The second type is called RLintA-M and concerns relevant links defined between an attribute and a method of the same class. Finally, the third type is called RLextA-A and concerns relevant links defined between attributes of different classes.

Relevant Link RLintA-A

A RLintA-A link leads to remove irrelevant attributes in the case where the initial schema has all the attributes associated with an object. Otherwise, it allows the determination of relevant attributes that are not included in the initial schema.

A RLintA-A link binds two sets of attributes of a class, possibly reduced to a singleton. This RL is defined by the following elements: the class concerned by the RL (c_i), the set of source attributes (A_s) and the set of target attributes (A_t). These two sets, A_s and A_t, are disjoint. We say that the source attributes semantically determine the target attributes. Table 4 presents the formal definition of RLintA-A.

This definition means that in the initial schema where it exists the set of (source) attributes $\{a_{s1},..., a_{sn}\}$, this RL adds in the final schema, the set of (target) attributes $\{a_{t1},..., a_{tm}\}$. Note here that all the attributes are not necessarily involved in this RL (e.g., key attribute). In fact, the key is not taken into account because keys operate at the functional dependency level in a database schema design.

As an example, from an agronomic point of view, a user who already has tree species, the phytosanitary status of the trees represents relevant information. Therefore, we define a RLintA-A link between two attributes of the Tree class (e.g., "species" and "phytosanitaryStatus"). This RL is defined by: -A_A> (Tree, species) = phytosanitaryStatus.

Operational complexity depends on the implementation of the database. The database administrator is in charge of defining such a complexity. It may be constant as defined for example in a framework of a relational database with sets materialization of source and target attributes in the same relation (the same tuple). It can be higher due to the third normal form, with or without index(es) in the second (third) relations. The complexity may be changed out of computed attributes (i.e., the complexity depends on the complexity of the evaluation).

Table 4. Formal definition of RLintA-A.

Let $c_i = (A_i, M_i, K_i)$ be a class where $c_i \in C$. Let $A_s = \{a_{s1},..., a_{sn}\}$ be a set
of attributes. Let $A_t = \{a_{t1},..., a_{tm}\}$ be a set of attributes
<u>Pre-conditions</u>: $A_s \subseteq A_i, A_t \subseteq A_i, A_s \cap A_t = \varnothing, K_i \not\subset A_s, K_i \not\subset A_t$
A RL of **RLintA-A (-A_A->)** type between A_s and A_t is defined by:
-A_A> : C x A x .. x A -> A x .. x A
-A_A-> $(c_i, a_{s1},..., a_{sn}) = (a_{t1},..., a_{tm})$

Relevant Link RLintA-M

A RLintA-M link leads to additional attributes in the final schema, which are results of methods.

A RLintA-M link binds a set of attributes of a given class (possibly reduced to a singleton) to a method of the same class. This set of attributes represents the source attributes and belongs to the initial schema. The result of the method (here a function) will be added to the final schema as a new attribute.

This RL is defined by the following elements: the class concerned by this RL (c_i), the set of source attributes (A_s), the method (m_k), its parameter list (P). Table 5 presents the formal definition of RLintA-M.

This definition means that in the initial schema where it exists the set of attributes $\{a_1,..., a_n\}$, this RL adds in the final schema an attribute that denotes the result of the method m_k $(a_{|A|+1})$. Note here that several links can possibly call the method m_k.

The recursive definition of the set A, with respect to the creation of attributes denoting the link, leads to establish (recursively or not) links between methods. One of these methods is considered as a source attribute. The origin of the recursion is necessarily, by the definition of the link, in the set of non-calculated attributes.

As an example, from a social point of view, we illustrate this type of RL between the attribute "population" and the method "density (integer, real)" of the City class. This RL is defined by: A_M-> (City, population, density

Table 5. Formal definition of RLintA-M.

Let $c_i = (A_i, M_i, K_i)$ be a class where $c_i \in C$. Let $A_s = \{a_{s1}, ... , a_{sn}\}$ be a set of attributes . Let $m_k \in M_i$ be a method. Let $|A|$ denote the cardinality of the set of attributes A. Let $a_{|A|+1}$ denote the result attribute. Let D be a domain. Let P be defined as P: D P | M (P) P | ε
<u>Pre-conditions:</u> $A_s \subseteq A_i$, $K_i \not\subset A_s$
A RL of **RLintA-M (-A_M->)** type between A_s and $a_{|A|+1}$ is defined by:
-A_M->: C x A x ... x A x M x (P) -> A
$(c_i, a_1, ... , a_n, m_k, (p_1, ... , p_q)) = a_{|A|+1}$
<u>Post-condition:</u> $A = A \cup \{a_{|A|+1}\}$

(population, surface (SR))) = density. The source attribute "population" semantically determines the method "density (integer, real)". This method has the following parameters: the attribute "population" and the result of the method "surface (SR)". The target attribute, here called "density", is provided as a new attribute in the final schema as a conventional attribute. It can thus be directly provided in complement of data schema information given by the WMS/WFS operations GetFeature or GetFeatureInfo.

Operational complexity concerns the complexity of the applied methods. In the case where a method has parameters that are results of methods, operational complexity will be aligned with the complexity of these methods.

We emphasize here that there are two approaches to the determination of methods. The first approach is to use external services. WPS (OGC 2007b), defines the meaning of rules to set up and to run a geo-processing as a Web service. The user executes in this case a geospatial operation, with all necessary data. The server starts the process and informs the user of its progress. An architectural alternative to this approach is given in Grosso et al. (2009) or Stollberg and Zipf (2007) with an orchestration of services. The second approach is internal and is linked to the database. Our proposal is oriented towards the second approach. These two approaches are complementary in the sense that the method could be the implementation of a Web service request (e.g., a WPS). The Web service is an encapsulation of the method and the method or the WPS are hidden to final users.

Relevant Link RLextA-A

A RLextA-A links leads to additional attributes in the final schema, which are retrieved from one or more classes.

A RLextA-A link binds two sets of attributes (possibly reduced to a singleton) of different classes. The classes involved in these links may be of relevance related among others, through the inheritance relationship, composition or be completely independent in the conceptual data schema. This RL is defined with a spatial operator in order to make an independent connection between different classes of the data model. It therefore represents a semantic spatial link.

RLextA-A is defined by the following elements: the class concerned by the RL (named of source class), the set of source class attributes, the target class, the relevant target class attributes, the spatial query involving the source class and target class. Table 6 presents the formal definition of RLextA-A.

This definition means that in a schema where exists the set of attributes $\{a_{s1},..., a_{sn}\}$ of a class c_i, this RL adds in the final schema the set of (target) attributes $\{a_{t1},..., a_{tm}\}$ of the class c_j.

Table 6. Formal definition of RLextA-A.

Let $c_i = (A_i, M_i, K_i)$ be a class where $c_i \in C$. Let $c_j = (A_j, M_j, K_j)$ be a class

where $c_j \in C$. Let $A_s = \{a_{s1}, .., a_{sn}\}$ be a set of attributes. Let $A_t = \{a_{t1}, ..,$

$a_{tm}\}$ be a set of attributes. Let REQSPAT defined as REQSPAT: OP

(ReqSpatP, ReqSpatP)

ReqSpatP : Geometry | OP (ReqSpatP ReqSpatF)

ReqSpatF: ',' ReqSpatP | ε

<u>Pre-conditions</u>: $A_s \subseteq A_i$ and $A_t \subseteq A_j$ and $K_i \not\subset A_s$

A RL of **RLextA-A (e-A_A->)** type between A_s and A_t is defined by:

e-A_A-> : C x REQSPAT x A x .. x A -> C x A x .. x A

$$(c_i, \textbf{reqspat}, a_{s1}, .., a_{sn}) = (c_j, a_{t1}, .., a_{tm})$$

Example 1 (1..* linked classes): From a tourist point of view, a user may want to eat in a restaurant near a point of interest. It would be appropriate to determine the types (e.g., international, French, Italian) and addresses of Restaurants within a reasonable distance. The administrator defines a RL between the spatial attribute "interestPoints" of the City class and alphanumeric attributes "typeRestaurant" and "addressRestaurant" of the Restaurant class. This RL is defined by: e-A_A-> (City, reqspat, interestPoints) = (Restaurant, typeRestaurant, addressRestaurant), where reqspat is a spatial query. This query is defined by the application of the $\sqrt{}$ operator: $\sqrt{}$(interestPoints, SR < 200) on the spatial attributes "interestPoints" of the City class and "SR" of the Restaurant class.

Example 2 (independent classes): From a tourist point of view, a hotel may have a park located in a wider green area. It would be appropriate to determine the name and type of this green space. The administrator defines a RL between the spatial attribute "park" of the Hotel class and the alphanumeric attributes "nameGS" and "typeGS" of the GeenSpace class. This RL is defined by: e-A_A-> (Hotel, reqspat, park) = (GreenSpace, nameGS, typeGS), where reqspat is a spatial query. This query is defined by the application of the $\subset$ operator: $\subset$(park, SR) on spatial attributes Park of the Hotel class and SR of the GreenSpace class.

Operational complexity is defined with two stages. The first stage concerns the complexity of the spatial operator. In a second stage, this complexity can be modified by the research of the potential instance(s) for this link. A value depends on the physical implementation of the database. For example, a specialization can be translated into a set of relational joins or a set of projections and unions.

We emphasize that the goal here is not to propose a standard solution like the OGC Table Joining Service (TJS) (OGC 2010a). TJS offers a method of data research, access and use from multiple sources and, dynamically, to supply the databases, to perform analysis. Indeed, we use the same schema of database to determine the relevant attributes.

Synthesis

The concept of relevant link (RL) presented in this section widely extends the functional dependence. They take into account more semantics such as the ones that may exist between attributes and between attributes and methods. The main advantage of this proposal lies on semantic definitions. Information will be processed with links of relevance and not only from their graphical representations.

Figure 7 presents the relationships between the various concepts of our meta-model formalized here in UML.

This diagram indicates the following rules:

- A class can have the role of source class and/or of target class within a RL.
- A class is composed of attributes, methods, and possibly RL.
- An attribute can be spatial or alphanumeric.
- An attribute can have the role of a source attribute or of a target attribute within a RL.
- An attribute has a single type.
- A method has a list of parameters and a return value defined by a type.
- A parameter can be an attribute or the result of a method.
- A constraint (pre-conditions and/or post-conditions) may be associated with an RL.
- A RL can be typed by RLintA-A, RLintA-M or RLextA-A.
- A RLintA-A establishes a link between two sets of attributes of a class.
- A RLintA-M is built with a set of attributes and methods of a class, the parameter list of this method and its result. A typed attribute denotes the result.
- A RLextA-A can bind two sets of attributes of two different classes.
- A RLextA-A does or does not involve a spatial query.

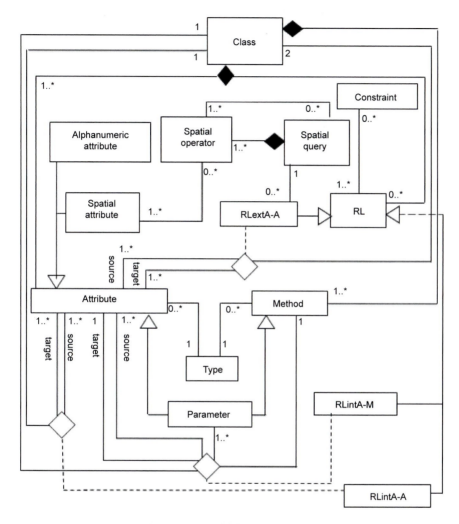

Fig. 7. Meta-model formalized in UML.

- A spatial query is composed of one or more spatial operators.
- An operand of a spatial operator is a spatial attribute or a spatial query.

In the following, we present the classification and the determination of relevant RL for GetFeatureInfo and GetFeature operations. These operations will provide relevant attributes from the database schema.

Determination of Relevant Links

At this stage, the problem is not the connection with the Data Base Management System but the identification and the choice of relevant attributes among the database schema. They should satisfy the better relevance criteria. The identification is therefore equivalent to determine the Relevant Links (RL). In that sense, we propose to classify the RL in order to facilitate the search.

Several formalisms exist to handle a classification, such as graph-based or logic-based paradigms for example (Bedel et al. 2007). We propose to use a lattice of concepts for the classification of RL. Foerster et al. (2010) and Messai et al. (2008) modify the lattice structure while analyzing end-users' query. In contrast to this approach, we propose starting from the data model of GetFeature or GetFeatureInfo data model to add or to delete attributes in order to define the data model associated with end-users' queries. This approach is similar to a conventional viewed mechanism in relational database.

Conventional applications in Information Systems have the knowledge of the database schema (eventually with the formalism of views). In our context, applications use a Web service that accesses to a distant database. Users do not have an a-priori knowledge of the database schema nor the database administrators have an a-priori knowledge of end-users' needs. Therefore a view mechanism cannot be provided. Nevertheless, data base administrators have a very good knowledge of the database schemas that are offered to external queries.

We proposed the concept of Relevant Links that provides a semantic relationship between database schemas and users. Starting with a user query, we propose to offer a semi-automatic design of a result database schema. The Data Base Administrator is no longer in charge of providing static external schemas (i.e., views in relational databases) for the set of queries. A middleware on the clients' computers is in charge of providing a dynamic "view". This operation is transparent to end-users. The "view" is relevant to an end-user's point of interest and is designed according to the operational complexity (s)he is willing to accept.

This section is organized into four parts. In the first part, we present the formal definition of lattice of concepts. In the second part, we present the links of this lattice and the available RL. In the third part, we present the extended formal definition of a RL. Then, in the fourth part, we present the determination of RL.

Lattice Formalization

We propose a classification built over a partial ordered structure named lattice of concepts (Wille 1982; Nauer and Toussaint 2009). This structure is induced by an environment (context) E, has a set of properties, T, a set of structures, L, and a binary relationship, I, named incidence of E, such as $I \subseteq T \times L$.

Let $L' \subseteq L$, be a set of structures. The set of common properties to the structures of L' is $X = \{x \in T \mid \forall y \in L', (x, y) \in I\}$.

Let $T' \subseteq T$, be a set of properties. The set of structures with all the properties in T' is $Y = \{y \in L \mid \forall x \in T', (x,y) \in I\}$.

A concept is formally represented with a couple (X, Y). Let F be the set of the concepts, relevant to an environment E. Let f_1 be (X_1, Y_1) and f_2 be (X_2, Y_2) in F, f_1 is subsumed by f_2 (noted $f_1 \leq f_2$) if $X_1 \subseteq X_2$ or from the dual point of view $Y_2 \subseteq Y_1$. $(F, \leq)$ is a complete lattice, named lattice of concepts linked to the formal environment E.

The lattice of concept we propose to use, references to a classification tool of the RL as an ontology of domain. This lattice describes concepts involved in a specific domain in order to take into account the different links between the components of this domain. Numerous works deal with the modeling (in the ontology sense) such as (OGC 2008) with CityGML. CityGML, recognized as a norm, provides a classification for 3D modeling of a town. Emgard and Zlatanova (2008) propose a generic meta-model for a 3D representation as a complement to CityGML.

In comparison with these works, we propose a meta-base (instead of a meta-model) with the formal representation of elements, rules and constraints that regulate the creation of a database (but not the application domain).

Within the lattice, a concept represents, from a semantic point of view, the level of interpretation for a group of data. This level is relevant for a specific user.

Let $E = (T, L, I)$ be a formal context. T is a set of themes. L is a set of Relevant Link (RL). I is an incident function of E such as $I \subseteq T \times L$ and $(t, l) \in I, t \in T$ and $l \in L$. That means that theme "t" is a property of the Relevant Link "l".

Let t be a theme such as $t \in T$. The set of RL sharing this theme is L' such as $L' = \{l \in L / (t, l) \in I\}$.

A formal concept is a theme shared by a set of RL. It is formally represented by a couple (t, L') where $t \in T$ and $L' \subseteq L$.

Extended Definition of Relevant Links

To provide a classification of Relevant Links (RL), we propose to extend the definition of a RL (given in the Data Conceptual Schema section). Let N denote the set of natural integers. The extensions are: an identifier (i.e., id, such as $id \in N$), a theme (i.e., $t \in T$) and a complexity θ associated with the process in charge of obtaining data. This complexity is data base implementation dependent.

Notation: Let n be a variable defining the quantity of data (e.g., the number of tuples in a database). Let T be a set of themes and $\theta = \{O(1), O(\log(n)), O(n), O(n^2), O(>)\}$ a set of complexity (Papadimitriou 1993).

The formal definition of RL typed by RLintA-A becomes:

-A_A->: N x T x θ x C x A x ... x A → A x ... x A

The formal definition of RL typed by RLintA-M becomes:

-A_M-> : N x T x θ x C x A x ... x A x M x (P) → A

$\quad$ (id, t, o, c_i, a_1, ... a_u, m_k, (p_1, ..., p_q)) = $a_{|A|+1}$

The formal definition of RL typed by RLintA-A becomes:

e-A_A-> : N x T x θ x C x REQ-SPAT x A x ... x A → C x A x ... x A

$\quad$ (id, t, o, c_i, request-spatial, a_1, ... a_u) = (c_j, a_v, ..., a_m)

Application of a lattice to Relevant Links

We propose to use a lattice of concepts, named here, lattice of reference, to classify Relevant Links. The Data Base Administrator is in charge of defining a theme to a Relevant Link using this lattice. Figure 8 presents an example of a (reference) lattice of concepts in our example. This lattice is shared by the set of applications that wish to use the enhancement service (in order to keep unchanged the signature of the normalized web services).

$\quad$ To illustrate the formal context E, we rely on a set of RL and themes presented in Table 7. Symbols are used to provide a better reading. An example of a formal context is presented Table 8.

$\quad$ Figure 9 presents the associated lattice of concept for a formal context E. This lattice is named Instantiated Lattice of Concepts. It corresponds to a sub set of the lattice of concepts and is specific to a geographical data server.

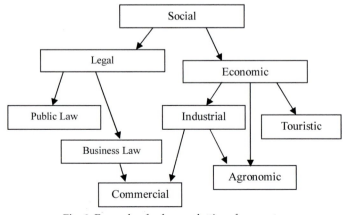

Fig. 8. Example of reference lattice of concepts.

Table 7. Designation of RL.

Designation of RL	Symbol
-A_A->(Tree, species) = phytosanitaryStatus	l1
-A_M->(City, population, density, (int, surface(RS))) = density	l2
e-A_A->(City, √, interestPoints) = (Restaurant, typeRest, adressRest)	l3
e-A_A->(Hotel, ⊂, park) = (GreenSpace, nameGS, typeGS)	l4
-A_A->(City, urbanArea) = mayor	l5
-A_A->(Country, population) = GNP	l6
-A_M->(City, urbanArea, percentUrban, (urbanArea, surface(RS))) = percentUrban	l7

Table 8. Example of formal context E.

	l1	l2	l3	l4	l5	l6	l7
Social	0	true	0	0	0	true	0
Legal	0	0	0	0	true	0	0
Economic	0	0	0	0	0	0	true
Touristic	0	0	true	true	0	0	0
Agronomic	true	0	0	0	0	0	0

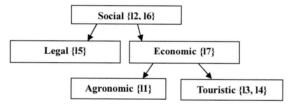

Fig. 9. Instantiated lattice of concepts.

Determination of Relevant Links

We consider in this section that a user query is based on two components: a temporal component (i.e., service time) and a semantic one.

About the temporal component, the user is invited to define the acceptable level. (S)he can accept the standard data provided by WMS and WFS services or can request for enhanced data. (S)he first indicates the maximum acceptable complexity to get these data (e.g., logarithmic, linear, quadratic). The complexity is widely linked to the database structure on which distant end users cannot interact. Her/His level of acceptation depends on her/his center of interests and may vary during the spatial analysis process. The temporal component is defined for each theme and provides an Operational Lattice of Concepts. This lattice is a sub-set of an Instantiated Lattice of Concepts and contains RL that are compatible with the temporal complexity for this theme. In the event of a change during the spatial analysis process, a new Operational Lattice of Concepts is generated.

The semantic component represents the position in the lattice. The user specifies her/his semantic requirement: asking enhanced data for a specific theme or indicating a theme and applying a closure operator on the graph modeling the lattice. This allows, for each query, the localization of the treatment(s) associated with the dynamic construction of a view into the end-users' side. This construction is completely transparent to geographic data servers.

In order to simplify the presentation, let us consider that the Instantiated Lattice of Concepts (Fig. 7) also corresponds to the Operational Lattice of Concepts. The application of the GetFeatureInfo or GetFeature operations will be applied to an instance (of a map or an object in a given map). The problem is now transformed into the definition of the data model associated with this operation.

The GetFeatureInfo operation (from a WMS service) will require the traversal of links typed by RLintA-M and by RLextA-A in order to define relevant attributes to be added to the initial schema. The standard GetFeatureInfo operation provides the whole schema associated with the result. The RLintA-A typed links are useful to determine attribute(s) to be taken off from the initial schema. To take into account the typed links RLintA-M and RLextA-A, the definition of a theme is essential. Based on the RLintA-M link, the definition of the theme "Economic" will lead to only traverse the link {"L7"}. The definition of the theme "Economic" with recursive application will lead to traverse the links {"L7"} * {"L3", "L4"}. Based on the RLintA-A link, the definition of the theme "Economy" will lead to only traverse the link {"L6"}. The source and destination attributes involved in this link and the source attributes of the links typed RLintA-M and RLextA-A will be considered as relevant and will be kept in the final schema. All other attributes will be removed (except the key attribute(s) if this last one is not already present in the final schema).

Table 9 illustrates an example of interactions of RL in a query based on the class "Town" with the theme "Economic".

To deal with the GetFeature operation (from a WFS service), we provide two approaches. In the first approach, we suggest automatically generating a schema from the user query and to limit the result to the theme and the key attributes. In this case, the GetFeature operation will lead to the traversal of RLintA-A, RLintA-M or RLextA-A typed links. Within the same definition of a theme (e.g., "Economic"), links {"L6", "L7"} will be traversed. The final schema will be composed of the different attributes involved in the traversed links. In a second approach, named standard, the user selects attributes that should be involved in the final schema. The GetFeature operation will require the traversal of links typed "RLintA-A". This traversal is performed

Table 9. Example of GetFeatureInfo query.

Id Town	urbanArea	percentUrban	interestPoints	TypeRest	Address Rest

Economic   Touristic

if the source belongs to the required attributes and at least an attribute of the destination does not belong to the required final attributes in order to define attributes to be added to the final schema. Within the same definition of a theme (e.g., "Economic") other links {"L6", "L7"} will be traversed under the same conditions for the source attributes. The process will be similar to a GetFeatureInfo operation in the case of a closure from a given theme. Let us mention that for the GetFeature operation, the schema contains relevant attributes.

Implementation

This proposition has been implemented with open source tools. Figure 10 sums up technical choices. They are compatible with norms and open standards that allow interoperability between different systems (applications, services and clients). PostgreSQL is used with its extensions: PostGIS and PgRouting for graph manipulations. Geoserver is used as a cartographical server since it provides raster and vector supports. Several databases can be handled with this server.

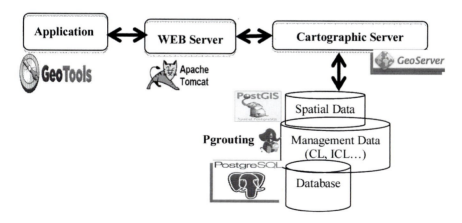

Fig. 10. Technical tools.

Easily integrated in the Java Eclipse platform, Geotools is the basic component of some GIS tools (i.e., UDig). Its modular architecture and its use to develop Geoserver, leads us to use it as a basic component for cartographic applications.

From the database server side, interfaces to model RL have been realized. From the application server, using a database schema provided by a distant server, an application has been developed to enhance attribute information and to display them. A presentation of these interactions is available in (El Ghouasli et al. 2012).

Figure 11 presents the specific application interactions. An application interacts with a middleware. The first step is the connexion that will provide the Instantiated Lattice of Concepts (ICL) from the server side using WFS services. The second step is a dynamic step that adapts the received ICL to users' needs. The Operational Lattice of Concepts (OLC) is now available. Since this step is dynamic, it can be performed several times within the same working session. The third step is an access demand. Using the RL, the received (G/X)ML files from WMS/WFS services are extended or reduced depending on users' needs. The application handles these (G/X)ML files as if they were original ones provided by WMS/WFS calls.

The lattice of concepts is shared by the different applications. The Operational Lattice is specific to each database server depending on available data. The modeling of each lattice is similar since they cover the same modeling (here an acyclic graph). A conventional database can take charge of this modeling.

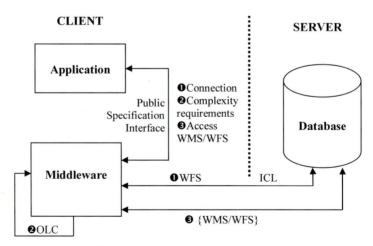

Fig. 11. Specific applications interactions.

Conclusion

To facilitate data and service interoperability, the Open Geospatial Consortium (OGC) proposes web services such as Web Map Service (WMS) and the Web Feature Service (WFS). These services allow a given system to access data from distributed data servers as soon as they both respect a common interface. These interfaces are based on operations such as for example GetCapabilities, GetFeatureInfo, and GetFeature.

These services present some weaknesses for the GetFeatureInfo operation of WMS and GetFeature operation of WFS from the semantical point of view. The DataBase Administrator of a distributed service does not know, by definition, the clients' needs (here end-users or applications using these services). The clients do not know the database schema (respectively, logical and external). Therefore, it is widely desirable to offer a mechanism close to the notion of a dynamic view in order to provide a better relevance to users' needs. This mechanism requires an enhancement of the database conceptual model from the server side.

The proposed extension defines Relevant Links (RL) between two sets of attributes belonging to the same class, between attributes and methods (i.e., functions) of the same class and finally between two sets of attributes belonging to two different classes (linked or not by constructors of the conceptual data model). Methods are viewed as attributes. In fact, users ignore the nature of available data (simple or computed). The goal is to provide a data model associated with the result of a query designed upon users' needs. The proposition must not modify the specifications of OGC norms for web services.

Relevant links, associated with classes, give a semi-automatic enhancement of the database schema provided with a result of a WMS or WFS call for GetFeatureInfo or GetFeature operations. A classification, based on a lattice, automatically gives the better enhancement of the external database schema provided with a result.

Perspectives of this work is to provide a metric evaluation of performances with real size applications focusing on the database server aspect without taking into account the network parameters (e.g., bandwidth) that are uncontrollable and not relevant for this proposal.

References

ANSI-SPARC Architecture. http://en.wikipedia.org/wiki/ANSI-SPARC_Architecture.

Bedel, O., S. Ferré, O. Ridoux and E. Quesseveur. 2007. GEOLIS: a logical information system for geographical data. Int. J. of Geomatics and Spatial Analysis, Hermes, France. 17: 371–390.

Brodie, M.L. and J.W. Schmidt. 1982. Final report of the ANSI/X3/SPARC, DBS-SG relational database task group, SIGMOD Record, Volume 12, Issue 4.

Bucher, B. and S. Balley. 2007. A generic prepocessing service for more usable geographical data processing services. http://www.agile-online.org/ Proc. of the 10th AGILE Conf. on Geographic Information Science, Aalborg, Denmark.

Donaubauer, A., A. Fichtinger, M. Schilcher and F. Straub. 2007a. Model driven approach for accessing distributed spatial data using web services—Demonstrated for cross-border GIS applications. Proc. 23rd Int. FIG congress, 2006, Munich, Germany. http://fig.net/pub/fig2006/techprog.htm.

Donaubauer, A., F. Straub and M. Schilcher. 2007b. mdWFS: a concept of web-enabling semantic transformation. Proc. of the 10th AGILE Conf. on Geographic Information Science, Aalborg, Denmark. http://agile.gis.geo.tu-dresden.de/web/index.php/conference/proceedings/proceedings-2007.

El Ghouasli, M., K. Jabri, N. Souissi and M. Mainguenaud. 2012. Improvement of cartographic information for the WMS and WFS Web services. 3rd International Conference on Multimedia Computing and Systems (ICMCS'12), May 10–12, Tangier, Morocco.

Emgård, K.L. and S. Zlatanova. 2008. Design of an integrated 3D information model. pp. 143–156. In: V. Coors, M. Rumor, E.M. Fendel and S. Zlatanova (eds.). Urban and regional data management: UDMS annual 2007. Taylor & Francis Group, London, UK.

Foerster, T., L. Lehto, T. Sarjakoski, T.L. Sarjakoski and J. Stoter. 2010. Map generalization and schema transformation of geospatial data combined in a Web Service context. Computers Environment and Urban Systems, Elsevier. 34(1): 79–88.

Grosso, E., A. Bouju and S. Mustière. 2009. Data integration GeoService: a first proposed approach using historical geographic data. J.D. Carswell et al. (eds.): W2GIS 2009, LNCS 5886, 103–119. Springer-Verlag Berlin Heidelberg.

Messai, N., M.D. Devignes, A. Napoli and M. Smaïl-Tabbone. 2008. Many-valued concept lattices for conceptual clustering and information retrieval, 18th European Conference on Artificial Intelligence (ECAI), IOS Press, Frontiers in Artificial Inteligent and Applications. 127–131M. Gallab et al. (eds.).

Nauer, E. and Y. Toussaint. 2009. CreChainDo, an iterative and interactive Web information retrieval system based on lattices. Int. J. of General System. (38–4) 363–378. Taylor and Francis (ed).

OGC, web. 2005. OpenGIS web services architecture description, Open Geospatial Consortium. http://www.opengeospatial.org/standards/bp.

OGC, web. 2006. OpenGIS Web Map Service (WMS) implementation Specification, Open Geospatial Consortium. http://www.opengeospatial.org/standards/wms.

OGC, web. 2007a. OpenGIS Geography Markup Language (GML) encoding standard, Open Geospatial Consortium. http://www.opengeospatial.org/standards/gml.

OGC, web. 2007b. OpenGIS Web Processing Service (WPS) implementation Specification, Open Geospatial Consortium. http://www.opengeospatial.org/standards/wps.

OGC, web. 2008. OpenGIS City Geography Markup Language (CityGML) encoding standard, Open Geospatial Consortium. http://www.opengeospatial.org/standards/citygml.

OGC, web. 2010a. Georeferenced Table Joining Service (TJS) implementation standard, Open Geospatial Consortium. http://www.opengeospatial.org/standards/tjs, 2010.

OGC, web. 2010b. OpenGIS Web Feature Service (WFS) implementation specification, Open Geospatial Consortium. http://www.opengeospatial.org/standards/wfs.

Papadimitriou, C. 1993. Computational complexity. Addison-Wesley.

Staub, P. 2007. A model-driven web geature service for enhanced semantic interoperability. OSGeo J. (3): 38–43. http://journal.osgeo.org/index.php/journal/issue/view/28.

Staub, P., H.R. Gnaegi and A. Morf. 2008. Semantic interoperability through the definition of conceptual model transformations. Transactions in GIS. 12–2: 193–207.

Stollberg, B. and A. Zipf. 2007. OGC Web processing service interface for web service orchestration aggregating geo-processing services in a bomb threat scenario. J.M. Ware and G.E. Taylor (eds.): W2GIS, LNCS n°4857, 239–251. Springer-Verlag Berlin Heidelberg.

Ullman, J.D. 1995. Principles of Database and Knowledge Base Systems—classical database systems, Computer Science Press.

Wiemann, S., L. Bernard, P. Wojda, P. Milenov, V. Sagris and W. Devos. 2012. Web services for spatial data exchange, schema transformation and validation as a prototypical implementation for the LPIS quality assurance. Int. J. of Spatial Data Infrastructures Research. (7): 66–87.

Wille, R. 1982. Restructuring lattice theory: an approach based on hierarchies of concepts. Formal Concept Analysis, LNCS n°5548, 314–339. Springer.

CHAPTER 7

Robust Workflow Systems + Flexible Geoprocessing Services = Geo-enabled Model Web?

Carlos Granell

Introduction

In January 2013, a United Nations (UN) Committee of Experts on Global Geospatial Information Management (GGIM) delivered a revised document on the "Future trends in geospatial information management: the five to ten year vision".[1] This document identified future directions in the use of geospatial data, the envisioned role of the national mapping agencies, private and voluntary sectors, and technology trends among others. Most of the experts and visionaries coincided in the permeability between geospatial and mainstream technologies. They foresaw an increasing adoption of geospatial technologies into mass-market technologies, and vice versa, classical geospatial developments will be absorbing and exploiting emerging technologies such as Big Data, Internet of Things, Linked Data and Cloud computing.

European Commission–Joint Research Centre, Institute for Environment and Sustainability, Digital Earth and Reference Data Unit (H06)–TP 262, Via e. Fermi 2749, 21027 Ispra (VA), ITALY.
Email: carlos.granell@jrc.ec.europa.eu

[1] http://ggim.un.org/docs/meetings/2ndHighLevelForum/UN-GGIM%20Future%20Trends%20Paper%20-%20Version%202.0.pdf

The overall aim of the present book fits nicely into the earlier UN GGIM vision, and surely other chapters will be reporting on hybridizations of these technologies such as Linked Geo Data and Cloud GIS. In this chapter we also adhere to this vision and report on the interplay between geoprocessing services and workflow systems as a new way to leverage geo-enabled workflows on the Web to support a vast array of scientific activities such as environmental modeling, simulation, and data visualization. Our exposition, though, is driven by the research question in the chapter title in the form of mathematical equation. We are questioning whether the robustness of workflow systems in conjunction with the flexibility exposed by geoprocessing web services *may* lead to the geo-enabled Model Web, which essentially is a novel approach that treats models as services and combination of models as composition of web services.

This chapter begins by briefly exploring the concept of modeling in geosciences which notably benefits from advances on the integration of geoprocessing services and workflow systems. In "The State of the Art Section", we provide a comprehensive background on the technology trends we treat in the chapter. On one hand we deal with workflow systems, categorized normally in the literature as scientific and business workflow systems (Barga and Gannon 2007). In particular, we introduce some prominent examples of scientific workflow systems—Kepler, Taverna, and VisTrails—and descriptive languages for modeling business workflows, namely BPMN and WS-BPEL. On the other hand, we visit the notion of geoprocessing web services—mainly interfaced by the Web Processing Service and Web Service Description Language specifications. After the current state of the art, we identify recent works that merge geoprocessing service technologies with scientific and business workflow systems to varying environmental modeling scenarios. Based on the earlier analysis, in "Towards the Geo-enabled Model Web Section" we introduce the notion of Geo-enabled Model Web to shape geo-enabled workflows on the Web, and raise some challenges that need to be addressed to promote robust and flexible tools for realizing the Model Web vision. Finally, some concluding recommendations are formulated.

The Role of Modelling for Geosciences

Modeling is a powerful tool in geosciences to understand the Earth and our environment. A model basically mimics and simplifies a natural system or a part of it (e.g., climate, ecosystems, watershed, and atmosphere), as Figure 1 illustrates. The analysis of complex natural phenomena, such as the examples above, requires the combination of several models that may even span over multiple disciplines (biodiversity, water, ocean, agriculture, etc.). The notion of Integrated Modeling (IM) captures exactly this. Rotmans

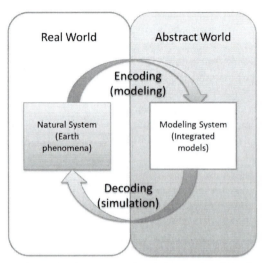

Fig. 1. Relationship between real and abstract (modeling) world.

and van Asselt (2001) defined IM as the process of structuring and sharing new knowledge that emerges from the interrelation of different constituent models in complex socio-economic-environmental scenarios. One of the common objectives in IM is to support simulation, anticipating potential impacts, policy and the decision-making processes (McIntosh et al. 2008). It is widely assumed that an individual model cannot be sufficient to handle with the complexity and the large quantity of parameters and variables needed in real-life, multi-dimensional scenarios.

In geosciences and environmental fields, IM is widely recognized as a proven mechanism to explore a given environmental problem or scenario, i.e., the combination of models will eventually depend on the environmental phenomenon being investigated (Parker et al. 2002; Jakeman and Letcher 2003). For example, integrated models for simulating crop growth and predicting watershed runoff will certainly rely on the combination of distinct constituent models. So, environmentalists and modelers demand novel approaches, technologies and tools that allow them to reuse, share and assemble different models together to address similar environmental issues (Voinov and Shugart 2013; Laniak et al. 2013b). In this chapter, we examine distinct strategies for IM, namely, workflow systems and geoprocesing services, identify synergies and their benefits, and eventually suggest a novel and powerful strategy exemplified by the Model Web vision.

The State of the Art

Integrated modeling (IM) has been an active research line during the last two decades in environmental and geosciences fields. From the

Goodchild et al. (1993) book on environmental modeling and Geographic Information Systems (GIS) and Abel et al. (1997) paper on integrating modeling for environmental management information systems, to the recent Environmental Modelling and Software's thematic issue on "the Future of Integrated Modelling Science and Technology" (Laniak et al. 2013a), there have been proposed lots of methodologies, strategies and approaches to deal with IM projects. In the 1990s, first approaches were focused on the integration of environmental models into desktop-based GIS tools. Existing GIS packages were containers to embed environmental models on such desktop tools. In the next decade, with the advent of the Internet and Web technologies, modeling strategies shifted from ad hoc, desktop-based solutions to distributed computing and web services technologies. Multiple middleware frameworks emerged to cope particularly with IM but still tailored to specific disciplines and fields (e.g., hydrology and ecology). Last years, though, have been particularly active with the emergence of new computing paradigms (e.g., cloud computing) and novel approaches such as Model Web and resource-oriented architectures which altogether promise platforms and tools to readily enable cross-domain IM.

In the following we present a state-of-the-art on technologies, systems and frameworks for supporting IM, i.e., the integration of various models. For doing so, we particularly focus on the concepts of workflow and services which drive scientific and business workflow systems and geoprocessing web services respectively.

Scientific Workflow Systems

Deelman et al. (2009) introduced the concept of workflow as the "activity of defining the sequence of tasks needed to manage a business or computational science or engineering process". A more specific definition would be a set of analytical tasks that may involve different processing tasks such as data access and querying, data analysis and processing, and data visualization. Workflow designers often specify a control flow (e.g., sequence, forks, switch, joins, etc.) and data flow (how outputs of a preceding task connect to the inputs of subsequent task) in order to structure the order of the required steps or activities.

In the scientific context, most tasks consist of the acquisition, manipulation, documentation, and processing of large amounts of scientific data, as well as the execution of computationally intensive analysis and simulations (Ludäscher and Goble 2005). Scientific workflows may be seen as a description of the combination of the previous scientific tasks to address data-centric applications. Such descriptions are then managed by scientific workflow systems, which are able to interpret and execute every single task specified within a scientific workflow.

Scientific workflow systems and component-based modeling frameworks (Argent 2004; Argent et al. 2006) share some commonalities. Both typically support dataflow-oriented workflows that can be computationally expensive, thereby data-intensive use-case scenarios characterize both sorts of systems.

Ludäscher and Goble (2005) however highlight annotation as the differentiator element, i.e., the ability to document a workflow in form of descriptions of the results of the workflow execution, inputs data sets used, restrictions and post-conditions after execution, and so on. Scientific workflows are often more annotation-intensive because by definition they pursue reproducibility (Jasny et al. 2011; Mesirov 2010) so that an annotated workflow with useful information can be replicated and reproduced in other scenarios later on. Reproducibility thus requires detailed context metadata and data provenance information (Cohen-Boulakia and Leser 2011). For instance, a common pattern in environmental sciences is to perform the same set of analytical tasks several times but changing the input data sets in each run. Take the example of a watershed runoff model: while the logic of the model itself remains invariable, the results change because the model is fed with distinct time-series data sets in each run. That is, control flow definitions are not as relevant and critical as data sets used in each execution. In these cases, annotation techniques (e.g., context metadata, provenance, traceability) allow modelers to document properly scientific workflows along with data sets used and produced, and to create accurate records of scientific experiments and procedures in order to be reproduced or analyzed later. The immediate benefit from workflow annotations is the ability to reuse others' workflows in a confident and reliable way, as scientists can reproduce and compare same scientific workflows applied to distinct use-case scenarios, leading to more transparent and reproducible science (Mesirov 2010).

In the workflow literature there is not a unique solution but dozens of scientific workflow systems specialized in different scientific domains and disciplines (Yu and Buyya 2005; Deelman et al. 2009). This vast array of systems suggests that there is no a single workflow system to handle with the diversity of scientific workflows, nor is there a common data model for all scientific workflows. All of this implies a lack of interoperability between workflow management systems (Elmroth et al. 2010). Given that, we focus in this section on widely-used scientific workflow systems and particularly in their support for web service technologies. Recent reviews on scientific workflow systems (Yu and Buyya 2005; Rahman et al. 2011) also include the scientific workflow systems analyzed below but complemented here by having a close look at the interplay between geo-computation and geoprocesing web services capabilities.

Kepler

Kepler[2] (Ludäscher et al. 2006) is a modeling, execution, and deployment framework for scientific workflows across different domains. Kepler is used in various kinds of projects such as biological sciences, DNA sequences, and geosciences and earth observation among others.[3]

The main components in Kepler are *directors*, *actors*, and *parameters* as well as *channels* and *ports*.[4] A scientific workflow in Kepler consists of a combination of actors. An actor can be regarded as an analytical task. Kepler follows an actor-oriented programming interface (Ludäscher et al. 2006) as a methodology to turn analytical tasks into Kepler-compliant actors. That is, an actor provides a customized interface to support distinct types of processing tasks such as file management actors, control actors, display actors, GIS actors, and so forth, which come pre-defined when installing the Kepler framework. Like a function signature or method, each actor has parameters that allow users to customize the behavior of a particular kind of actor. For instance, a pre-defined file management actor may be customized to locate and unzip a given data file in a specific format.

Data communication between actors is carried out via ports. Each actor in a Kepler workflow can expose one or more ports used to consume (input ports) or produce (output ports) data in order to communicate with other actors within the workflow. The data-flow connection between one actor port and another actor port is called a channel. Finally, a director is meant to control the execution of a workflow. Every workflow must have a director that defines whether the set of actors are executed for example in sequence or in parallel. Roughly speaking, directors specify the control flow of a scientific workflow.

The ability of reusing a part of or an entire workflow is managed via a specific type of actors. *Composite actors*, or sub-workflows, wrap some atomic actors for performing more elaborated operations. As an entire workflow can be a composite actor, it can be re-used by other Kepler workflows.

The Kepler framework also contains a graphical user interface and an execution engine to help scientists to edit, manage and execute scientific workflows. Scientists can create their own scientific workflows by simply dragging and dropping actors onto a design area, where actors can be customized, combined and executed.

---

[2] http://kepler-project.org
[3] https://kepler-project.org/users/projects-using-kepler
[4] Getting Started Guide: https://kepler-project.org/users/documentation

Taverna

Together with Kepler, Taverna[5] (Oinn et al. 2004) is one of the most widespread scientific workflow systems. It is aimed to design and execute scientific workflows, used extensively across a range of life science projects such as gene and protein annotation, proteomics, and so forth (Oinn et al. 2006).

The main components in Taverna are *processors, data links* and *coordination constraints*. A Taverna workflow is a combination of processors, i.e., each individual step that transforms a set of data inputs into a set of data outputs. The notion of processor in Taverna is conceptually similar to an actor in Kepler, though, while actors encompass the idea of adapters or wrappers, processors are more aligned with the idea of "service clients" (Hull et al. 2006). Each type of processor in Taverna calls to remote or local services. For example, the processor *arbitrary WSDL type* executes a request to an operation from a web service described in WSDL format (we visit this later in Section "Web Service Description Language"). Other examples of processor types are *local processor type* (e.g., a local Java function) and *nested workflow type* (a nested workflow processor). Alternately, if the set of pre-defined processor types does not support an external component interface, a developer can either transform the interface of that component into an interface supported natively by Taverna or develop its own processor type.[6]

Processors in Taverna workflows are described in a XML-based language called SCUFL (Simple Conceptual Unified Flow Language). It is a high-level language to describe the data flow interactions between different processors, even if a processor is of *nested workflow type* (i.e., a sub-workflow), in a Taverna workflow.

Apart from processors, a SCUFL-based workflow description also contains data links and coordination constraints. Data links facilitate data communication between two processors, which in turn may be of nested workflows type, thereby enabling data communications between workflows. A coordination constraint links two processors and controls their execution. It represents a kind of control flow pattern, especially in the case of two processors connected with no direct data dependency between them. However, in most cases, the level of control flow is defined by the very data flow connections between two processors imposed by data links components.

Taverna encompasses a suite of open source tools made up of the Taverna Engine (used for enacting workflows) that powers the Taverna

---

[5] http://www.taverna.org.ok
[6] Taverna 2.x Documentation: http://www.taverna.org.uk/documentation/

Workbench (the desktop client application) and the Taverna Server (which allows remote execution of workflows). Like in Kepler, the main function of the workbench itself is to enable the construction and editing of scientific workflows. The main mechanism for interaction with the workflow is through the Explorer View, which presents a tree view interface showing the types of processors available.

VisTrails

VisTrails[7] is an open-source scientific workflow and provenance management system that is focused on data visualization (Callahan et al. 2006a). VisTrails pays special attention to the management of data and metadata of a *visualization output*, i.e., the result or output of a workflow execution in VisTrails. Examples of visualization-centric applications are medical imaging and physics simulation,[8] even though it is also being applied to the geospatial and environmental fields like in the Observations Network on Environmental Change in South Africa (SAEON).[9] Apart from the emphasis on data exploration and visualization, the other key value of VisTrails is the extensive set of capabilities for capturing and managing data provenance, including the provenance of the visualization process (i.e., workflow) and its resulting data sets.

The main components in VisTrails are *modules*, *ports*, *connections*, and *vistrails*.[10] As a visualization output is central in VisTrails, a workflow is regarded as a sequence of steps used to generate such a visualization output. Each step is called *module* in VisTrails terminology. Each module has input and output ports, that is, input and output parameters. Modules are linked together to define data communication between modules by connecting output to input ports. In addition, VisTrails can manage several versions of the same workflow, called *vistrails*, by recording past steps or changes done in generating a particular visualization product (Callahan et al. 2006b).

Third-party components can be added into VisTrails by defining and developing custom modules. Application developers must develop a set of Python classes following a pre-defined packages structure and respecting some code conventions. The collection of Python classes is bundled together in a VisTrails package which may then be used as a built-in module in the VisTrails framework.

VisTrails is architected as a set of interrelated tools (Callahan et al. 2006a,b). Users can create and edit a workflow using the VisTrail Builder

---

[7] http://www.vistrails.org
[8] http://www.vistrails.org/index.php/Projects_using_VisTrails
[9] http://www.saeon.ac.za/
[10] VisTrails Documentation: http://www.vistrails.org/index.php/Downloads

user interface. The tool is able to capture every user-made modification on the workflow in form of provenance data. As a result the user is able to trace back to previous versions of the same workflow. A VisTrails workflow, both the set of operations and provenance data, is serialized in XML which is saved in a VisTrails Repository. Execution is controlled by the VisTrails Cache Manager, the VisTrails Log keeps a log of each workflow execution, and the VisTrails Player invokes the appropriate functions from the visualization libraries within VisTrails.

Business Workflow Languages

Scientific workflows and business workflows share the term workflow, understood as an array of tasks to manage a certain process (Deelman et al. 2009). In principle, both types of workflows share a lot of commonalities such as modularity, encapsulation, and reuse of tasks. Differences, however, come in practical terms. For instance, Barga and Gannon (2007) stated that entire or part of scientific workflows are often constructed by scientists themselves, who often are not technologically as skilled as software engineers who are behind business workflows. As commented earlier, scientific workflows are typically data-centric and dataflow-oriented, as opposed to task-centric and control-flow oriented as defining characteristics of business workflows (Ludäscher and Goble 2005). These differences suggest that scientific workflows are more dynamic in nature because they reflect evolving scientific needs in terms of large data sets, which require robust and sophisticated scientific workflow systems (Section "Scientific Workflow Systems").

Having said this, while scientific workflows were the focus in the previous section, in this section we treat business workflows that are materialized by means of business workflows languages (BWL). Note that we use the term BWL throughout the chapter only for grouping descriptive languages for modeling business workflows. Interested readers may find a comprehensive survey on business workflows standards in Mili et al. (2010).

Business Process Model and Notation

The Object Management Group (OMG) recently released version 2 of the Business Process Model and Notation[11] (BPMN) (OMG 2011) that standardizes a set of graphical symbols with well-defined semantics (e.g., task, data input, parallel flow, etc.) to visually specify abstract processes. By abstract we refer to a set of high-level tasks, procedures and steps to achieve a business goal or process, avoiding at this stage technical and implementation details.

---

[11] http://www.bpmn.org

In this context, the primary goal of BPMN is to provide a graphical notation (Fig. 2) that is readily understandable by all business users, from the business analysts, who create the initial abstract processes, to the technical developers, who are responsible for implementing these processes, and finally, to the business people, who manage and monitor the life cycle of these processes. Thus, BPMN creates a standardized bridge for the gap between the business process design and process implementation (Chinosi and Trombetta 2012).

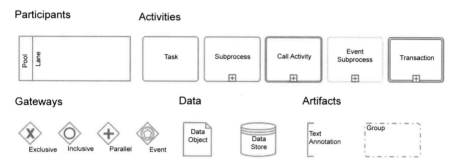

Fig. 2. Examples of BPMN notation and symbols (Source: http://www.camunda.org).

Web Services—Business Process Execution Language

Under the Organization for the Advancement of Structured Information Standards (OASIS),[12] WS-BPEL (Jordan and Evdemon 2007) specification has turned into de facto business process description language for which a process is defined as a workflow of web services. WS-BPEL is a description language composed of variables and operations. These operations have a strong support to handle XML messages (e.g., SOAP[13] envelopes, i.e., an XML communication protocol specification under the World Wide Web Consortium (W3C))[14] and are able to call various remote web services, synchronize results, transform them to be packaged as inputs for subsequent web services in the workflow, and store intermediate results in variables.

Figures 3 and 4 illustrate the same toy example of WS-BPEL. Figure 3 shows a graphical representation of a WS-BPEL process composed of one single web service. Figure 4 lists the XML description of the WS-BPEL process in Fig. 3. In this simple business workflow, only a *sequence* (line 23) activity encompasses a couple of *receive* (line 24) and *reply* (line 27) activities to coordinate the flow of messages with the web service.

---

[12] http://www.oasis-open.org/
[13] http://www.w3.org/TR/soap
[14] http://www.w3.org

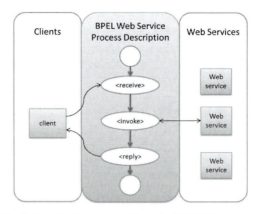

Fig. 3. Graphical representation of a WS-BPEL process.

```
01 <?xml version = "1.0" encoding = "UTF-8" ?>
02 <process name="HolaMundo" targetNamespace="http://xmlns.oracle.com/HolaMundo"
03      xmlns="http://schemas.xmlsoap.org/ws/2003/03/business-process/"
04      xmlns:client="http://xmlns.oracle.com/HolaMundo">
05
06 <!-- PARTNERLINKS: List of services participating in this BPEL process -->
07 <partnerLinks>
08    <partnerLink name="client" partnerLinkType="client:HolaMundo"
09          myRole="HolaMundoProvider"/>
10 </partnerLinks>
11
12 <!-- VARIABLES: List of messages and XML documents used within this BPEL process -->
13 <variables>
14   <variable name="inputVariable"
15          messageType="client:HolaMundoRequestMessage"/>
16   <variable name="outputVariable"
17          messageType="client:HolaMundoResponseMessage"/>
18 </variables>
19
20 <!-- ORCHESTRATION LOGIC: Set of activities coordinating the flow of messages across the
21            services integrated within this business process -->
22
23 <sequence name="main">
24   <receive name="receiveInput" partnerLink="client"
25          portType="client:HolaMundo" operation="process"
26          variable="inputVariable" createInstance="yes"/>
27   <reply name="replyOutput" partnerLink="client" portType="client:HolaMundo"
28          operation="process" variable="outputVariable"/>
29 </sequence>
30 </process>
```

Fig. 4. XML description of the WS-BPEL process in Fig. 3.

WS-BPEL is supported by various workflow enactment engines either commercial like ActiveVOS[15] or open source based implementations such like Apache ODE.[16] Interested readers can find more information on the primer document[17] and in Vasiliev (2007).

---

[15] http://activevos.com/products/activevos/overview

[16] Apache ODE (Orchestration Director Engine), http://ode.apache.org/

[17] http://docs.oasis-open.org/wsbpel/2.0/Primer/wsbpel-v2.0-Primer.htm

XML Process Definition Language

The XML Process Definition Language (XPDL)[18] is an XML format standardized by the Workflow Management Coalition (WfMC)[19] aimed at interchanging business process definitions between different workflow products by defining an XML schema for specifying the declarative part of business workflows. The XPDL format is mostly used for exchanging BPMN diagrams since it has been specifically designed to store all aspects of a BPMN diagram, both the graphic notation and the semantics of a BPMN process. This distinguishes XPDL from WS-BPEL because the latter focuses exclusively on the executable aspects of the process. WS-BPEL, in fact, does not contain elements to represent the graphical aspects of a business process. Therefore, XPDL has been widely adopted as a common standard interchange format for BPMN diagrams (Chinosi and Trombetta 2012).

Service-based Modeling

Like a task in scientific and business workflows, a service is an abstraction unit that allows modelers to share and encapsulate any piece of functionality. In service-based modeling, building blocks such as data and processes are distributed over a network and exposed as services available and accessible via standard interfaces on the Web (Alonso et al. 2004). In such a context, service-oriented architectures (SOA) are usually adopted in the development of collaborative, distributed web applications based on interoperable services. Friis-Christiensen et al. (2009) define SOA as "an open and interoperable environments based on reusability and standardized components and services". Web services represent autonomous, platform-independent entities that can be described, published, discovered, and eventually composed into service compositions to support the development of rapid, interoperable and massively distributed applications (Papazoglou et al. 2007; Papazoglou and van den Heuvel 2007).

Web services, as an implementation of a service, can perform functions that range from answering simple requests to executing sophisticated analysis and business processes. Indeed, any piece of code deployed in backend systems and applications can potentially be transformed and exposed as a web service. One of the required steps to convert a backend component into a web service is to describe its public service interface so that potential clients can connect and interact with it.

---

[18] http://www.xpdl.org
[19] http://www.wfmc.org

Web Service Description Language

The Web Service Description Language (WSDL)[20] (Christensen et al. 2001) is a W3C standard for describing the public interface of web services. It is an XML-based language to determine the functional properties of a service such as its method signatures, input and output messages, details of the transport protocol used (endpoint, SOAP envelope, etc.). In short, a WSDL document contains all of the public operations or methods offered by a given service (Fig. 5).

A WSDL document is not intended to be read by humans but to support interoperable machine-to-machine interactions. Client applications can automatically gather from a WSDL document the required information (e.g., data types for input and output messages, communication protocol used) to query and interact with the corresponding web service. That is the way how some scientific workflow systems like Taverna are able to turn WSDL-based web services into workflow tasks.

```
01 <definitions
02    name="HolaMundo"
03    targetNamespace="http://xmlns.oracle.com/HolaMundo"
04    xmlns="http://schemas.xmlsoap.org/wsdl/"
05    xmlns:plnk="http://schemas.xmlsoap.org/ws/2003/05/partner-link/"
06    xmlns:client="http://xmlns.oracle.com/HolaMundo">
07
08    <types>
09      <schema xmlns="http://www.w3.org/2001/XMLSchema"
10                       xmlns:plnk="http://schemas.xmlsoap.org/ws/2003/05/partner-link/"
11              xmlns:client="http://xmlns.oracle.com/HolaMundo">
12          <import namespace="http://xmlns.oracle.com/HolaMundo" schemaLocation="HolaMundo.xsd"/>
13      </schema>
14    </types>
15
16    <message name="HolaMundoRequestMessage">
17      <part name="payload" element="client:HolaMundoProcessRequest"/>
18    </message>
19
20    <message name="HolaMundoResponseMessage">
21      <part name= "payload" element="client:HolaMundoProcessResponse"/>
22    </message>
23
24    <portType name="HolaMundo">
25      <operation name="process">
26        <input message="client:HolaMundoRequestMessage"/>
27        <output message="client:HolaMundoResponseMessage"/>
28      </operation>
29    </portType>
30
31    <plnk:partnerLinkType name="HolaMundo">
32      <plnk:role name="HolaMundoProvider">
33        <plnk:portType name="client:HolaMundo"/>
34      </plnk:role>
35    </plnk:partnerLinkType>
36 </definitions>
```

Fig. 5. WSDL description of the WS-BPEL process in Fig. 4.

---

[20] http://www.w3.org/TR/wsdl

Geoprocessing Web Services

The term geoprocessing encompasses a computational operation to manipulate geospatial data. Examples are network analysis and coordinate transformation, thematic classification, and geocoding to name a few. Geoprocessing web services extend geoprocessing capabilities according to the SOA principles (Zhao et al. 2012). In such a context, the Open Geospatial Consortium (OGC)[21] has defined a set of web service interfaces, communication protocols, and data encodings for discovering, accessing and visualizing geospatial data as well as executing geoprocessing operations on the Web in a standard way (Lee and Percivall 2008). Further readings relating OGC service specifications applied to varied application domains are well documented in the literature (Reichardt 2010; Foerster et al. 2011; Zhao et al. 2012).

One of these OGC service specifications is the Web Processing Service (WPS) specification (Schut 2007). The OGC WPS specification defines a service interface to expose computational operations and algorithms as a web service. In contrast to traditional desktop-based geoprocessing implementations, WPS-enabled services are flexible and remotely accessible algorithms available and inherit the benefits from service-based modeling and web services (Kiehle 2006).

The notion of process is central to WPS services (Granell et al. 2008). To run a process implies the following steps: (i) to find geospatial data required by the process, (ii) to initiate the process, (iii) to control the output, and finally (iv) to make the results available to the client.

To follow the above steps an instance of a WPS service must offer three public methods to the client applications. These methods can be called using either HTTP-GET in combination with key-value pair (KVP), or HTTP-POST and XML documents. Its *getCapabilities* and *describeProcess* methods offer service and process metadata respectively, while the *execute* method triggers a concrete process. Figure 6 shows how a client communicates with a WPS service instance through the above three types of requests.

Emerging Approaches for Geo-enabled Workflows

This section outlines recent works that combine the two factors of the left part of the equation from the chapter title, workflows systems and geoprocessing web services, to lead to a kind of geo-enabled (scientific and business) workflows. Such experiments and proof-of-concept tools for

---

[21] http://www.opengeosatial.org

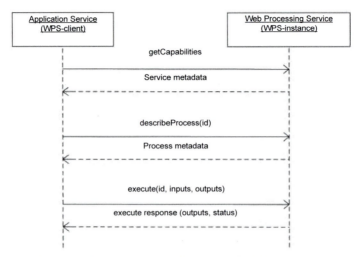

Fig. 6. Interaction between client applications and a WPS service instance.

geo-enabled workflows shift from the left part of the equation into the right one on the way towards the concept of Model Web. Finally, we summarize the main points that characterize geo-enabled workflows.

Geoprocessing Services and BWL

Business workflow languages seen earlier—BPMN, XPDL, and WS-BPEL—play complementary roles to model business workflows or processes. While BPMN is essentially a graphical notation to build abstract workflows—a set of high-level tasks to achieve a certain goal—XPDL and WS-BPEL are aimed to describe executable workflows—the specification of low-level implementation details such as calling web services, gathering their response, and handling data transformation operations. Rather than being disjoint, however, these standards are complementary. This means that proper combinations of BPMN and XPDL/WS-BPEL may lead to flexible and scalable approaches to cover the entire lifecycle of business workflows (Fig. 7).

BPMN is an excellent means for creating and maintaining workflows at high level of abstraction, where implementation details are hidden to workflow modelers (top-right Fig. 7). User requirements can be easily incorporated in BPMN diagrams in a graphical basis, which enable the sharing and understanding of the workflow context and requirements by a wider community. By using BPMN diagrams, the intended meaning and goal of workflows is shared between stakeholders, who are not necessarily software developers.

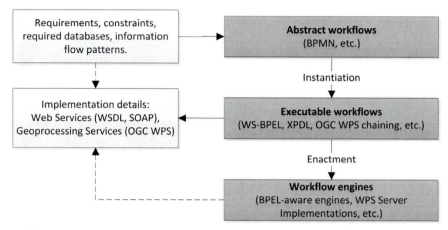

Fig. 7. Phases involved in modeling business workflows and how they relate to web services and geoprocessing web services.

WS-BPEL, however, is a description language for defining a concrete set of tasks associated with web services in conjunction with data dependences among such tasks. WS-BPEL workflow descriptions (e.g., Fig. 4), which point to particular web services used within a workflow and specify lots of details such as input and output data sets (mid-right Fig. 7), are executed by WS-BPEL-complaint workflow engines (bottom-right Fig. 7). Finally, XPDL acts as the glue standard capable of transforming BPMN diagrams into executable WS-BPEL descriptions together.

On the other hand, WSDL and WPS specifications are widely used for describing service interfaces in mainstream web services and geoprocesing services respectively (Schade et al. 2012). Their use has led to so-called WSDL-based or WPS-based services, respectively. The OGC web services have been developed in parallel with the evolution of mainstream web services defined by the stack of protocols and specifications under the W3C and OASIS. The dichotomy between WSDL and OGC interfaces has led often to disjoint solutions where both types of services are mostly incompatible when they are used together in business processes.

This dualism is demonstrated through many examples that assemble web services, either WSDL- or WPS-based, into geospatial service compositions. On one side, Li et al. (2010) describe how GRASS[22] commands can be exposed as WSDL-based web services to be orchestrated by WS-BPEL engines. On the other side, Kiehle (2006), Granell et al. (2010) and Maué et al. (2011) are just a few examples aiming at integrating and sharing WPS-based geoprocessing services.

---

[22] Geographic Resources Analysis Support System: http://grass.osgeo.org/

The previous examples are still isolated solutions in terms of using only WSDL-based or WPS-based services. To address the lack of interoperability between these types of services, the OGC WPS specification encourages the use of application profiles and supports pointing to associated WSDL descriptions. In particular, explicit support for WSDL descriptions is a way to make OGC-based services in the closed geospatial domain available to mainstream, mass-market web services (mainly based on WSDL), so that hybrid solutions might be suitable. In this context, a WSDL document can describe both an entire WPS service and each single contained process. In the first case, the capabilities document can include a *WSDL* element pointing to the corresponding WSDL description file (URI). The WSDL document describes the interface of the entire WPS service instance. In the second case, each contained process description can have its own WSDL element that points to a specific WSDL description file, which provides the functional signature of the WPS process. Schade et al. (2012) discussed on these WSDL-WPS integration cases.

In March 2011, Lopez-Pellicer et al. (2012) checked the availability of WPS-based services deployed on the Web. The results however were disappointing and showed scarce support for WSDL, which seriously limits interoperability with mass-market WSDL-based web services and also with BWL and scientific workflow systems that explicitly support WSDL-based services.

Some incipient projects are changing this trend though. Yu et al. (2012) provide WSDL-based descriptions for various kinds of OGC services, including WPS. These OGC services were orchestrated and executed using a BPEL-compliant engine, called BPELPower, for a variety of environmental modeling use cases. This represents a step further, since OGC services are being orchestrated through WS-BPEL process descriptions. The authors, however, introduced some enhancements into the BPELPower engine to be able to interpret the singularities of geospatial-enabled workflows.

Indeed, there have been identified in the literature some restrictions that make geoprocessing services hard to integrate with existing business process languages (e.g., WS-BPEL) and mass-market web services (e.g., WSDL-based services). First, current workflow engines provide limited support for geospatial schemas and sometimes even the parsing and use of geography markup language (GML)[23] may not be always possible. Because of the inherent complexity of many OGC schemas in terms of complicated relationships and recursive type definitions (Tamayo et al. 2012), business workflow engines are not designed for managing such complexity. Business workflows usually handle simple data items (orders, client records, etc.)

---

[23] http://www.opengeospatial.org/standards/gml

between workflow tasks. Similarly, WSDL parsers also offer poor support for managing complex data type definitions as part of geospatial elements schema (see some examples in Tamayo et al. (2012)). Second, geospatial workflows are often data-intensive and may require long processing times in execution and retrieving inputs data sets, as well as asynchronous calls. These and other advanced features are not often part of business workflow engines (Barga and Gannon 2007).

Geoprocessing Services and Scientific Workflow Systems

This section explores the ability of the aforementioned scientific workflow systems to interact with distributed web services in general, and geoprocessing web services in particular, in compliance with the vision of geo-enabled scientific workflows (Altintas et al. 2011).

Kepler-based Geo-enabled Workflows

As earlier commented in Section "Scientific Workflow Systems", Kepler provides the concept of actor to enable the addition of any algorithm or component in a scientific workflow (Ludäscher et al. 2006). Users select components, data sources and remote web services via specialized actors (e.g., *Web Service* actor for WSDL-based services) to include them in Kepler workflows. In this line, Pratt et al. (2010) explored how Kepler workflows can be exposed as OGC WPS-based services. However, the authors found several limitations because of the way inputs and output parameters are declared. For example, Kepler uses the particular *Object* type for capturing complex structures, which basically is a flexible container to allow dynamic type binding, whereas all of the elements necessary (mime types, schemas, etc.) to declare the WPS parameter types must be known at design time. Similarly, Barseghian et al. (2010) suggested the potential integration of OGC Sensor Web Enablement (SWE)[24] standards within Kepler workflows, so that scientists can benefit from these sensor web standards and protocols to access to disparate environmental observations and measurements (Bröring et al. 2011).

Apart from supporting web services technology, Kepler includes distributed computing technologies to permit scientists to share their data and workflows with other scientists and to use data and analytical workflows. It is recently expanding to support cloud computing capabilities. For example Wang et al. (2009) integrated Kepler and Hadoop,[25] an open-source implementation of MapReduce[26] computation model to

---

[24] http://www.opengeospatial.org/ogc/markets-technologies/swe
[25] http://hadoop.apache.org/
[26] http://labs.google.com/papers/mapreduce.html

perform parallel and scalable computations for data-intensive business and scientific applications. The authors proposed the implementation of specific MapReduce actors to support MapReduce applications into Kepler scientific workflows.

Taverna-based Geo-enabled workflows

Taverna provides a processor of type *arbitrary WSDL type*, which is able to call a single operation of a WSDL-based web service (Section "Taverna"). Indeed, a simple way to integrate external services into Taverna is just by providing a WSDL interface for these services. In the case of services in the life sciences domain, Taverna provides a specific mechanism via the BioCatalogue[27] tool, a registry for life sciences web services that is accessible and discoverable from Taverna. One registers a web service in the BioCatalogue and immediately it is made reachable from Taverna through the corresponding processor type. Bhagat et al. (2010) described the advantages of using BioCatalogue against WSDL-based services, as for example the ability to annotate a service at several dimensions (functional, profile, provenance, etc.) and the use of lightweight response formats such as JSON.[28]

Regarding the support of OGC services, de Jesus et al. (2012) have recently presented a WPS server implementation (based on pyWPS)[29] that is able to automatically generate WSDL descriptions from WPS-based services. In their tool, such WSDL descriptions are then used in conjunction with a Taverna WSDL processor type to invoke the counterpart geoprocesing web service from Taverna workflows. Rather than extending Taverna by developing a new kind of *OGC WPS processor type*, de Jesus et al. (2012) took the approach to augment WPS services with WSDL descriptions which are natively supported in Taverna.

It is worthwhile to note that Taverna workflows are mostly addressed to bioinformatics activities.[30] Indeed, one of Taverna's key values is the availability of bioinformatics services already integrated in the platform (Deelman et al. 2009). For example, Hull et al. (2006) described how bioinformatics can use hundreds of web services without prior programing knowledge. Taverna provides a built-in mechanism for discovery and access to a wide range of web service repositories (Hull et al. 2006) such as the aforementioned BioCatalogue. Kawas et al. (2006) exploited the extension capacity of Taverna by implementing a plug-in named BioMobu, whose

---

[27] http://www.biocatalogue.org/
[28] JavaScript Object Notation: http://www.json.org
[29] https://github.com/geopython/PyWPS
[30] http://www.taverna.org.uk/introduction/taverna-in-use/

aim is to facilitate the discovery and access to BioMOBY services,[31] which are a kind of mediators (Wiederhold 1992) to ease the exchanging and transformation of data among resources from distinct genomic and biology data repositories. Li et al. (2008) described how Taverna can be used for composing R[32] and Matlab[33] scripts for quantitative data analysis as long as these are exposed as web services.

VisTrails-based Geo-enabled workflows

As mentioned in Section VisTrails is also being used in the geospatial and environmental domains. For example, the Observations Network on Environmental Change in South Africa (SAEON) is using a geospatial extension (EO4VisTrails) to VisTrails as a service chaining utility for geospatially-enabled scientific workflows (McFerren et al. 2012). The EO4VisTrails extension provides access to spatial databases and some OGC web services which can then be integrated into VisTrails workflows for geospatial analysis and visualization. The use of VisTrails behind these applications is likely motivated by the full support of visualization libraries for data-intensive scientific applications (Santos et al. 2009; Silva et al. 2011).

Summary

From the previous examples in Section "Emerging Approaches for Geo-enabled Workflows", the encapsulation of WPS-based geoprocessing services through WSDL descriptions has been the widely adopted approach to boost interoperability and reusability of geoprocessing services both in BWL and scientific workflow systems. Hereby encapsulation becomes the natural way to reuse web services and geoprocesing services in flexible business workflows and robust scientific workflow systems. In summary, WPS-based services are being integrated in geo-enabled workflows in the following ways:

- WPS services in conjunction with business workflow languages, chiefly only with WS-BPEL. As WS-BPEL works mainly with WSDL-based web services, WPS services should be realigned to this standard in order to benefit from existing tools and engines, and penetrate into the mass-market web service community. For example Yu et al. (2012) have implemented a geospatially-aware WS-BPEL engine that in reality manages WPS services but wrapped as WSDL-based services.

---

[31] http://biomoby.open-bio.org/index.php/for-developers/
[32] http://www.r-project.org/
[33] http://www.mathworks.com/products/matlab/

- WPS services in conjunction with scientific workflow systems. Like in BWL, WPS services are mostly wrapped in WSDL descriptions to be part of scientific workflows. In this context de Jesus et al. (2012) have presented a proof-of-concept implementation that allows WPS services wrapped again as WSDL-based services to be incorporated into Taverna workflows, exploiting then all of the Taverna tools available to manage and execute scientific workflows.

Furthermore, we can conclude that the current trend is to encapsulate OGC services as WSDL services, that is, the provision (e.g., via links) of WSDL-based descriptions of OGC services. This way, WSDL-based OGC services can be readily used in business workflow engines and scientific workflow systems whereas their internal implementations remain unchanged. Nevertheless, the success of this approach depends strongly on the nature of the geoprocessing services themselves. As earlier commented, the level of complexity of some geospatial schema may be a limitation at runtime, because target systems (e.g., Taverna, WS-BPEL engine) may not properly deal with some geospatial singularities. Furthermore, future developments in scientific workflow systems (e.g., Taverna, VisTrails) or business workflow engines (e.g., BPELPower) may address these geospatial requirements (complexity schemas, long processing times, etc.) and consequently increase up reusability and interoperability between OGC and mass-market services, and eventually lead to the so-called geo-enabled scientific workflows (Altintas et al. 2011).

Towards the Geo-enabled Model Web

Sharing models would promote collaboration and experience exchange between research teams which would eventually result in much better informed decisions to handle environmental issues. The set of technologies, frameworks and research works analyzed in the previous sections proceed in this direction. They pursue the reuse of pieces of functionality in form of web services to create geo-enabled business and scientific workflows. At this point, we wonder whether such geo-enabled workflows may shape IM. Can environmental models, like a watershed runoff model be transformed into WSDL-based web services or WPS-based services? Can scientific workflows emulate combinations of environmental models? In this section we discuss these questions and identify open issues that need to be addressed to bridge the Model Web to its full potential.

What is the Model Web?

The Model Web vision promotes the idea of having distributed networks of interoperating models capable of communicating each other using web services interfaces (Geller and Turner 2007; Geller and Melton 2008). It fosters reusability and sharing of models with the goal that altogether can answer more questions than an individual model. The Model Web, however, must be understood as a vision—a high-level concept—rather than a particular architecture or implementation. Indeed, as a vision, it may be presented through different architectures and implementations.

Nativi et al. (2013) defines the Model Web as the "World Wide Web for models", and so it makes a lot of sense to build the Model Web upon similar principles. The authors propose the following four principles upon which the Model Web vision sits:

- Open access: anybody has the freedom to create, share and access models on the Web.
- Minimal barriers to entry: model providers and model users are not constrained by difficult mechanisms to publishing and accessing models on the Web. The Model Web encourages the use of open standards to describe and publish models that are publicly accessible on the Web.
- Service-driven approach: Model access is driven by the notion of services, i.e., in the form of web services.
- Scalability: The network of models can grow without being impeded by implementation and design factors.

Simply put, the Model Web vision would embrace models exposed as web services and integrated models as the result of composing these web services by adhering to the SOA principles (Papazoglou et al. 2007; Papazoglou and van den Heuvel 2007; Friis-Christiensen et al. 2009), i.e., the result of one web service (model) can be reused by the next service (model).

Some examples can help us to materialize the vision of the Model Web in practical terms. Dubois et al. (2013) presented *eHabitat*, a WPS-based service for ecological modeling for computing the likelihood of finding ecosystems with equal properties to those specified in a user request. When used in conjunction with other web services such as those providing data on species occurrences, bird distribution maps and data on climate changes, the eHabitat WPS service becomes a key tool for ecological forecasting and IM in general (Skøien et al. 2013). This is an example of a compound web service that gives proxy to other backend web services through a common service interface (WPS) and reuses the results of these services. That's an example of the principle of service-driven approach.

Feng et al. (2011) and Castronova et al. (2013) also worked on the idea of sharing models as web services being compliant with the WPS service interface. Both, however, go one step further by developing approaches to bring these models as web services into workflows that represent complex environmental modeling scenarios such as wetland ecosystems services and watershed runoff prediction, respectively.

In contrast to the previous examples where models are exposed as pure WPS-based services, Jones et al. (2012) proposed a WSDL-based implementation for publishing models on the Web built internally on the WPS service interface which can be later integrated with scientific workflow systems such as Taverna and Kepler. In this sense, this work closely relates with the approaches for geo-enabled scientific workflows described in the previous section.

These examples are a first attempt towards the Model Web vision, where open networks of models exposed as web services are ready to be shared and to support IM by allowing these services to talk to each other directly using standard services interfaces. Yet, the above examples are still far from the ideal vision of Model Web because integrated models have some intrinsic characteristics that should be taken into account in practical terms. In particular, the following issues are barriers to realizing the vision of a useful Model Web and require further research: (i) the variability of types of models; (ii) the increasing complexity of chained models; (iii) the existence of multiple services interfaces for exposing models; and (iv) the wide range of infrastructures and implementations towards realizing the Model Web.

The idea of augmenting the Model Web with the notion of reproducibility is a logical next step to modelers and decision makers. Essentially, IM is seen as an example of scientific practice that seeks to reproduce modeling results to be compared with historical data and verify their validness. Therefore, reproducibility implicitly requires sustainable development practices about enduring existing infrastructures, platforms and tools, and re-using existing models and resources (Granell and Schade In press). Indeed, models are reused every time they are reproduced. All of this poses an ambitious challenge: one thing is to reuse a model, and the other is to (v) reproduce the execution of such a model again by others. The latter implies a large set of requirements that must be taken into account. For example, the access to the same input data sets used in that model, the access to notes or documentation regarding the execution of that model, access to historical results so that new results may be assessed, as well as technical notes and configurations to recreate the same execution environment. In the rest of the section, we delve into these five open issues and sketch some possible solutions.

Not One but Multiple Types of Models

WSDL-based services and WPS-based services are designed to perform the same computation every time they are invoked by a client application. This is true for example to run a spatial intersection process between two given geometries and to geocode a place name. In these cases, a geoprocessing service takes input data, uses this data to perform a computation, encodes the result into XML, and finally sends it back to the client. In contrast, environmental models often consist of several phases (e.g., initialization, run, and finish) and must be capable of performing different types of computations in each phase. Unlike environmental models, geoprocesing services are often time-independent and therefore do not require storing and referencing previous calculations, which implies that WPS-based services are stateless.

Is the Model Web in conjunction with geoprocesing services and workflow systems able to capture not only one type of model but multiple types of models such as time-dependent, stochastic, empirical and so on? To this respect, Castranova et al. (2013) recently showed that it is possible, under certain circumstances, to adapt WPS-based services for encoding time-dependent model computations and maintaining model state. Nevertheless, innovative approaches are yet required to bring different computations models together.

Not One but Multiple Perspectives of Complexity

As introduced earlier, a model basically simplifies a natural system or a part of it. Therefore, IM should be regarded as a simplification process that abstract from complex interactions among natural systems. Nevertheless, most of available models cannot be easily linked with each other when assembling IM chains due to several reasons (Granell et al. 2013b). Here, we overview some of them.

First, the expectation that the Model Web can deliver integrated models by simply linking existing models as web services can lead to meaningless models. Link to the previous issue on variability of types of models, some environmental models are regarded as time-dependent models and empirical models that rely strongly on calibration data. That is, "as more components are brought together, the calibration of the whole integrated model becomes only more difficult" (Voinov and Shugart 2013). It is less thus evident how current web service technologies and workflow systems can handle calibration data tied to models (see Section "Not One but Multiple Reproducible Resources").

Second, semantics and formats of model input/output data sets (Feng et al. 2011) along with multiple spatial resolutions and scales (Voinov and Shugart 2013) may difficult IM. Again, it is less clear how to exchange data between models with distinct level of spatial resolutions and scales.

Third, accessing models without proper supporting tools may be seen as an obstacle at first sight by non-expert users (Skøien et al. 2013) thereby breaking the Model Web principle on lowering entry barriers. For example, when putting services together we need easy-to-use tools to allow us to assemble these WPS-based services together and to understand the new context led by the service composition (see Section "Not One but Multiple Infrastructures").

Not One but Multiple Service Interfaces

It seems unrealistic to assume that every single environmental model fits into WSDL and WPS service description. Models by nature are heterogeneous and are often tightly tied to the singularities of each discipline (e.g., biodiversity models, watershed models, ecological models), which by large exceed the capabilities of these service description services. For example, several authors have found several limitations in matching data structure between models and service description specifications because of the way inputs and output parameters are declared. Inputs parameters in models are much richer than those supported by WSDL and WPS specifications. Although the modeling community requires the development of standards for describing and publishing environmental models in suitable form for automated access and integration (Laniak et al. 2013b), in practical terms, however, no silver bullet exists and the description of models are being proposed to vertebrate sharing and integration of models within each discipline. These proposals range from describing biomedical (de Bono et al. 2011) and biodiversity models (Endresen and Knüpffe 2012; Constable et al. 2010) to ecology (Schmolke et al. 2010) and climate modeling (Parsons 2011).

The wide diversity of descriptions of models raises the question whether such models can interact across disciplines. Some recent works have addressed this limitation. One approach relies on the brokering pattern (Buschmann et al. 1996). A brokering component would implement the required business logic to allow clients for discovering and accessing heterogonous models from distinct disciplines. It extends the traditional SOA architecture (Papazoglou and van den Heuvel 2007) by introducing an "intelligent" component to consume heterogeneous models in a transparent way and interact with them using a single and well-known endpoint, i.e., the brokering component (Nativi et al. 2011). In addition, if brokering components are well designed to turn into reusable building blocks, various brokers can be aggregated and work cooperatively to deliver meaningfully functionality and connect increasingly more models and sources across disciplines (Vaccari et al. 2012; Díaz et al. 2012).

An alternative to the brokering approach is to design a common interface that abstracts from service description specifications and other specific APIs to access and interact with backend models in a uniform manner. To this respect, Granell et al. (2013a) recently discussed on design practices and the implementation recommendations to build resource-based interfaces for environmental models based on the capabilities of HTTP as application protocol and the REST architectural principles (Fielding 2000), which aligned nicely to the Model Web principles mentioned earlier.

Not One but Multiple Infrastructures

As commented earlier the Model Web is a vision which may materialize through multiple infrastructures and implementations. Assuming that each discipline sits on its thematic information infrastructure, such as health information infrastructures (Kamel Boulos 2004; Barret et al. 2011) and geospatial information infrastructures (Wright and Wang 2011), the question raised here again is how to bridge such infrastructures together and to facilitate the sharing and integration of models as services deployed on distinct information infrastructures.

The term Virtual Research Environments (VREs) attempts to support multidisciplinary research teams to work collaboratively by managing the increasingly complex range of tasks involved in carrying out research at both small and large scales. Voss and Procter (2009) define VRE as a means to support the integration of resources like environmental models throughout the life cycle of an IM project. VREs encompass information infrastructures, collaborative tools and technologies needed by researchers to do their daily research activities, interact with other researchers, and to enable vertical and horizontal integration of resources available both locally and remotely. VREs are sometimes also known as research or scientific infrastructures (Bernard et al. 2013) and cyber-infrastructures (Yang et al. 2010). Regardless of the particular label, the same goals and objectives are behind the scenes.

The concept of VRE does not should thought of as a unique research platform but multiple VREs interconnected addressed to different purposes, similarly to the vision of multiple Digital Earths addressed to user needs (Craglia et al. 2012; Goodchild et al. 2012). This of course has implications into the geo-enabled scientific workflows, in which each web service in the workflow may belong to distinct institutions or individuals and be deployed in distinct information infrastructures. Enabling collaboration across information infrastructures is not a desire but a must.

Apart from supporting generic research tasks such as data management and analysis (horizontal view), VREs should also incorporate the context in which those tools and technologies are used, i.e., specific characteristics, settings and execution environments for each research discipline (vertical view). For example the sharing, management and execution of geo-enabled scientific workflows in geosciences distinguishes from the bio-genomics field in terms for example of data tools and methodologies used. So, VREs are information infrastructures to enable the sharing and (vertical and horizontal) integration of resources needed by different stakeholders (e.g., scientists, policy makers, modelers, etc.) throughout the lifecycle of a project (e.g., IM project).

Yet, open issues require further research such as to provide customized VREs, i.e., the ability to incorporate the specific characteristics of each discipline, and the exploitation of cloud computing paradigm. For the latter, recent works are making good progress on effective geoprocessing computation by exploring the interplay between geoprocessing services and cloud computing platforms (Yang et al. 2011; Wen et al. 2013; Yue et al. 2013), as well as novel strategies for standardized reusable geoprocessing packages that can be moved closer to data sources to alleviate exchanging large amounts of data sets (Müller et al. In press). All in all, these works pursue to bring pieces from distinct research infrastructures together.

Not One but Multiple Reproducible Resources

The Model Web highlights models which may be executed again and again under similar conditions and settings. For doing so, it is necessary to record in sufficient detail the contextual information, also known as data provenance, not only for a given model but also for any related resource so that model can be fully reproduced later on. So, is Model Web through scientific workflows concepts and web service technologies prepared to capture model reproducibility on the Web?

The notion of *aggregation* is widely used in some scientific workflow systems to bring together related resources into a logical unit. Aggregation by definition implies reusability, i.e., an aggregation simply reuses the resources it contains. For example, a geo-enabled scientific workflow can aggregate varied types of resources such as workflow descriptions, log records, inputs datasets and related publications and presentations reporting on the execution results of that workflow.

Complementary to aggregation, data provenance allows users to document each contained resource within an aggregation so as to determine the usability, reliability and even uncertainty of such a aggregation as a whole (Di et al. 2013; Rotmans and van Asselt 2001; Bastin et al. 2013). Uncertainty management is a must in IM because contained models may potentially utilize data sets from uncontrolled and unverified sources. Recent works have experimented data provenance techniques applied to geoprocessing web services and compositions of web services (Yue et al. 2010; Yue et al. 2011; Jones et al. 2012), which pave the way towards the applicability of data and model provenance to the Model Web vision. Nevertheless, the pending question is how to map and implement provenance and aggregation in conjunction with the concepts of reliability and uncertainty into Geo-enabled Model Web vision.

Concluding Remarks

Modeling in environmental and geosciences fields is a data-intensive science that increasingly requires joint research and efforts toward a sharing and better understanding of multi-disciplinary resources (data, metadata, results, workflows, etc.) by the modeling community (e.g., decision makers, policy makers, modelers, scientists) involved in IM activities. Ad-hoc IM solutions may work for specific environmental issues but others do not benefit from these modeling results. On the other hand, a unique IM solution seems to be unaffordable because of the great variety of environmental problems, data formats, stakeholders, models, perspective and social aspects, and needs involved in modeling activities.

In this context, the chapter first reviewed a set of technologies that, when properly combined, could become essential ingredients for leveraging the next generation of IM frameworks and tools. In particular scientific workflow systems, BWL, and geoprocessing web services have been put on the table to analyze altogether how they can work collaboratively to address a common goal. The integration of geoprocesing web services into workflow systems opens a new range of possibilities to cope with IM since the resulting geo-enabled workflows are characterized by the robustness of well-proven workflow execution environments along with the ability of geoprocessing web services to meet particular needs.

In the second part of the chapter, we set out the concept of Geo-enabled Model Web as a novel approach to leverage IM on the Web. Models are seen as services which can be composed, discovered and integrated as web services do. We examined the role of current workflow systems and geoprocessing web services in this settings, and identified open issues and challenges that certainly need further research in the immediate future to truly realizing the Geo-enabled Model Web vision (Nativi et al. 2013), i.e., models envisioned as a networks of reusable services deployed on an ecosystem of inter-related research information infrastructures capable to handling with multi-disciplinary and ever-increasing environmental issues, along the lines of the next generation of Digital Earth applications (Goodchild et al. 2012).

References

Abel, D.J., K. Taylor and D. Kuo. 1997. Integrating Modelling Systems for Environmental Management Information Systems. SIGMOD Record. 26(1).

Alonso, G., F. Casati, H. Kuno and V. Machiraju. 2004. Web Serivces: Concepts, Architectures and Applications. Springer.

Altintas, I., D. Crawl, C.J. Crosby and P. Cornillon. 2011. Scientific workflows for the geosciences: An emerging approach to building integrated data analysis systems. pp. 237–250. In: G.R. Keller and C. Baru (eds.). Geoinformatics—Cyberinfrastructure for the Solid Earth Sciences. Cambridge University Press.

Argent, R.M. 2004. An overview of model integration for environmental applications-components, frameworks and semantics. Environmental Modelling and Software. 19(3): 219–234.

Argent, R.M., A. Voinov, T. Maxwell, S.M. Cuddy, J.M. Rahman, S. Seaton, R.A. Vertessy and R.D. Braddock. 2006. Comparing modelling frameworks—A workshop approach. Environmental Modelling and Software. 21: 895–910.

Bhagat, J., F. Tanoh, E. Nzuobontane, T. Laurent, J. Orlowski, M. Roos, K. Wolstencroft, S. Aleksejevs, R. Stevens, S. Pettifer, R. Lopez and C.A. Goble. 2010. BioCatalogue: a universal catalogue of web services for the life sciences. Nucleic Acids Research. 38(suppl. 2): W689–W694.

Barga, R. and D. Gannon. 2007. Scientific versus business workflows. pp. 9–18. In: I. Taylor et al. (eds.). Workflows for e-Science. Springer-Verlag, Berlin.

Barrett, M.A., T.A. Bouley, A.H. Stoertz and R.W. Stoertz. 2011. Integrating a one health approach in education to address global health and sustainability challenges. Frontiers in Ecology and the Environment. 9(4): 239–245.

Barseghian, D., I. Altintas, M.B. Jones, D. Crawl, N. Potter, J. Gallagher, P. Cornillon, M. Schildhauer, E.T. Borer, E.W. Seabloom and P.R. Hosseini. 2010. Workflows and extensions to the Kepler scientific workflow system to support environmental sensor data access and analysis. Ecological Informatics. 5(1): 42–52.

Bastin, L., D. Cornford, R. Jones, G.B.M. Heuvelink, E. Pebesma, C. Stasch, S. Nativi, P. Mazzetti and M. Williams. 2013. Managing uncertainty in integrated environmental modelling: The UncertWeb framework. Environmental Modelling and Software. 39: 116–134.

Bernard, L., S. Mäs, M. Müller, C. Henzen and J. Brauner. In press. Scientific geodata infrastructures: challenges, approaches and directions. International Journal of Digital Earth. http://dx.doi.org/10.1080/17538947.2013.781244.

Bröring, A., J. Echterhoff, S. Jirka, I. Simonis, T. Everding, C. Stasch, S. Liang and R. Lemmens. 2011. New Generation Sensor Web Enablement. Sensors. 11: 2652–2699.

Buschmann, F., R. Meunier, H. Rohnert, P. Sommerland and M. Stal. 1996. Pattern-oriented software architecture, volume 1: A system of patterns. John Wiley & Sons Ltd.

Callahan, S.P., J. Freire, E. Santos, C.E. Scheidegger, C.T. Silva, H.T. Vo. and T. Huy. 2006a. VisTrails: visualization meets data management. N Proc. of the 2006 ACM SIGMOD international conference on Management of data. New York, USA. pp. 745–747.

Callahan, S.P., J. Freire, E. Santos, C.E. Scheidegger, C.T. Silva and H.T. Vo. 2006b. Managing the Evolution of Dataflows with VisTrails. In Proc. of the 22nd International Conference on Data Engineering Workshops. Atlanta, USA.

Castronova, A.M., J.L. Goodall and M.M. Elag. 2013. Models as web services using the Open Geospatial Consortium (OGC) Web Processing Service (WPS) standard. Environmental Modelling and Software. 41: 72–83.

Chinosi, M. and A. Trombetta. 2012. BPMN: An introduction to the standard. Computer Standards & Interfaces. 34(1): 124–134.

Christensen, E., F. Curbera, G. Meredith and S. Weerawarana (eds.). 2001. Web Services Description Language (WSDL) 1.1. The World Wide Web Consortium (W3C). http://www.w3.org/TR/wsdl.

Cohen-Boulakia, S. and U. Leser. 2011. Search, Adapt, and Reuse: The Future of Scientific Workflows. SIGMOD Record. 40(2): 6–16.

Constable, H., R. Guralnick, J. Wieczorek, C. Spencer, A.T. Peterson et al. 2010. VertNet: A New Model for Biodiversity Data Sharing. PLoS Biology. 8(2): e1000309.

Craglia, M., K. de Bie, D. Jackson, M. Pesaresi, G. Remetey-Fülöpp, C. Wang, A. Annoni, L. Bian, F. Campbell, M. Ehlers, J. van Genderen, M. Goodchild, H. Guo, A. Lewis, R. Simpson, A. Skidmore and P. Woodgate. 2012. Digital Earth 2020: towards the vision for the next decade. International Journal of Digital Earth. 5(1): 4–21.

de Bono, B., R. Hoehndorf, S. Wimalaratne, G. Gkoutos and P. Grenon. 2011. The RICORDO approach to semantic interoperability for biomedical data and models: strategy, standards and solutions. BMC Research Notes. 4: 313.

de Jesus, J., P. Walker, M. Grant and S. Groom. 2012. WPS orchestration using the Taverna workbench: The eScience approach. Computers & Geosciences. 47: 75–86.

Deelman, E., D. Gannon, M.S. Shields and I. Taylor. 2009. Workflows and e-Science: An overview of workflow system features and capabilities. Future Generation Computer Systems. 25(5): 528–540.

Di, L., P. Yue, J. Gong and M. Zhang. 2013. Geoscience Data Provenance: An Overview. IEEE Transactions on Geoscience and Remote Sensing. 51(11): 5065–5072.

Díaz, L., C. Granell, J. Huerta and M. Gould. 2012. Web 2.0 Broker: A standards-based service for spatio-temporal search of crowd-sourced information. Applied Geography. 35(1–2): 448–459.

Dubois, D., M. Schulz, J. Skøien, L. Bastin and S. Peedell. 2013. eHabitat, a multi-purpose Web Processing Service for ecological modeling. Environmental Modelling and Software. 41: 123–133.

Elmroth, E., F. Hernández and J. Tordsson. 2010. Three fundamental dimensions of scientific workflow interoperability: Model of computation, language, and execution environment. Future Generation Computer Systems. 26(2): 245–256.

Endresen, D.T.F. and H. Knüpffe. 2012. The Darwin Core extension for genebanks opens up new opportunities for sharing genebank datasets. Biodiversity Informatics. 8: 12–29.

Feng, M., S. Liu, N.H. Euliss, Jr., C. Young and D.M. Mushet. 2011. Prototyping an online wetland ecosystem services model using open model sharing standards. Environmental Modelling and Software. 26(4): 458–468.

Fielding, R.T. 2000. Architectural Styles and the Design of Network-based Software Architectures. Ph.D dissertation, University of California (Irvine). http://www.ics.uci. edu/~fielding/pubs/dissertation/top.htm.

Foerster, T., B. Schäffer, B. Baranski and J. Brauner. 2011. Geospatial Web Services for Distributed Processing: Applications and Scenarios. pp. 245–286. In: P. Zhao and L. Di (eds.). Geospatial Web Services: Advances in Information Interoperability. Hershey, IGI Global.

Friis-Christiensen, A., R. Lucchi, M. Lutz and N. Ostländer. 2009. Service chaining architectures for applications implementing distributed geographic information processing. International Journal of Geographical Information Science. 23(5): 561–580.

Geller, G.N. and W. Turner. 2007. The model web: a concept for ecological forecasting. In IEEE International Geoscience & Remote Sensing Symposium (IGARSS 2007). Barcelona, Spain. pp. 2469–2472.

Geller, G.N. and F. Melton. 2008. Looking Forward: Applying and Ecological Model Web to assess impacts of climate change. Biodiversity. 9(3&4): 79–83.

Goodchild, M.F., B.O. Parks and L.T. Steyaert. 1993. Environmental Modeling and GIS. Oxford University Press, Oxford.

Goodchild, M.F., H. Guo, A. Annoni, L. Bian, K. de Bie, F. Campbell, M. Craglia, M. Ehlers, J. van Genderen, D. Jackson, A.J. Lewis, M. Pesareri, G. Remetey-Fülöpp, R. Simpson, A. Skidmore, C. Wang and P. Woodgate. 2012. Next-generation Digital Earth. Proceedings of the National Academy of Sciences. 109(28): 11088–11094.

Granell, C., L. Díaz and M. Gould. 2008. Distributed Geospatial Processing Services. pp. 1186–1193. In: M. Khosrow-Pour (ed.). Encyclopedia of Information Science and Technology, Second Edition. IGI Global, Hershey.

Granell, C., L. Díaz and M. Gould. 2010. Service-oriented applications for environmental models: reusable geospatial services. Environmental Modelling and Software. 25(2): 182–198.

Granell, C., L. Díaz, S. Schade, N. Ostländer and J. Huerta. 2013a. Enhancing Integrated Environmental Modelling by Designing Resource-Oriented Interfaces. Environmental Modelling and Software. 39: 229–246.

Granell, C., S. Schade and N. Ostländer. 2013b. Seeing the forest through the trees: A review of integrated environmental modelling tools. Computers, Environment and Urban Systems. 41: 136–150.

Granell, C. and S. Schade. In press. Environmental Informatics for Sustainable Development. Encyclopedia of Information Science and Technology (3rd Edition). Hershey, IGI Global.

Hull, D., K. Wolstencroft, R. Stevens, C. Goble, M.R. Pocock, P. Li and T. Oinn. 2006. Taverna: a tool for building and running workflows of services. Nucleic Acids Research. 34: W729–W732.

Jakeman, A.J. and R.A. Letcher. 2003. Integrated assessment and modelling: features, principles and examples for catchment management. Environmental Modelling and Software. 18(6): 491–501.

Jasny, B.R., G. Chin, L. Chong and S. Vignieri. 2011. Again, and Again, and Again. Science. 334(6060): 1225.

Jones, R., D. Cornford and L. Bastin. 2012. UncertWeb Processing Service: Making Models Easier to Access on the Web. Transactions in GIS. 16(6): 921–939.

Jordan, D. and J. Evdemon. 2007. Web Services Business Process Execution Language. http://docs.oasis-open.org/wsbpel/2.0/wsbpel-v2.0.pdf.

Kamel Boulos, M.N. 2004. Towards evidence-based, GIS-driven national spatial health information infrastructure and surveillance services in the United Kingdom. International Journal of Health Geographics. 3(1).

Kawas, E., M. Senger and M.D. Wilkinson. 2006. BioMoby extensions to the Taverna workflow management and enactment software. BMC Bioinformatics. 7: 523.

Kiehle, C. 2006. Business logic for geoprocessing of distributed geodata. Computers & Geosciences. 32(10): 1746–1757.

Laniak, G.F., A.E. Rizzoli and A. Voinov. 2013a. Thematic Issue on the Future of Integrated Modeling Science and Technology. Environmental Modelling and Software. 39: 1–2.

Laniak, G.F., G. Olchin, J. Goodall, A. Voinov, M. Hill, P. Glynn, G. Whelan, G. Geller, N. Quinn, M. Blind, S. Peckham, S. Reaney, N. Gaber, R. Kennedy and A. Hughes. 2013b. Integrated environmental modeling: A vision and roadmap for the future. Environmental Modelling and Software. 39: 3–23.

Lee, C. and G. Percivall. 2008. Standards-based computing capabilities for distributed geospatial applications. Computer. 41(11): 50–57.

Li, P., J.I. Castrillo, G. Velarde, I. Wassink, S. Soiland-Reyes, S. Owen, D. Withers, T. Oinn, M.R. Pocock, C. Golbel, S.G. Olivier and D.B. Kell. 2008. Performing statistical analyses on quantitative data in Taverna workflows: An example using R and maxdBrowse to identify differentially-expressed genes from microarray data. BMC Bioinformatics. 9: 334.

Li, X., L. Di, W. Han, P. Zhao and U. Dadi. 2010. Sharing geoscience algorithms in a Web service-oriented environment (GRASS GIS example). Computers & Geosciences. 36(8): 1060–1068.

Lopez-Pellicer, F.J., W. Rentería-Agualimpia, R. Béjar, P.R. Muro-Medrano and F.J. Zarazaga-Soria. 2012. Availability of the OGC geoprocessing standard: March 2011 reality check. Computers & Geosciences. 47: 13–19.

Ludäscher, B. and C. Goble. 2005. Guest Editors' Introduction to the Special Section on Scientific Workflows. SIGMOD Record. 34(3): 3–4.

Ludäscher, B., I. Altintas, C. Berkley, D. Higgins, E. Jaeger, M. Jones, E.A. Lee, J. Tao and Y. Zhao. 2006. Scientific workflow management and the Kepler system. Concurrency and Computation: Practice and Experience. 18(10): 1039–1065.

Maué, P., C. Stasch, G. Athanasopoulos and L.E. Gerharz. 2011. Geospatial Standards for Web-enabled Environmental Models. International Journal of Spatial Data Infrastructures Research. 6: 145–167.

McFerren, G., T. van Zyl and A. Vahed. 2012. FOSS geospatial libraries in scientific workflow environments: experiences and directions. Applied Geomatics. 4(2): 85–93.

McIntosh, B.S., C. Giupponi, A. Voinov, C. Smith, K.B. Matthews, M. Monticino, M.J. Kolkman, N. Crosman, M. van Ittersum, D. Haase, J. Mysiak, J.C.J. Groot, S. Sieber, P. Verweij, N. Quinn, P. Waeger, N. Gaber, D. Hepting, H. Scholten, A. Sulis, H. van Delden, E. Gaddis and H. Assaf. 2008. Bridging the Gaps Between Design and Use: Developing Tools to

Support Environmental Management and Policy. pp. 33–48. In: A.J. Jakeman et al. (eds.). Environmental Modelling, Software and Decision Support. Elsevier, Amsterdam.

Mesirov, J.P. 2010. Accessible Reproducible Research. Science. 327(5964): 415–416.

Mili, H., G. Tremblay, G. Bou Jaoude, E. Lefebvre, L. Elabed and G. El Boussaidi. 2010. Business process modelling Languages: Sorting through the Alphabet Soup. ACM Computing Surveys. 43(1).

Müller, M., L. Bernard and D. Kadner. 2013. Moving code—Sharing geoprocessing logic on the Web. ISPRS Journal of Photogrammetry and Remote Sensing. 83: 193–203.

Nativi, S., M. Craglia, L. Vaccari and M. Santoro. 2011. Searching the New Grail: interdisciplinary interoperability. In Proc. of the 14th AGILE International Conference on Geographic Information Science (AGILE 2011), Utrecht, The Netherlands.

Nativi, S., P. Mazzetti and G.N. Geller. 2013. Environmental model access and interoperability: The GEO Model Web initiative. Environmental Modelling and Software. 39: 214–228.

Object Management Group (OMG). 2011. Business Process Model and Notation (BPMN), Version 2.0 http://www.omg.org/spec/BPMN/2.0/.

Oinn, T., M. Addis, J. Ferris, D. Marvin, M. Senger, M. Greenwood, T. Carver, K. Glove, M.R. Pocok, A. Wipat and P. Li. 2004. Taverna: a tool for the composition and enactment of bioinformatics workflows. Bioinformatics. 20: 3045–3054.

Oinn, T., M. Greenwood, M. Addis, M.N. Alpdemir, J. Ferris, K. Glover, C. Goble, A. Goderis, D. Hull, D. Marvin, P. Li, P. Lord, M.R. Pocock, M. Senger, R. Stevens, A. Wipat and C. Wroe. 2006. Taverna: lessons in creating a workflow environment for the life sciences. Concurrency Computation: Pract. Exper. 18: 1067–1100.

Papazoglou, M.P., P. Traverso, S. Dustdar and F. Leymann. 2007. Service-Oriented Computing: State of the Art and Research Challenges. Computer. 40(11): 38–45.

Papazoglou, M.P. and W.J. van den Heuvel. 2007. Service oriented architectures: approaches, technologies and research issues. The VLDB Journal. 16: 389–415.

Parker, P., R. Letche, A. Jakeman, M.B. Beck, G. Harris, R.M. Argent et al. 2002. Progress in integrated assessment and modelling. Environmental Modelling and Software. 17(3): 209–217.

Parsons, M.A. 2011. Making data useful for modelers to understand complex Earth systems. Earth Science Informatics. 4: 197–223.

Pratt, A., C. Peters, S. Guru, B. Lee and A. Terhorst. 2010. Exposing the Kepler Scientific Workflow System as an OGC Web Processing Service. In International Environmental Congress on Environmental Modelling and Software Society (iEMS 2010), Otawa, Canada.

Rahman, M., R. Ranjan, R. Buyya and B. Benatallah. 2011. A taxonomy and survey on autonomic management of applications in grid computing environments. Concurrency and Computation: Practice and Experience. 23(16): 1990–2019.

Reichardt, M. 2010. Open standards-based geoprocessing Web services to support the study and management of hazard and risk. Geomatics, Natural Hazards and Risk. 1(2): 171–184.

Rotmans, J. and M. van Asselt. 2001. Uncertainty management in integrated assessment modelling: towards a pluralistic approach. Environmental Monitoring and Assessment. 69(2): 101–130.

Santos, E., L. Lins, J. Ahrens, J. Freire and C. Silva. 2009. VisMashup: Streamlining the Creation of Custom Visualization Applications. IEEE Transactions on Visualization and Computer Graphics. 16(6): 1539–1546.

Schade, S., N. Ostländer, C. Granell, M. Schulz, D. McInerney, G. Dubois, L. Vaccari, M. Chinosi, L. Díaz, L. Bastin and R. Jones. 2012. Which Service Interfaces fit the Model Web? In Proc. of 4th International Conference on Advanced Geographic Information Systems, Applications, and Services (GEOProcessing 2012), Valencia, Spain. pp. 1–6.

Schmolke, A, P. Thorbek, D.L. DeAngelis and V. Grimm. 2010. Ecological models supporting environmental decision making: a strategy for the future. Trends in Ecology and Evolution. 25(8): 479–486.

Schut, P. 2007. OpenGIS Web Processing Service 1.0.0. OpenGIS Standard 05–007r7. Open Geospatial Consortium Inc.

Silva, C.T., E.W. Anderson, E. Santos and J. Freire. 2011. Using VisTrails and Provenance for Teaching Scientific Visualization. Computer Graphics Forum. 30(1): 75–84.

Skøien, J.O., M. Schulz, G. Dubois, I. Fisher, M. Balman, I. May and É.Ó. Tuama. 2013. A Model Web approach to modelling climate change in biomes of Important Bird Areas. Ecological Informatics. 14: 38–43.

Tamayo, A., C. Granell and J. Huerta. 2012. Measuring Complexity in OGC Web Services XML Schemas: Pragmatic Use and Solutions. International Journal of Geographical Information Science. 26(6): 1109–1130.

Vaccari, L., M. Craglia, C. Fugazza, S. Nativi and M. Santoro. 2012. Integrative Research: The EuroGEOSS Experience. IEEE Jounral of Selected Topics in Applied Earth Observations and Remote Sensing. 5(6): 1603–1611.

Vasiliev, Y. 2007. SOA and WS-BPEL: Composing Service-Oriented Architecture Solutions with PHP and Open-Source ActiveBPEL. Packt Publishing.

Voinov, A. and H.H. Shugart. 2013. 'Integronsters', integral and integrated modelling. Environmental Modelling and Software. 39: 149–158.

Voss, A. and R. Procter. 2009. Virtual research environments in scholarly work and communications. Library Hi Tech. 27(2): 174–190.

Wang, J., D. Crawl and I. Altintas. 2009. Kepler + Hadoop: A General Architecture Facilitating Data-Intensive Applications in Scientific Workflow Systems. In Proc. of the 4th Workshop on Workflows in Support of Large-Scale Science (WORKS09) at Supercomputing 2009 (SC2009) Conference.

Wen, Y., M. Chen, G. Lu, H. Lin, L. He and S. Yue. 2013. Prototyping an open environment for sharing geographical analysis models on cloud computing platform. International Journal of Digital Earth. 6(4): 356–382.

Wiederhold, G. 1992. Mediators in the Architecture of Future Information Systems. IEEE Computer. 25(3): 38–49.

Wright, D.J. and S. Wang. 2011. The emergence of spatial cyberinfrastructure. Proceedings of the National Academy of Sciences. 108(14): 5488–5491.

Yang, C., R. Raskin, M. Goodchild and M. Gahegan. 2010. Geospatial Cyberinfrastructure: Past, Present and Future. Computers, Environment and Urban Systems. 34(4): 264–277.

Yang, C., M. Goodchild, Q. Huang, D. Nebert, R. Raskin, Y. Xu, M. Bambacus and D. Fay. 2011. International Journal of Digital Earth. 4(4): 305–329.

Yu, J. and R. Buyya. 2005. A Taxonomy of Scientific Workflow Systems for Grid Computing. SIGMOD Record. 34(3): 44–49.

Yu, G.E., P. Zhao, L. Di, A. Chen, M. Deng and Y. Bai. 2012. BPELPower—A BPEL execution engine for geospatial web services, Computers & Geosciences. 47: 87–101.

Yue, P., J. Gong and L. Di. 2010. Augmenting geospatial data provenance through metadata tracking in geospatial service chaining. Computers & Geosciences. 36(3): 270–281.

Yue, P., Y. Wei, L. Di, L. He, J. Gong and L. Zhang. 2011. Sharing geospatial provenance in a service-oriented environment. Computers, Environment and Urban Systems. 35(4): 333–343.

Yue, P., H. Zhou, J. Gong and L. Hu. 2013. Geoprocessing in Cloud Computing platforms—a comparative analysis. International Journal of Digital Earth. 6(4): 404–425.

Zhao, P., T. Foerster and Y. Peng. 2012. The Geoprocessing Web. Computers & Geosciences. 47: 3–12.

CHAPTER 8

Architecture for Including Personalization in a Mobile GIS via Semantic Web Techniques

Laia Descamps-Vila, Jordi Conesa, Antoni Pérez-Navarro*
and Joan Casas

Introduction

Smartphones are nowadays a common device with millions of users around the world. They can execute thousands of applications that support any interest of users: productivity, mail access, internet navigation, entertainment, tourism, etc. These applications transform the experience of using a smartphone to something similar to use a pocket extension of a personal computer. They are not simply PC software ported to smartphones, but applications that have been created from scratch to be adapted to these devices in order to take full advantage of the possibilities of smartphones. Among these applications we find geographical applications, like Google Maps (Google 2012) or OSMAnd (2012), that allow users to have access to a map anywhere and at anytime and to be positioned and located thanks to the GPS capacities. Some applications even take advantage of the compass and accelerometer and are able to find out what the user is looking at or

IT, Multimedia and Telecommunications Department, Universitat Oberta de Catalunya, Rambla Poblenou, 156, 08018 Barcelona, Spain.
Emails: ldescamps@uoc.edu; jconesac@uoc.edu; jcasasrom@uoc.edu
* Corresponding author: aperezn@uoc.edu

where the user is moving to. Other applications may also offer navigation functionalities, able to guide the user from one point to another.

Tourism is one of the most popular sectors of applicability of geographical applications and some tourism applications are able to (Gavalas et al. 2013): 1) find out where the user is; 2) guide the user to a point of interest; 3) identify what the user is looking at; and 4) provide some information about the point of interest the user is looking at. Therefore, with these functionalities smartphones become an up to date guide wherever the users go and true Recommender Systems (RS).

And, how do these geographical applications work? There are several categories of RS (Gavalas et al. 2013, and references therein), regarding their target applications, the knowledge used, the way they formulate recommendations and the algorithms they implement: collaborative filtering; content-based filtering; knowledge-based filtering; demographic filtering; matrix factorization; hybrid RSs.

The approach presented in this work is a hybrid RD that uses primarily knowledge-base filtering, but corrected by content-based filtering. According to the classification from Burke (Burke 2002), the system presented here is weighted in a first approximation and switched depending on the current situation.

Regarding the performance, many of actual approximations take advantage of the capacities of the smartphone (see the review from Gavalas et al. 2013), since they use the GPS and/or the network. However, when the task to perform is costly (computationally or by its memory requirements), these applications tend to delegate such task to a remote server (see for example Noguera et al. 2013). Examples of these delegated tasks are the download of the map of the zone, the retrieval of information of the points of interest the user is looking at, or the calculus of a route from one point to another, or some extra information which is computationally very costly. Therefore, most geographical applications need a cellular network to work properly.

Unfortunately, the constant need for a Internet connection poses several drawbacks that make the application not usable in all conditions:

- 3G coverage is low in most of the territory: although 3G covers the most populated areas and usually covers 90% of the population, isolated areas are not covered.[1] This would make geographical applications useless in isolated zones, despite the interest tourists may have in these areas.
- Battery consumption: batteries of smartphones last only for few hours when using the cellular network connection, making applications useless for longer routes.

---

[1] See for example de zones covered by Orange in Spain: most of the main cities are covered, but not small villages or the country. http://movil.orange.es/cobertura-orange/mapa-de-cobertura.html

On the other hand, current tourism applications have another inconvenience: the amount of information provided may overwhelm users. In places with several Points Of Interest (POIs) to visit, it is important to choose the most relevant point of interest for the user. The user has to choose where to go, and this choice is even more difficult when the user has a very limited time. But once the POI is chosen, the user has to face a great deal of information (pictures, comments and descriptions, etc.): and decide which ones are relevant. Therefore, the tourism experience becomes a "play of choices", with a cost of choice that: 1) makes the user loses his or her time; and 2) makes the user experience unsatisfactory when the choice is not the most suitable for him or her (Gavalas et al. 2013).

Summarizing, actual geographic applications have several limitations: the lack of a cellular network connection in some zones, the low battery duration and the necessity to filter the useful information for each user (personalization).

In order to overcome those drawbacks, we propose that a geographical application should satisfy the following requirements:

1. Ability of working anywhere and at anytime. This can be achieved by creating an application able to work completely off-line: *a Geographic Information System (GIS) within a smartphone.*
2. Ability of preparing a visit suitable for the user, giving a satisfactory user experience: *personalization.*

Thus, the research question dealt with in this chapter is: how can we build a touristic GIS that works anywhere, anytime, consumes less battery and provides personalized information for each tourist? To answer this question we have to satisfy the aforesaid requirements.

To satisfy the first requirement, *working off-line*, it is mandatory to execute all the processes and to store all the relevant data in the smartphone. Thus, we will need to adapt the software to smartphones.

To satisfy the second requirement, *personalization*, we need to provide the user only with the suitable information to make a satisfactory user experience. This goal can be achieved by using semantic web techniques. But, once more, these techniques have to work within the smartphone itself.

This chapter faces the research question by designing a touristic GIS architecture that satisfies the stated requirements. The result is Itiner@, which is a proof of concept of the presented architecture. The presented architecture has the fundamentals of personalization and allows to build a Context-Aware Recomender System (CARS, see Adomavicius and Tuzhilin 2011) through semantic information.

The main contribution of the chapter is to show the process used and the drawbacks found: 1) to develop geographic applications able to work off-line; 2) to manage semantic information in a smartphone; and 3) to combine geographic and semantic information within a smartphone. We show and test several technologies to satisfy these three tasks. Answering the third is the main goal of the subject. A proof of concept, Itiner@, is built to show the viability of the architecture presented.

The chapter is structured as follows: Section GIS in Smartphone explains the importance of GIS in current smartphone applications and the state of the art of GIS techniques implemented by current geographic applications. Section Semantic Web defines the concepts of Semantic Web. Section Expected System details the design of the system that should satisfy the requirements of the project. Section Tests and Discussion tests and discusses the use of GIS and Semantic web technologies to achieve the requirements presented in Section Expected System. Section Proof of Concept: Itiner@ presents the final architecture of the application, comparing the functionalities with state of the art features, and describes the conducted proof of concept, Itiner@. Finally, the chapter ends with the conclusions.

GIS in Smartphone

Our world has become more dynamic and mobile than ever. The advancement of technologies and communications has changed the ways that people and businesses conduct their day-to-day activities. Most daily actions are done through smartphones, such as: checking traffic before going to work, consulting if the public transport is on time, to receive a discount coupon from a nearby tea shop while going to work, to find the location of a restaurant and to know which is the best route to get there, etc. The underlying technology that powers all these *actions* is GIS. Thus, GIS are very important in modern society and their functionalities are used in multiple fields.

Before Android and iOS operative systems, GIS were run mainly in PC and servers, and mobile applications were just extensions of the desktop counterparts and their use were restricted to professionals or very specific users. But now, due to the technological development of mobile devices (tablets and smartphones), the operations with geographic data are executed on portable devices and, since these devices have become very popular, geographical operations have become available for most people.

The main advantage of mobiles over computers is that the user can carry the device with him/her. Hence, smartphones allow continuous access to information, anywhere at any time.

In addition, smartphones have location-aware technologies, namely, technologies that allow knowing the location of the device continuously, like GPS. Therefore, smartphones add a useful aspect of mobility, which gives added value compared to the computers because most of actions may be related with the user location.

Importance of GIS in LBS

Everything happens in some place in a single time. Because of this, geographically referenced data is important, and thanks to the actual mobile technology, currently it is widely used and it is becoming the key to better decision-making in business and activities of daily living related to location.

A study by the Pew Internet and American Life Project (Mashable 2012) has determined that about 74% of smartphone users use location services for information on what is around them; and one out of five (18%) checks for local businesses such as restaurants or theatres. Thus, we could say that people are interested in things near their location. This was already stated in the year 1970 by Tobler, who presented his first Law (Tobler 1970): "everything is related to everything else, but near things are more related than distant things".

To satisfy the need to know what is happening near the user, services that use his or her position to deliver information have been developed, which are called Location Based Services (LBS). GIS provides the basic tools that make LBS functionally possible (Frank et al. 2004). Thus, LBS systems can be considered as specialized GIS (Virrantaus et al. 2001). In addition, the paper of McKee et al. (2002) explains that the introduction of LBS has also further increased the importance of GIS within today's society. From a business-to-business perspective LBS enables more effective "real time" data collection, improving data base accuracy and ultimately delivering a more efficient service to the end customer. For all these reasons, location and the knowledge about the things around the user are becoming of great interest and make appear *location-aware systems for geographic applications*.

Location-Aware Systems for Geographic Applications

The increasing number of smartphones equipped with multiple sensors and the ability to use different connection networks, have opened several possibilities to the deployment of novel and exciting context-aware mobile systems (Adomavicius and Tuzhilin 2011; Poveda et al. 2010). On the other hand, the use of LBS within smartphones provides the geographical context to the applications. Therefore, the combination of smartphones, GIS technologies and LBS adds a new context, *location-aware systems*.

Location-aware systems provide information on the user's environment in a given time and position, which can help people to get aware about their environment. Hence, the deployment of these systems in a tourism application could provide a lot of benefits. These systems deployed in smartphones result in location-based information, that, for example, allow the user to search for the nearest restaurant or store; provide information on items of interest in a city; give step by step navigation to a particular location; or report the estimated travel time to reach a store based on the current traffic conditions. These functionalities can be seen in Google Maps (Google 2012), OsmAnd (OsmAnd 2012), or Foursquare (Crowley and Selvadurai 2009). All of them offer information on the smartphone while considering the location of the user.

GIS Functionalities Necessary for Location-Aware Systems

As aforesaid, location-aware systems must implement multiple GIS techniques. If GIS capabilities are fully available, location-aware systems will be much more efficient for end-users (Virrantaus et al. 2001).

Focusing on the literature (Marble and Peuquet 1990), the features that provide a common GIS can be summarized in four points: 1) data input system that collects spatial data, 2) data storage system that allows a quickly retrieval of data, 3) data analysis system that allows mathematical operations over data, and 4) data reporting system that is capable of displaying all or part of the data.

These features provide the following functionalities of a GIS in a location-based application:

1. Storage and management of spatial data.
2. Tools to perform geographic operations. Tourism applications for smartphones in a real-world scenario use GIS methods such as proximity analysis, distance measurement and path finding, which is also called routing.
3. Displaying of cartographic data to facilitate interaction with geographic data. For instance, the application must display maps and text data and when the user touches the screen of the smartphone, the device must be aware of the coordinates on the map.

GIS features available in current applications

In order to achieve GIS functionalities, there are libraries that provide them as extensions of software or spatial databases. For example, Java Topology Suite (JTS-Topo-suite 2012) provides a suite of functions and spatial predicates and PgRouting (PgRouting 2012) is an extension of PostgreSQL (postgreSQL 2012) database which provides routing functionalities. However, these libraries are designed for computers and are inappropriate for devices with the memory and CPU features of a smartphone.

Thus, some questions arise: how can we add GIS functionalities to current location-based applications? Since the common GIS libraries are suitable only for desktop environments, what GIS features are actually implemented in current location-based applications?

Although it has been a great progress with location-based applications, we still miss some important features, especially regarding the step by step navigation, working off-line.

Most smartphone applications perform routing operations through servers, where most of operations are pre-calculated to be more efficient (see for example, Noguera et al. 2013). When the user is off-line, Google Maps for Android has a re-routing function. If the user starts the route on-line and the server connection drops, the application is still able to guide the user to the right path. However, the user still needs a server for the initial draw. Another application, called OsmAnd (OsmAnd 2012) has routing off-line still in an experimental stage and works only for short distances.

It is important to note, that a routing system is not a tourist application, since the user has to choose the places to visit. A true tourist application should have to be able to provide customized information to the user, according to his or her geographic location, as well as his or her personal preferences, and all these processes should run off-line. Such a system does not still exist. How can we build it?

There are several solutions for the routing part that work in desktop and off-line environments. But how personalization can be addressed is lacking in them. Our initial approximation is by using the techniques of Semantic Web and ontologies.

Semantic Web

The large amount of data in the Web, their centralization and lack of standardization makes very difficult the task of the agents in providing complete and updated information to the user. The selection of information relevant to the user gains more importance when smartphones come into play.

Therefore, in order to manage the information available on the Web it is essential to locate, process and integrate all relevant information. Since most of the information on the Web has been generated without any control or organization, it is necessary to use technologies that allow computers to process all available information from a semantic point of view for classification and subsequent use. Thus arises the Semantic Web concept, defined by Berners-Lee (Berners-lee et al. 2001), whose objective is to provide a well-defined meaning to information, allowing computers to *understand* and then *use* the information available on the Web.

One of the main benefits that arise from the Semantic Web is the ability to integrate data from different sources in order to obtain more accurate search results and to be able to automate complex tasks. To do so, computers must have access to structured collection of information and sets of inference rules that can be used to conduct automated reasoning (Berners-lee et al. 2001). To achieve these requirements, Semantic Web languages and ontologies have been developed for knowledge representation, and Semantic Web reasoners have been introduced for inferring knowledge.

Ontologies are the cornerstone of the Semantic Web and are used for organizing data, improving a search or data integration. They can be defined as conceptual models of reality written in a language interpreted by a program, which represent concepts and relationships of a specific domain. They can acquire or share knowledge about a system and use this knowledge for a specific purpose, such as designing or developing software, designing a database or understanding how a specific system works (Guarino 1998).

On the other hand, Semantic Web allows inferring new information creating new relations between data. This is done using Semantic Web reasoners, which are programs that infer logical consequences from a set of explicitly asserted facts or axioms, and typically provides automated support for reasoning tasks such as classification, debugging and querying (Dentler et al. 2011).

Through reasoners it is possible to infer new information from ontologies (Bry and Marchiori 2005). The inference rules are commonly specified by means of an ontology language, and often a description language. For instance, using an ontology that describes simple preferences of a user allows inferring more sophisticated preferences. Schickel-Zuber and Faltings (2006) presented examples of the benefits of ontologies in recommender systems. Bradley et al. (2000) used an ontology to build a personalized search engine that increased classification accuracy by more than 60% and Middleton et al. (2004) used ontological relationships between topics of interest to infer other topics of interest, which might not have been browsed explicitly.

Therefore, using ontologies and reasoners in the appropriate context, it is possible to develop a system able of reasoning, in a minor extent, as a person, what we can call an *intelligent* system. This could be especially useful to filter the big quantity of data available on the Web taking into account user interests and geographic location.

Current Ontologies

To implement technologies of the Semantic Web in a location-aware system for tourism application, it is necessary to define the ontologies of every domain involved in the project and consider some rules for personalization of data, defining forms of reasoning that might be needed to provide the required data (Bry and Marchiori 2005).

Since the topic of this chapter is focused on a location-based application that offers touristic routes considering user preferences, the ontologies described in this section deal with the domains related with touristic location-based services and the user preferences; i.e., our goal is to offer the tools to build a CARS. These domains and associated ontologies are:

Tourism. Harmonise Ontology (Ontologyan 2011) contains around 200 concepts and properties, which are mainly focused on accommodation, events and activities, gastronomy, monuments and places of interest. The ontology Cruzar (Mínguez et al. 2010) has been developed to facilitate the calculus of touristic routes within the city of Zaragoza according to the profile and context of each user. DERI e-Tourism Ontology is an ontology created to support a web portal that supports touristic searches using Semantic Web technologies (Siorpaes 2005).

Time. TimeOntology (W3C 2006) provides a vocabulary for expressing facts about topological relations among instants and intervals, together with information about durations, and about date time information.

User Information. FOAF (W3C 2010) is used to represent user information, tastes, preferences and even their participation and interventions in social networks.

Mobile devices. WURFL (ScientiaMobile 2011) provides information on thousands of mobile devices and their characteristics. Delivery Context ontology (W3C working group 2010) allows describing the characteristics of mobile devices (hardware, software and network information of the data connection it uses). Mio! Network Ontology (Poveda et al. 2010) defines a network of ontologies to represent knowledge related to the context of a user, including information like: where is the user, which are his or her preferences and information about his or her device. Such information would be useful to filter what information makes sense to send to each user, what is the best delivery format for each user and what kind of information is not available in the user mobile device.

Note that with mobile ontologies we mean ontologies that allow defining the information of the internal sensors of mobile devices, but also information about their other hardware, capabilities and functionalities. The goal is to use the ontology information not only to locate the user, but also to allow proactive recommendations. A proactive system like that would be able to detect in which data format the information should be delivered to a user according also to the limitations of the cell phone. This system would be able to deliver the information in an audio format when the battery of the mobile device is low, avoiding the use of the screen and therefore saving battery.

Linked Data. Today, perhaps the main problem of current tourist applications is interoperability. A possible solution to interoperability problems could be the use of well-known ontologies adopted by the majority of the community, such as the ones within the Open Linked Data (OLD) project (Bizer et al. 2009). OLD is an initiative to interconnect the information from the web in a way that is understandable to computer programs. A set of ontologies are defined, where each ontology describes the information structure of a different domain. However, the power of OLD is not in their ontologies, but in their interrelationships. Ontologies are related to each other, allowing programs to navigate from one concept to other concepts, either in the same domain or in different domains. For example, the system links LinkedGeoData (LGD) ontology, which contains geographical data from OpenStreetMap, with DBpedia, which contains data on Wikipedia.

Semantic Languages

Once we know how to represent an ontology from a conceptual point of view, we need to know the language used to represent it. As ontologies are conceptual schemas focused on computers, they should be written in a language that a program should be able to understand. Today, the most used ontology languages are Resource Description Framework (RDF) (W3C 2004) and Web Ontology Language (OWL) (W3C 2007).

RDF/OWL

RDF is a standard model for data interchange on the Web. This language is based on *Triples* that represent expressions of the kind subject-predicate-object (*who* does *what* with *something*). For example, a triple could be "Jack has a house", where "Jack" is the *who*, "has" is the *what* and "a house" is the *something*. Triplets can be semantically processed by machine agents and most of the current Semantic Web applications apply reasoning using such triplets (Stojanovic et al. 2002). OWL language extends RDF with the possibility to express additional constraints.

RDF and OWL languages use different standard terms to represent concepts and relationships of ontologies. Firstly, there are *Classes*, which provide an abstraction mechanism for grouping resources with similar characteristics. Secondly, there are the *instances*, which are the individuals in the class. Thirdly, there are the properties that relate information and can be defined as: *Object Properties*, which link individuals to individuals, and *Datatype Properties*, which link individuals to data values. These terms are very important, because for every ontology we have to define every single aspect interesting to the knowledge domain by using these terms consistently.

For example, in the particular case of a tourism system, the basic concepts of a city should be described. From an ontology point of view, every one of these concepts is a class. Figure 1 shows part of an ontology used for a tourism domain. The classes of the ontology could be a *Point of interest*, a *Resource*, a *Spatial point*, etc. Classes may also be specialized using subclasses. For instance, a subclass of *Point of interest* class could be a *Historic point* and a subclass of the Historic point could be a *Monument* (represented in orange color in Fig. 1). On the other hand, an instance of *Monument* class may be *SagradaFamilia* and an instance of *Resource* class may be a picture: *Photo* (represented in purple color in Fig. 1). An Object Property *hasResource*, allows to link the *SagradaFamilia* instance with the *Photo* instance, indicating what is shown in the picture. Then, the *name* is a data property that is used to relate the *SagradaFamilia* instance with a String value that indicates the name of the monument.

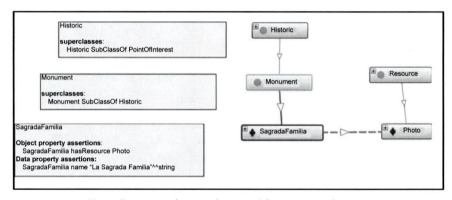

Fig. 1. Fragment of an ontology used for a tourism domain.

Color image of this figure appears in the color plate section at the end of the book.

SPARQL

Data that represents ontology information of Fig. 1 is stored in RDF triplets within a triplestore, which is a specific database built for the storage and retrieval of triples. Since in a relational database, triplestores can be queried by using a query language, which is called SPARQL (W3C 2012). SPARQL is the equivalent of SQL in relational databases, but limited to query operations. As an example, the following lines show a SPARQL query over the ontology of Fig. 1.

PREFIX owl: <http://www.w3.org/2002/07/owl#>
PREFIX xsd: <http://www.w3.org/2001/XMLSchema#>
PREFIX rdfs: <http://www.w3.org/2000/01/rdf-schema#>
PREFIX rdf: <http://www.w3.org/1999/02/22-rdf-syntax-ns#>

> *PREFIX : <http://www.itinera.cat/ontology/itinera core.owl#>*
> *SELECT ?resource*
> *WHERE {:SagradaFamilia :hasResource ?resource}*

The query asks for the *resources* related with the *Sagrada Familia* instance. In the case of Fig. 1, the result of this query would be the instance *Photo*.

GeoSPARQL

Spatial information can also be tagged as triplets in RDF data, resulting in what is called RDF spatial data. In this case, the difference is that RDF semantic data is tagged with geographical position. Thus, tools are required to perform spatial operations, define spatial relations and infer deductions over geographic RDF data, which are different from non-spatial RDF data.

Therefore, the question is: does SPARQL language support geospatial data? The answer is it does not. Without spatial reasoning, the value of the spatial context and geographic data is limited. In order to meet this need GeoSPARQL language has been introduced (Battle and Kolas 2011).

GeoSPARQL language is a standard from the Open Geospatial Consortium (OGC) for querying and reasoning geospatial data represented in RDF format, from simple points of interest to complex geospatial data sources. The main advantage is that it provides the ability to query and filter the relationships between geospatial entities (Battle and Kolas 2011).

Reasoners

Besides ontologies, the other important mechanisms of Semantic Web are the reasoners. As already mentioned, a reasoner infers logical consequences from a given knowledge base (Mishra and Kumar 2010). There are multiple reasoners to work with semantic data depending on the language they are written, the logic or the algorithms they use. The most well-known are FaCT++ (Tsarkov and Horrocks 2006), Pellet (Clark and Parsia 2011) and HermiT (University of Oxford)[#]. For a deeper benchmark about reasoners see Dentler et al. (2011).

In order to understand the benefits of using a reasoner, we follow the previous example of the ontology in Fig. 1. In that case, the ontology declares that *La Sagrada Familia is A Monument* and a *Monument is Historic*. Then, a Semantic Web program with a reasoner can add the statement *La Sagrada Familia is A Historic* to the set of relationships, although it was not part of the original data. This is only a simple example of inference of knowledge, but we can use the reasoners for more complex tasks.

[#]http://www.cs.ox.ac.uk/projects/HermiT/

Frameworks

In order to use semantic data in an application, it is necessary to deploy the Semantic Web concepts described previously in an application. To this end RDF frameworks have been introduced. These frameworks provide facilities to build Semantic Web applications that use ontologies and reasoners. To filter out all the existing frameworks, we focus on those which have the generic tools necessary to develop a Semantic Web system and they provide: 1) support for RDF and OWL languages, 2) support for SPARQL specification, 3) tools to store RDF data with the ability to scale to large datasets, and 4) inference capabilities for ontologies.

Desktop Frameworks

Two frameworks that accomplish these requirements and are well-established for the Semantic Web community are Jena (Carroll et al. 2004) and OpenRDF Sesame (Aduna 2012) .

Jena is an open-source Java framework that provides a programmatic environment for RDF, OWL and SPARQL languages and offers support for several reasoners. Regarding storage systems, it offers two different persistent modules: SDB and TDB. SDB stores information in a transactional database, using JDBC interface to access the most well-known databases; and TDB stores the information in a pure-Java dataset that offers non-transactional functionality.

OpenRDF Sesame is an extensible Java framework and web server for RDF parsing, storage and SPARQL querying. Sesame has a memory-based and a disk-based RDF store and RDF inferences. The framework is fully extensible and configurable regarding storage mechanisms, inferences and query languages.

Spatial Reasoning

Therefore, both frameworks, Jena and OpenRDF Sesame, have support for RDF/OWL languages, offer query engines for RDF data and provide inference functionalities. But, do they support to deal with RDF spatial data?

Similar to relational databases that have extensions to work with spatial data,[2] Semantic Web frameworks have extensions to work with RDF spatial data. Thus, Jena can use GeospatialWeb (KONA 2010) as spatial extension

---

[2] For example, PostGIS spatial database provides spatial objects for storing and querying geographic information in PostgreSQL database.

to Jena RDF databases and OpenRDF Sesame may use uSeekM component (Sahara 2013), a library that contains spatial predicates and indexes for Sesame based RDF databases.

In order to provide geospatial reasoning and querying in a triplestore, the implementers must define an ontology for representing spatial objects and additional query predicates for retrieving these spatial objects (Battle and Kolas 2011).

GeospatialWeb offers spatial predicates such as: 1) a *Within* function that returns whether a spatial element of an ontology is contained in a geographic area, 2) a *Nearby* function that returns geographic elements of an ontology that are near to a given latitude and longitude and, 3) a *Distance* function that returns the distance between two geographic elements defined in an ontology.

uSeekM library provides indexing, efficient search and computations on geometries integrated into the SPARQL query language and supports all OpenGIS Simple Features geometries (such as Point, Line, Polygon) and functions (such as *Within, Intersects, Overlaps, Crosses*) standardized in the OGC GeoSPARQL standard. In addition, uSeekM extends existing triplestores with new types of indexes (R-Tree, Quadtree or Geohash). It also has a PostgisIndexer module that uses a PostGIS database to build indexes.

Note that these semantic spatial libraries provide similar mathematical operations that a GIS provides in a location-aware system, as described in the GIS Functionalities Necessary for Location-Aware Systems section, but with the advantage that they allow to work with RDF data and to integrate with all the Semantic Web technologies.

Smartphone Frameworks

Desktop Semantic Web frameworks are very powerful and offer a complete range of functionalities. However, as Zander et al. (Zander and Schandl 2010) show, these heavy-weight systems cannot be deployed on typical smartphones because of their limited memory and processing capacities. Desktop frameworks are developed for powerful server computing infrastructures, which are able to process huge amounts of data. However, they are not appropriate for a smartphone environment, which has lower CPU power. Hence, Jena and Sesame must be modified to work on a smartphone.

Jena has a port of different components to Android Operative System (OS). The Android version comprises the core of Jena, called Androjena (Group 2010), the SPARQL query engine-ARQ, called ARQoid, and the non-transactional database storing system-TDB, called TDBoid. However, Jena SDB component is not ported to Android. SDB uses relational databases as back-end and the databases supported by SDB are not available in smartphone OS. Thus, SDB cannot be used as a triplestore for Androjena.

There is also a project called android-sparql (Sesame 2012) developed for Android OS, which is an extension of OpenRDF. It can be used to create and populate a triplestore on an Android device, as well as a web application with a simple implementation of the SPARQL protocol. Even though it is possible to store data on a triplestore in the device, SPARQL queries require Internet connection to perform the queries over the triplestore. Thus, it is not useful for the application we are looking for, because it does not accomplish the requirement of working offline.

Finally, there is the MobileRDF (Hedenus 2008) open source project. It is a lightweight implementation of RDF developed for Java ME that offers a simple and easy-to-use API. It has simple functionalities like RDF parsing and serialization. Java ME is a Java platform designed for embedded systems, thus MobileRDF could be used as a library in an Android application. The main drawback is that MobileRDF does not offer inference or more complex RDF tasks.

Integration of Semantic Web in GIS

Summarizing, the building blocks of Semantic Web are: 1) semantic data, which is represented as triplets in the standard model, called RDF; 2) ontologies, which contains information about the domains of interest by representing the related data, relationships between data, domain constraints and inference rules; 3) reasoners that provide functions to infer new knowledge via inference and validate RDF instanced data via ontologies; and 4) query language, that allows to retrieve data from ontologies.

Therefore, using geographic ontologies and frameworks able to reason over RDF spatial data, we can integrate GIS and Semantic Web functionalities. However, current state of the art of GIS does not support completely semantics (Van Hage et al. 2010). Thus, the integration between semantic and spatial context is a current focus of study nowadays (Descamps-Vila et al. 2012a; Hoekstra et al. 2010; Van Hage et al. 2010). On the other hand, several initiatives are working to incorporate Semantic Web that is an important research focus (see for example, Ordnance Survey 2013).

The use of semantic data together with geographic functions in smartphones would allow, for example, creating tourism guides like the mobile cultural guide of Van Aart et al. (2010), developing systems to create and share tourism information like CsxPOI (Max 2009); creating urban computing demonstrators, such as the one of Milano city (Della Valle et al. 2010), or prototypes of DBPedia mobile (Becker and Bizer 2012). Currently, the drawback of all these applications is they use a server to perform semantic operations.

When dealing with offline services in mobile environments, it is very important to filter, in advance, the information of interest by the users in order to minimize the storage of the required information and to maximize the users' satisfaction with the provided information. To perform such personalization it is necessary to be aware of the context in which the user is placed. In our particular case, the context may be defined by the user preferences, the user-traveling configuration (how the user is traveling, whether the user travels with friends or family, the time of the route, the place where the route will be done, etc.). By using semantic information it is possible to define the contexts the user can found and what information is more relevant for each possible context. For example, using a single ontology, it would be easy to define what points of interest are relevant when traveling with children, and then they can be used to propose routes for travelers who travel with family. In addition, using semantics may provide other benefits, such as the possibility of reusing existent and standard information and the ability of improving the proposed system only by extending the current ontologies with new information.

Hence, we propose to create a tourism application that integrates Semantic Web and a location-aware system in a smartphone working locally, off-line.

Expected System

At this point we have analyzed the current state of the art of the technologies needed to answer the research question and have seen that there is currently no solution that fits the project requirements described in Section "Introduction": the creation of a location-based mobile application that works offline and provides the user's personalization. This section proposes a system to integrate these missing functionalities in a smartphone system.

The current missing features in a location-aware system are: 1) to offer personalized results to every single user by filtering the information and automatically adapt the filtered information to each user according to her or his profile and preferences, 2) to provide GIS functionalities, and 3) to execute both 1) and 2) in a smartphone without Internet connection. As a case study and proof of concept a tourism application is designed.

To offer personalization of routes and work without Internet connection a system may use the following Semantic Web technologies:

- Ontology for each domain of the project (user preferences, tourism, time and historic of routes) that will be the core of the RS.
- Framework to store and manage RDF data in smartphone.
- Spatial reasoning to create a system that performs simple geographic operations over spatial RDF data.

Note that, a tourism application must integrate additional functionalities, such as displaying maps and performing routing. Thus, there are additional requirements that must be taken into account for tourism applications that should be executed offline: 1) cartographic data (raster data) must be stored, managed and displayed in the smartphone, 2) vector map data must be stored and managed in the smartphone, and 3) a routing algorithm to guide the user through one point of interest to another over the map must run locally on the smartphone.

It is important to take into account that our goal is to develop an end user application; therefore the application has to run fast and efficiently in order to offer a satisfactory user experience.

A location-aware system integrated in a tourism application must address all these requirements to achieve the goal of the project. To achieve these requirements, the following section tests and discusses some technologies described in Section "GIS in Smartphone" and Section "Semantic Web".

Tests and Discussion

To know how to build an application with the requirements of previous section, this section details the ontology created from scratch, the tests to find out whether it is feasible to deploy semantic web technologies in a smartphone system, and create an algorithm to personalize the information provided to each user.

Ontology

The ontology presented in this chapter is designed with the objective of representing the necessary information in order to support personalized LBS that assist users in their touristic routes, i.e., to build a CARS. The personalization is based on user preferences, the place to visit, the time zone of the visit and the limitations of the smartphone to use. Therefore, we need ontologies that address the following domains: tourism, temporal context (to take into account opening hours, user temporal preferences, etc.), personal preferences, smartphones' limitations and information from previous routes (this item can help to better know the user and to better fit her/his experience to her/his true preferences (see Ono et al. 2009). The ontology is created from scratch, but part of data is linked with ontologies of OLD (Fig. 2). The ontology allows easy exchange of data with other related applications/systems as it is aligned with LinkedGeoData, WGS_84 and TimeOntology.

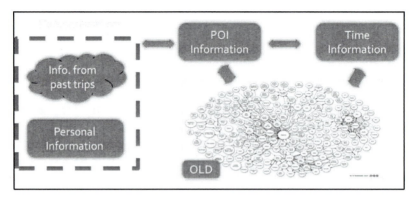

Fig. 2. Diagram of ontologies relationships.

Points of interest ontology

The POI ontology (Fig. 3) contains information about the points of interest. A point of interest is any spatial element that includes both monuments and tourist attractions. Some of the POI-related classes have been reused from other ontologies, such as the SpatialThing class, which comes from the ontology WGS_84 (W3C 2003). Also, the specialization of POI is based on the categories of points of interest defined in the ontology LinkedGeoData (Lehmann and Hellmann 2009), facilitating the compatibility of our ontology with Open Street Maps and Linked Data.

But the definition of a POI does not provide interesting information of the POI to the user. The information related to every single POI that can be of interest to users is specified by external resources like images, videos, textual information, maps, etc. These resources provide extra information to the user, and enhance the travel experience.

Time

According to Gavalas et al. (2013) and Noguera et al. (2013) time is one of the items that should have to be taken into account by future application. Thus, one of the critical factors in tourism applications is the time factor, it is necessary to take into account issues such as opening hours of attractions, calendar aperture, the time to visit, etc. Therefore, it is necessary to have a representation that facilitates the management of time, dates and schedules. To store and operate efficiently with temporary variables, we reused concepts of TimeOntology. Figure 4 shows a fragment of the ontology that deals with the description of time and date intervals that are used to represent a POI schedule. The classes reused from the TimeOntology are colored in grey in the figure.

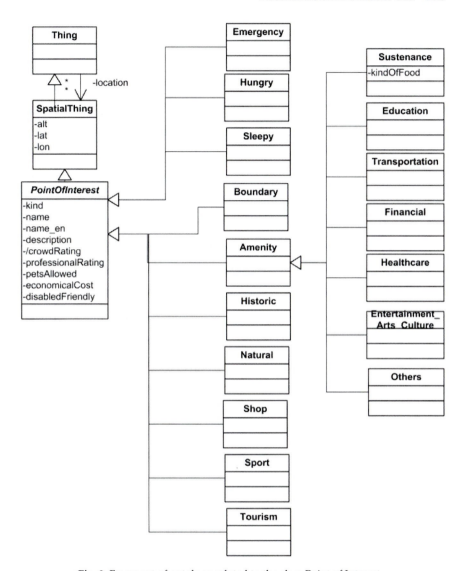

Fig. 3. Fragment of ontology related to the class Point of Interest.

User Preferences and Historical of Routes

Another important factor to consider in the ontology is the specification of the user preferences. These preferences should be used by the system to offer different routes to each user based on her/his preferences and the context in which she/he is. Figure 5 shows a fragment of the ontology that stores user preferences.

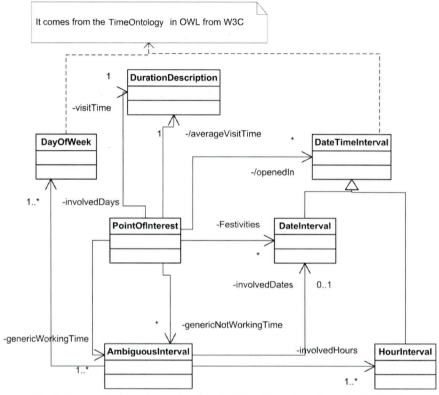

It comes from the TimeOntology in OWL from W3C

Fig. 4. Fragment of ontology related to the time dimension of points of interest.

A user can define preferences according to: 1) his/her interests in the type of POI to visit, 2) the expected duration of visits, 3) his/her economic restrictions, 4) the preferred visiting zone, 5) the size of the area to cover per visit, and 6) the users' companion, that is, if he/she is traveling with children, friends, pets or people with disabilities. Furthermore, the system also lets to set composite preferences, combining two or more different types of preferences (including composites). This would allow representing things like "when I travel alone I like to visit historical POIs, but when I go with children I prefer to go to infantile or zoo points of interest and finish the route before 5 pm." These composite preferences are specified by the concept *CompositePreference*.

The ontology allows representing explicit and implicit preferences. Explicit preferences are those entered into the system by the user. Implicit preferences are automatically extracted by analyzing all the routes performed by each user (both automatically and manually). Explicit preferences may be seen as what the user thinks, he or she likes and implicit preferences as what the user really likes. For example, someone may say that

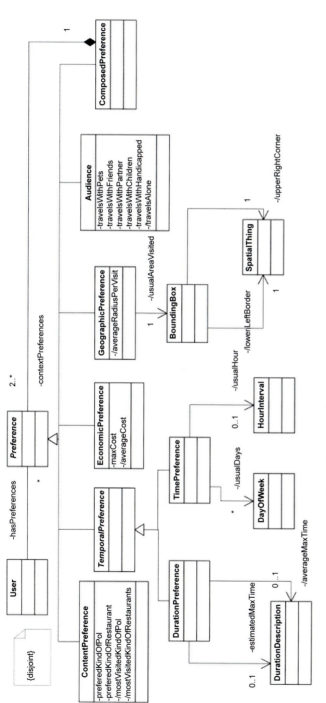

Fig. 5. Fragment of ontology related to user preferences.

she/he likes to visit Romanesque churches with her/his partner (explicit preferences) but actually she/he never visit Romanesque churches when has the opportunities and visit soccer stadiums (implicit preferences). Implicit preferences are easily identified in Fig. 6 since they are represented by derived (calculated) attributes and relationships and therefore they have a "/" at the beginning of their name.

Both types of preferences are taken into account to customize user routes. To find out the implicit preferences of each user, the ontology will represent historical information of the routes taken by users (Fig. 6).

It is important to note that relationships are established when ontologies and logical rules are designed. Ontologies already consider the different possibilities of the user context. What change over time are the values of such relationships as soon as the system gathers data from user history of routes. Thus, when the system generates new routes, it takes into account previously stored data from past users. Opinions and ratings of users are added as data in the points of interest ontology, and comments are displayed on the point of interest screen data.

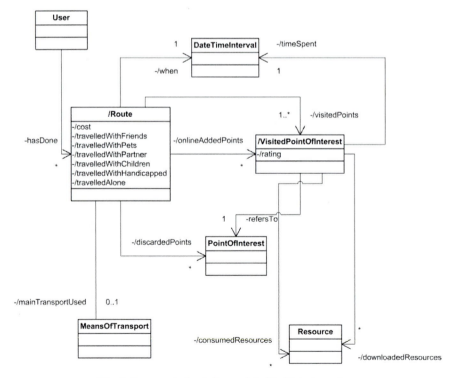

Fig. 6. Fragment of ontology related to historic routes.

Spatial RDF Data in Smartphone

This subsection proposes a framework to store and manage RDF data in smartphones and to use spatial reasoning to create a system that performs simple geographic operations over spatial RDF data.

In order to perform geographical queries over ontology data, first of all we import the GeospatialWeb library into Android OS and we integrate spatial predicates in the Androjena framework. Therefore, we create a system that contains Androjena plus ARQoid and TDBoid components and GeospatialWeb library. The system has been deployed into a prototype system for an Android smartphone.

Feasibility tests

To test whether the architecture is feasible, we implemented three different GeoSPARQL queries, the most common in a tourism application. The first one selects all the points of interest inside a given area (*within* function), the second one selects the nearest points from the user position (*nearby* function), and the third one selects the distance between two points of interest (*distance* function).

Figure 7 shows that it is feasible to integrate a GIS with spatial operations over RDF data. It shows the result of *within* query over the ontology: the points of interest within the region of *La Sagrada Familia* (a POI in Barcelona). The query returns all the POIs of LinkedGeoData from this area, which are displayed over a map.

Efficiency tests

The next step is to check whether such integration is efficient enough and scalable to be used in end-user applications. Therefore, we performed some queries with the GeoSPARQL *within* function done over several geographic ontologies, in order to evaluate the scalability of results. Additionally, we also perform tests with simple SPARQL queries, without spatial reasoning, in order to compare the results.

The smartphone used for the tests is an HTC Desire, which has Android 2.1 operative system, 1.0 GHz processor and 576 GB of RAM. Each test is repeated five times, in order to have an average of results, because they may be variable.

Results of SPARQL queries are displayed in Table 1 and results of GeoSPARQL queries are displayed in Table 2. Note that ontologies in Table 1 and Table 2 have different number of triplets, since it is not possible to use the same ontologies to query spatial and non-spatial data.

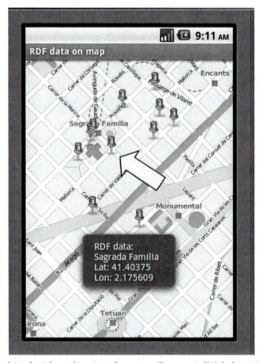

Fig. 7. Prototype of Android application that uses GeospatialWeb functions over ontology. *Color image of this figure appears in the color plate section at the end of the book.*

Table 1. SPARQL query results.

# triplets	SPARQL query time (s)
402	3.43
2075	5.05
5633	12.37
8444	12.53

Table 2. GeoSPARQL query results.

# triplets	GeoSPARQL query time (s)
147	2.63
1138	7.61
5880	18.50
6522	19.64

Figure 8 displays graphically the results of Table 1 and Table 2. It shows a comparison of the results of queries over spatial and non-spatial RDF data. Each point represents a single ontology, where X axis displays the number of triplets of each ontology and Y axis the query time.

The graphic shows that a system performing geographic operations over RDF data needs more time than a system without geographic queries for the same number of triplets. Furthermore, that difference increases as the number of triplets increases. Therefore, we can say that spatial reasoning considerably slows down the SPARQL queries.

Both results show linearity: the more triplets the ontology has, the more time the query needs. Execution time for SPARQL queries over ontologies with 1000 triplets or less is acceptable but queries over larger ontologies are not as efficient enough for usable smartphone applications. A system that uses SPARQL queries is not scalable enough for a tourism application because a district of a city like Barcelona has more than 6000 triplets. Thus, a system with this architecture cannot achieve the required efficiency.

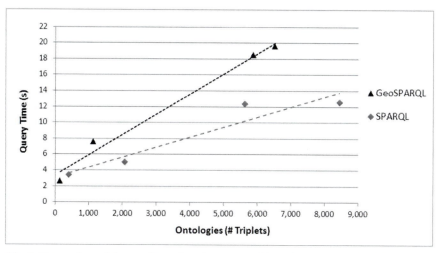

Fig. 8. Comparison of the results performing geographic and non-geographic RDF spatial queries over ontologies.

Alternative Architecture for End-user Applications

We achieved one of the goals of the paper: to develop a system that integrates GIS and semantic data using GeoSPARQL queries over ontology data in a smartphone locally. Nevertheless, the queries in such system are too slow to be used in an end-user application. Thus, it is necessary an alternative architecture to get an end-user real life application.

The main drawback we found is the inefficiency of SPARQL and GeoSPARQL queries. Therefore, in order to improve query time, we rewrite the conceptual model of the ontology to a relational database schema. In consequence, the instances of the ontology are inserted into the database as relational data. Then, we could use SQL to perform queries, which is far more efficient than SPARQL queries.

To create the database from the ontology, we created the ontologies in OWL format with triplets that represent the knowledge of each domain. Then, we imported the ontology model into a relational database schema. We tried to minimize the loss of semantics with a translation of the ontology model to a relational database: each class of the ontology becomes a table in the relational database. Therefore, the instances of the ontology are imported into the tables as relational data.

In this way, OWL reasoners cannot be used in the system because we implemented the ontology using a relational database representation. Therefore, the inference capabilities of reasoners cannot be used in the prototype. However, the system still has a clear and unambiguous definition of the touristic and geographic information and supports inference rules that allow taking profit of such information. In particular, the system has an ontology that stores the past trips of users (see Fig. 6) and some rules use this information to adjust the personalization values of each user according to her/his history, like tuning the average visit time of a point of interest according to the visit time of similar points of interest in the user past visits. Thus, even though moving the ontology to a relational model avoids using inference engines, the semantics of the concepts are kept and some basic inferences are still possible. In particular, most of the integrity constraints and derivation rules of the ontology have been kept after a direct translation of the RDF ontologies to the relational model, minimizing the loss of semantics. Note that the fact of not using reasoners is not only due to limitations of OWL/RDF frameworks in smartphones, but also due to technical limitations that make the use of these reasoners impracticable for real life smartphone applications.

Furthermore, it is important to take into account that personalization is one of the requirements to achieve. Therefore, we must filter data and suggest the best information for each different user. In order to do so, we design a personalization algorithm over the database.

Personalization Algorithm

Section "Spatial RDF Data in Smartphone" showed that it is inefficient to work with RDF data in a smartphone locally and proposed to work with ontology data exported to a relational database. Thus, to get personalization we designed a personalization algorithm with different rules that selects the best data for each user considering different preferences, taking also into account context, i.e., we propose a CARS. Thanks to this algorithm, we give different and personalized routes to each user.

In the following we present an overview of the algorithm. This algorithm uses mainly knowledge-base filtering, but corrected by content-based filtering. It is based on two key points:

1) It only offers viable routes, namely, the algorithm suggests a visit only when the user can carry it out. For example, it does not offer to visit a place that would be closed at the expected arrival time.

2) The proposed route does not necessarily suggest the maximum number of attractions to visit, because the visitor may want to visit fewer POIs, but mainly those that are more interesting to him/her. Thus, from a collection of viable routes, the algorithm will choose those most attractive to every single user.

Considering these two features, we designed an iterative algorithm. For each step we indicate, in parentheses, from which ontology the data is obtained:

1. From a group of POIs inside a region, a set of random routes doing combinations of those POIs is generated (*tourism ontology* —POI information, *personal ontology*—temporal and geographic preferences).

2. POIs are ordered by distance (*tourism ontology*—POI location)

3. Routes that are in one of the following cases, are discarded:

   • Case 1: the needed time to finish the route exceeds the available time for the visitor (*historical of routes* ontology—POI average length of visit)

   • Case 2: there is a POI closed when the user arrives to visit it (*time ontology*—POI timetable)

   • Case 3: there is no place to eat when the visitor wants to eat—the visitor decides which meals he/she wants to do (*time ontology*—POI timetable, *personal ontology*—content preferences)

4. Once there is a group of viable routes, the algorithm selects the most appropriate to the user needs considering the following parameters:

   a. POI average rating: given by other visitors (*tourism* and *historical of routes* ontologies)

   b. Companions of the user: according to who is travelling with the user different kind of POIs will be suggested. For instance, a night club has a higher rate when travelling with friends (*personal, tourism* and *historical of routes* ontologies)

   c. Kind of POIs preferred by the user (*personal* and *tourism* ontologies)

    d. Kind of food preferred by the user (*personal* and *tourism* ontologies)

    e. POI general importance rating. There are attractions that are of general interest to anybody, even when they are not the usual type of POI interesting to the visitor. For example, a monument such as *La Sagrada Familia* in Barcelona is considered essential for all users (*tourism* ontology)

Taking into account these parameters, a cost is assigned to each route. The higher the cost of a route, the more interesting it is. The cost is calculated with a function that takes into account every single point previously cited:

$$c_{route} = c_{average} + c_{intrinsic} + c_{type} + c_{meal} + c_{general}$$

Where:

- c_{route} is the cost function.
- $c_{average}$ is the average importance of every single POI and is obtained from the rates given by users.
- $c_{intrinsic}$ takes into account the intrinsic importance of the POI for the user and his or her situation (for example, a night pub weighs more for a user travelling with friends than for a user travelling with the family).
- c_{type} takes into account if the POI type is a type of POI that the user prefers (for example, if the user likes churches it gives more weight to churches that a POI not preferred by the user).
- c_{meal} takes into account if the POI offers the kind of food preferred by the user.
- $c_{general}$ takes into account the general importance of the POI, like *La Sagrada Familia* in Barcelona.

For a deeper explanation of the cost function, see (Descamps-Vila et al. 2012b).

Spatial Functions

As the research discarded to work with RDF data, it is not possible to use GeospatialWeb to perform geographic queries. Therefore, the geographic functions used by spatial reasoning are substituted by geographic functions designed by our own. Thus, we developed spatial operations such as nearby, distance and bounding box measurements as an external library.

As described in Section "Expected System", the expected system also requires a routing algorithm, because once the personalization algorithm has selected the best points of interest to visit, it is necessary to guide the user through one point to another.

We created a routing algorithm from scratch that works efficiently running locally in a smartphone device and it is scalable until distances up to 1 km. The algorithm uses a backtracking function with some heuristics that improve efficiency and reduce memory requirements, in order to work on a smartphone device. For a deeper explanation see (Descamps-Vila et al. 2012b).

Proof of Concept: Itiner@

This section presents the final product developed considering the architecture presented in Section "Tests and Discussion", which is an Android application, called Itiner@.

Final Architecture

The final architecture (see Fig. 9) combines GIS functionality with semantic data stored in a relational database.

We developed a location-aware application that provides an all-in-one system with the following features: 1) stores and manages spatial data locally; 2) displays cartographic data and facilitates interaction with

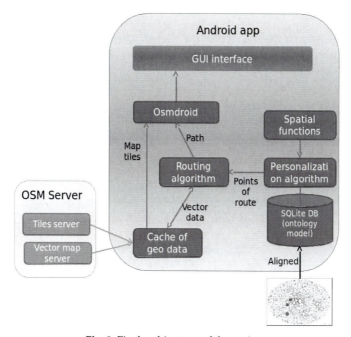

Fig. 9. Final architecture of the project.

Color image of this figure appears in the color plate section at the end of the book.

map data; 3) performs geographic operations; and 4) offers personalized information automatically considering geographic, user and temporal context.

First of all, we integrated the library of spatial functions described in the previous section with the personalization algorithm in order to be able to calculate distances and travel times. This algorithm uses the ontology model, which is stored in the smartphone SDcard as a SQLite database. The algorithm identifies the most attractive POIs for a given user and situation and sends them to the routing algorithm, which uses vector map data stored previously in the smartphone in order to provide a suitable path to go from one single POI to another.

In order to display geographic data, we used *osmdroid* library (OpenStreetMap 2012), which provides tools to work with OpenStreetMap data in an Android application. It allows displaying maps, getting the coordinates when the user touches the screen and drawing lines and images over the map. Hence, we implemented different functionalities to display and interact with geographic data in the device.

Firstly, we developed a cache of maps from scratch in order to store raster and vector map data downloaded from OSM server in order to avoid memory limitations. The reason is that map tiles are displayed as raster images with a high memory consumption. Since smartphone applications have very limited memory by application, every time a zoom or a scroll is done on the smartphone screen, the memory of the cache must be cleared. The created cache of maps for Itiner@ application avoids such problem.

Secondly, we implemented an additional layer over the map view in order to display the points of interest as clickable pushpins and to draw the path between these points as blue lines.

Thirdly, we implemented a tool that allows the user to touch screen of the smartphone to make the device aware of the coordinates of the map. By doing so, the user can inform the application about his/her current location when there is not GPS or 3G connection.

Note that the application connects to OSM Server to download the required tiles and vector data before starting the route. This task should be done when the user is at home or in a place with high bandwidth.

Proof of Concept

To show that the proposed theoretical approximation can be transferred to a real application, we achieved to develop a proof of concept application that could be the seed for a true *intelligent* travel guide able to work without network connection, that allows people to use it anywhere and adapted to every single user. Figure 10 shows some screenshots of the implemented Android application.

Fig. 10. Screenshots of Android pilot application. (a) Map with an example of a route. (b) User preferences. (c) Preferred POIs types. (d) Preferred food types.

Color image of this figure appears in the color plate section at the end of the book.

The first screenshot of Fig. 10 shows an example of a route in *Barcelona* city. The pushpins are the points of interest selected by the personalization algorithm, which are data extracted from *tourism ontology*. Green pushpin is the next point of interest the user will visit, the red flag represents a POI already visited and the grey pushpins represent the other points of the route. The blue line is the path calculated by the routing algorithm.

The other screenshots of Fig. 10 show some of the preferences of the user, which are the same detailed on *personal ontology* (Fig. 5).

Evaluation

To evaluate the proof of concept, 20 tests have been performed taking into account several roles and personal situations:

- Man alone
- Woman alone
- Family with babies
- Family with children
- Group of friends
- Young couple
- Middle age couple
- Elderly couple

We have also included roles with mobility problems and travelling with pets. Several starting times have also been taken into account: early in the morning, during the day and late in the night.

The test have been developed in the 22@ district of Barcelona, near some important touristic points like *La Sagrada Familia*, *Torre Agbar* and

Arc de Triomf, where there are available several bars, restaurants, cinemas, concert halls and theaters. People who conducted the tests already knew the city.

For each test, the prototype has generated a route that fits the preferences of the role. Therefore, the route has been performed and a set of questions have been asked to evaluate the system. In particular, after the test, the questions to answer were:

1. How would you rate the route?
    a. There is some place that you expected to visit and was not suggested?
    b. Did you visit any unexpected place?
2. The route was appropriate in time?
3. Did you stop to eat?
4. Did the route plan the place on the desired geographic area?
5. If so, how adequate was the place to eat?
6. Comments

Questions 1, 2, and 5 are answered with a 1–5 scale and marked with 5 minus 1 for every drawback. These questions are addressed to evaluate, respectively, the route, the time and the suggested restaurant. Questions 1a, 1b, 3 and 4 are YES/NO questions and question 6 is an open question. Figure 11 shows the results obtained.

From the evaluation rates it can be seen that most of the routes (95%) have a rate of 3 or more, but only 30% of routes are rated with 5. From the analysis of the comments we have seen that this is because *La Sagrada Familia*

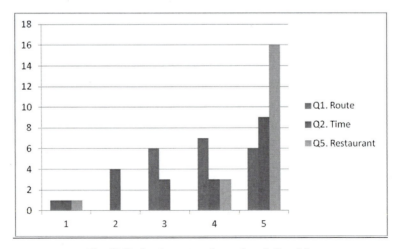

Fig. 11. Evaluation rates of questions 1, 2 and 5.
Color image of this figure appears in the color plate section at the end of the book.

did not appear as a point to visit in most of the routes. The reason is it takes a long time to visit *La Sagrada Familia*, and consequently, it weighs less than the addition of several other monuments with a shorter visit. On the other hand, *La Sagrada Familia* does not accept pets and thus people with pets could not visit it.

Regarding time, 75% of the routes have a rate of 3 or more, but 35% of routes have a rate of 2 or 3. The reasons are: elderly people and families with small children cannot walk so long as is required by the route proposed. Another drawback we found is that cinemas and theaters are not proposed taking into account the time when the shows start, and the system does not take into account the duration of the show. Therefore, some unsatisfactory routes have been generated.

The worst evaluation came from a role that decided to start the visit at 7 o'clock in the morning and, since there was nothing open at that time the system did not proposed any route.

The best evaluation came from the role of group of friends at night, and people alone.

Regarding restaurants, the ones proposed were also correct according to the results of the test.

It is important to note, also, that the system generates different routes for different roles. Therefore, the roles that, for example, "wanted to go shopping" found a route totally different than the roles that "wanted to visit monuments".

From this evaluation we see that the system is able to offer different routes to different people according to their preferences and personal situation. The system is able to create a route that offers restaurants, shopping centers, monuments, etc., taking into account how much time is needed to visit every single place. The provided routes have been obtained in less than 6 seconds, totally offline, in the smartphones where they have been tested. The combination of all these items with the proposed architecture is the main contribution of the application to the actual constellation of CARS.

As has been seen in the test, to increase the satisfaction of the user it is important to have the appropriate information in the database and different visiting times according to each role (child and a young couples do not take the same time to see the zoo for example). In addition, it is important to take into account that some monuments maybe need a shorter time to view them, than to visit them (we could go to view *La Sagrada Familia* and taking some pictures of the facade, but going inside requires more time); and also that some points of interest have a timetable (theaters or cinemas). Finally, walk velocity must be adapted to every kind of user, since elderly persons do not walk so fast as young persons.

The experiment also shows that a reasonable explanation to the user is necessary when a route is not provided. Such explanation should explain why the system has not been able to provide any route. Accordingly, the user can change some parameters to get a suitable route by changing the type of restaurant, the start time or performing other minor adjustments.

Comparison with Existent Systems

After the results of the evaluation, we showed that the proposed CARS offers personalized information to users considering their preferences and the context of the visit, while running off-line in the smartphone.

This represents an advance in research in the tourism and GIS area because currently there are systems that are not able to provide complete routes with personalized data without requiring server connection.

As explained in Section "GIS in Smartphone", there are location-aware systems like OsmAnd or Google Maps that provide offline functionalities. However, their off-line features are limited to: 1) provide simple routing from one POI to another, and 2) provide lists of POIs ordered by type or rating. Then, the applications display these data over a map. Therefore, these applications do not offer a complete experience for the user, because the filtering and the posterior decision of what places to visit must be done by the user. Table 3 presents the comparison of off-line functionalities of current context-aware systems.

Table 3. Comparison of off-line functionalities of current context-aware systems.

	Application		
	OsmAnd	**Google Maps**	**Itiner@**
Base Map	Raster maps	Raster maps	Raster maps
POI	List of POIs	List of POIs	POIs ordered to create a route
Data organization	Data are not organized with ontologies	Data are not organized with ontologies	Data are organized with ontologies
User preferences	Takes into account only one selection from the user	Takes into account one selection from the user	Takes into account multiple user preferences
Timetables	Does not take into account time	Does not take into account time	Takes into account the time for recommendations
Routing	Routing off-line	Only re-routing off-line	Routing off-line
Extra information	Text information from Wikipedia	Text and photos	Shows multimedia content

In conclusion, nowadays there are location-aware applications that are able to work off-line, however, there are not CARS such as Itiner@ system. These systems may be seen as digital maps or travel guides with a huge quantity of information that, in some cases, can overwhelm users.

On the other hand, the proposed system will behave more like a travel assistant that supports the users in the search of the points of interest that are suitable for them and can be visited in the time window available for users.

The main novelty of Itiner@ is that it becomes a complete travel assistant that can be carried over into the smartphone, which adapts to the user in time and place where multimedia content can be viewed. In addition, this guide offers data appropriate for the user needs.

Conclusions

The research presented in this chapter shows how to build a location-based/ tourism-oriented application for smartphone to be able to work off-line. It incorporates a Context-Aware Recommender System (CARS), able to offer personalized information for every single user regarding his/her preferences, time availability and companions.

This goal has been achieved by combining Semantic Web technologies to filter the information sent to the user with a GIS to manage geographic information. Thus, we have created a geospatial semantic system that manages spatial and non-spatial RDF-data and runs off-line in a smartphone. The system has been created using Androjena RDF framework with TDBoid and ARQoid components, together with some predicates of GeospatialWeb library ported to Android.

To show the feasibility of the proposal, a system has been implemented in an Android application. We have seen that such a system is not efficient and scalable enough to be used in end-user applications, because of the slowness of geographic SPARQL queries.

As an alternative system, we proposed to use a relational database instead of RDF ontology. Such change implies using SQL to query the data instead of SPARQL, improving the query time response greatly. The main drawback of such architecture is that it is not possible to use inference engines, and therefore, some inferences cannot be done.

However, with a good translation of the ontology to a relational model, most of the constraints and derivation rules can be maintained, minimizing greatly the loss of semantics. In addition, the efficiency improves greatly and allows satisfying all requirements proposed in this research. As an alternative to filter data, we developed a personalization algorithm that offers different routes to different visitors, taking into account time, location and preferences.

In this chapter we also presented the ontology created to represent the necessary information to support the touristic application. The ontology is aligned with LinkedGeoData, TimeOntology and WGS_84 ontologies.

The contribution of this chapter is to show the architecture to integrating a personalization system and a GIS in a smartphone application that runs entirely off-line. These features are still missing in most of the Android location-based applications. The research presented demonstrates that: a) it is possible to work with geographic data in a smartphone even without spatial databases being adapted to Android; b) it is possible to perform routing functionalities off-line in a smartphone. Additionally it is showed that, so far, it is not efficient to use Semantic Web technologies in an Android device. It is important to note that the system looks for a total tourist experience, taking into account restaurants, monuments, cinemas, theaters, shopping centers, etc. as well as the opening times, timetables and visiting times.

The chapter presented a proof of concept that shows the application works properly when data about POI is known, correct and complete. The proof of concept also shows that the visit time of a single POI may be variable according to the nature of the POI itself or the kind of user who is traveling, and so should be taken into account in a recommender system like ours.

As further work there are three main goals: to achieve efficient GeoSPARQL queries for bigger datasets, thus, to have a scalable system in order to develop the prototype with GeoSPARQL queries into a true end-user application; to improve the data model of POI to take into account different times to visit them and timetables; and to develop a true usable end-user application.

Acknowledgments

This work has been partially supported by the Ministerio de Industria under project AVANZA2 IST-020110-2009-442 and by the Ministerio de Educación y Ciencia and FEDER under project TIN2008-00444/TIN, Grupo Consolidado.

References

Aduna. 2012. Sesame. Available at: http://www.openrdf.org [Accessed September 19, 2012].

Adomavicius, G. and A. Tuzhilin. 2011. Context Aware Recommender Systems. In: Recommender Systems. Handbook, Springer Science, pp. 217–253.

Battle, R. and D. Kolas. 2011. GeoSPARQL: Enabling a Geospatial Semantic Web. Semantic Web–Interoperability, Usability, Aplicatibility 0(0): 1–17. Available at: http://semantic-web-journal.org/sites/default/files/swj176_1.pdf [Accessed March 11, 2013].

Becker, C. and C. Bizer. 2012. DBpedia Mobile: A Location-Aware Semantic Web Client. pp. 1–8. mes-semantics.com/wp-content/uploads/2012/09/Becker-Bizer-DBpediaMobile-Submission.pdf [Accessed February 14, 2014].

Berners-lee, T., J. Hendler and O. Lassila. 2001. The Semantic Web. Scientific American. 284(5): 34–43.

Bizer, C., T. Heath and T. Berners-Lee. 2009. Linked Data—The Story So Far T Heath, M. Hepp and C. Bizer (eds.). International Journal on Semantic Web and Information Systems. 5(3): 1–22. Available at: http://www.citeulike.org/user/omunoz/article/5008761 [Accessed October 28, 2013].

Bradley, K., R. Rafter and B. Smyth. 2000. Case-based user profiling for content personalisation. Adaptive Hypermedia and Adaptive Web. Available at: http://www.springerlink.com/index/rr7xq1mkqdx7ua8u.pdf [Accessed March 4, 2013].

Bry, F. and M. Marchiori. 2005. Reasoning on the semantic web: beyond ontology languages and reasoners. 2nd European Workshop on the Integration of Knowledge, Semantics and Digital Media Technology (EWIMT 2005). 317–321.

Burke, R. 2002. Hybrid Recommender Systems: Survey and Experiments. User Modeling and User-Adapted Interaction 12, 4, 331-370. DOI=10.1023/A:1021240730564 http://dx.doi.org/10.1023/A:1021240730564 [Accessed June 23, 2013].

Carroll, J.J. et al. 2004. Jena. In Proceedings of the 13th international World Wide Web conference on Alternate track papers & posters - WWW Alt. '04. New York, New York, USA: ACM Press: 74. Available at: http://dl.acm.org/citation.cfm?id=1013367.1013381 [Accessed March 29, 2012].

Clark and Parsia. 2011. Pellet. Available at: http://clarkparsia.com/pellet/.

Crowley, D. and N. Selvadurai. 2009. Foursquare. Available at: https://foursquare.com/.

Della Valle, E., I. Celino and D. Dell'Aglio. 2010. The Experience of Realizing a Semantic Web Urban Computing Application. Transactions in GIS. 14(2): 163–181.

Dentler, K., R. Cornet, A. ten Teije and N. de Keizer. 2011. Comparison of reasoners for large ontologies in the OWL 2 EL profile. Semantic Web. 2: 71–87.

Descamps-Vila, L., J. Conesa and A. Perez-Navarro. 2012a. How Can Semantic Be Introduced in GIS Mobile Applications: Expectations, Theory and Reality. In 2012 Fourth International Conference on Intelligent Networking and Collaborative Systems. IEEE: 477–482.

Descamps-Vila, L., J. Conesa and A. Perez-Navarro. 2012b. Implementation of High Complex Routing Algorithms in Mobile Devices: The Itiner@ Case. In 2012 Fourth International Conference on Intelligent Networking and Collaborative Systems. IEEE: 443–450.

Frank, C., D. Caduff and M. Wuersch. 2004. From GIS to LBS–An Intelligent Mobile GIS. IfGI prints: 1–11. Available at: http://www.geo.uzh.ch/~caduff/publications/Frank_GIDays04.pdf [Accessed February 4, 2013].

Gavalas, D., C. Konstantopoulos, K. Mastakas and G. Pantziou. 2013. Mobile recommender systems in tourism. Journal of Network and Computer Applications, 39: 319–333. Available at: http://dx.doi.org/10.1016/j.jnca.2013.04.006 [Accessed June 6, 2013].

Google. 2012. Google Maps Mobile. Available at: http://www.google.com/mobile/maps/ [Accessed May 9, 2012].

Group, J. 2010. Androjena. Available at: http://code.google.com/p/androjena/ [Accessed May 11, 2012].

Guarino, N. 1998. Formal Ontology and Information Systems. In IOS Press: 3–15.

Hedenus, M. 2008. MobileRDF. Available at: http://www.hedenus.de/rdf/index.html.

Hoekstra, R., R. Winkels and E. Hupkes. 2010. Spatial Planning on the Semantic Web. Transactions in GIS. 14(2): 147–161.

JTS-Topo-suite, 2012. Java Topology Suite. Available at: http://sourceforge.net/projects/jts-topo-suite/(accessed March, 26 2013).

KONA, 2010. GeospatialWeb. Available at: http://code.google.com/p/geospatialweb/.

Lehmann, J. and S. Hellmann. 2009. LinkedGeoData—Adding a Spatial Dimension to the Web of Data. The Semantic Web. 5823: 731–746.

Marble, D.F. and D.J. Peuquet. 1990. Introductory Readings In Geographic Information Systems, Taylor and Francis, London. Available at: http://books.google.com/books?hl=esandlr=&id=_y5Dk7NjEBoC&pgis=1 [Accessed January 17, 2013].

Mashable. 2012. More Smartphone Owners Use Location-Based Products. Available at: http://mashable.com/2012/05/11/location-based-services-study/ [Accessed January 23, 2013].

Max, B. 2009. Context-aware Collaborative Creation of Semantic Points of Interest as Linked Data. Thesis, Computer Science Degree Program, University of Koblenz-Landau.

McKee, F. 2002. The Importance of GIS in Modern Society and the effective partnering of Education and Industry. In 5th AGILE Conference on Geographic Information Science. Palma Mallorca. Available at: http://itcnt05.itc.nl/agile_old/Conference/mallorca2002/proceedings/dia25/Session_5/s5_McKee.pdf [Accessed January 17, 2013].

Middleton, S., N. Shadbolt and D. De Roure. 2004. Ontological user profiling in recommender systems. ACM Transactions of Information Systems. 22: 54–88. Available at: http://dl.acm.org/citation.cfm?id=963773 [Accessed March 4, 2013].

Mínguez, I., D. Berrueta and L. Polo. 2010. CRUZAR: An Application of Semantic Matchmaking to e-Tourism. Cases on Semantic Interoperability for Information Systems Integration: Practices and Applications. 255–271.

Mishra, R.B. and S. Kumar. 2010. Semantic web reasoners and languages. Artificial Intelligence Review. 35(4): 339–368.

Noguera, J.M., M.J. Barranco, R.J. Segura and L. Martínez. 2012. A mobile 3D-GIS hybrid recommender system for tourism. Information Sciences, 215, 37–52. Available at: http://dx.doi.org/10.1016/j.ins.2012.05.010 [Accessed June 20, 2013].

Ono, C., Y. Takishima, Y. Motomura and H. Asoh. 2009. Context aware Preference Model Based on a Study of Difference between Real and Supposed Situation Data. Proceedings of the 17 th International Conference on User Modeling, Adaptation, and Personalization (UMAP, 2009): 102–113.

Ontologyan, H. 2011. TheHarmoNET Ontology. Available at: http://euromuse.harmonet.org/web/guest/23.

OpenStreetMap. 2012. Osmdroid. Available at: http://code.google.com/p/osmdroid/ [Accessed May 9, 2012].

Ordnance Survey, 2013. Research Publications—GeoSemantics. Available at: http://www.ordnancesurvey.co.uk/oswebsite/partnerships/research/publications/semantics.html [Accessed June 24, 2013].

OsmAnd, 2012. OsmAnd. Available at: http://code.google.com/p/osmand/ [Accessed May 9, 2012].

PgRouting, 2012. PgRouting. Available at: http://pgrouting.org/[accessed March, 26 2013]. PostgreSQL, 2012.

PostgreSQL. Available at: http://www.postgresql.org/[accessed March, 26 2013].

Poveda Villalon, M., M.C. Suárez-Figueroa, R. García-Castro and A. Gómez-Pérez. 2010. A Context Ontology for Mobile Environments. In: Workshop on Context, Information and Ontologies - CIAO 2010 Co-located with EKAW 2010, October 11, 2010, Lisbon, Portugal.

Sahara, O. 2013. uSeekM. Available at: https://dev.opensahara.com/projects/useekm.

Schickel-Zuber, V. and B. Faltings. 2006. Inferring User 's Preferences using Ontologies. In proceedings of the 21st national conference on Artificial intelligence. 1413–1418.

ScientiaMobile. 2011. WURFL Ontology. Available at: http://wurfl.sourceforge.net/.

Sesame, O. 2012. Android-sparql. Available at: http://code.google.com/p/android-sparql/ [Accessed November 8, 2012].

Siorpaes, K. 2005. OnTour System Design. Available at: http://e-tourism.deri.at/ont/.

Stojanovic, L., A. Maedche, B. Motik and N. Stojanovic. 2002. User-Driven Ontology Evolution Management. Knowledge Engineering and Knowledge Management: Ontologies and the Semantic Web Lecture Notes in Computer Science. 2473: 285–300.

Tobler, W. 1970. A computer movie simulating urban growth in the Detroit region. Economic geography. 46: 234–240. Available at: http://www.jstor.org/stable/10.2307/143141 [Accessed January 18, 2013].

Tsarkov, D. and I. Horrocks. 2006. FaCT ++ Description Logic Reasoner : System Description. Automated Reasoning Lecture Notes in Computer Science. 4130: 292–297.

University of Oxford, Hermit. Available at: http://www.hermit-reasoner.com/.

Van Aart, C., B. Wielinga and W. van Hage. 2010. Mobile cultural heritage guide: location-aware semantic search. Knowledge Engineering and Management by the Masses. 257–271.

Van Hage, W.R., J. Wielemaker and G. Schreiber. 2010. The Space Package: Tight Integration between Space and Semantics. Transactions in GIS. 14(2): 131–146.

Virrantaus, K., J. Markkula, A. Garmash, V.Y. Terziyan, J. Veijalainen, A. Katasonov and H. Tirri. 2001. Developing GIS-Supported Location-Based Services. Proceedings of the 2nd International Conference on Web Information Systems Engineering (WISE'01), Kyoto, Japan. 66–75.

W3C, 2003. wgs_84 ontology. Available at: http://www.w3.org/2003/01/geo/.

W3C, 2004. Resource Description Framework (RDF): Concepts and Abstract Syntax, W3C. Available at: http://www.w3.org/RDF/.

W3C, 2006. Time Ontology. Available at: http://www.w3.org/TR/owl-time/.

W3C, 2007. OWL Web Ontology Language Reference. Available at: http://www.w3.org/TR/owl-ref/#OWLDocument.

W3C, 2010. FOAF Ontology. Available at: http://www.foaf-project.org/.

W3C, 2012. SPARQL. Available at: http://www.w3.org/TR/rdf-sparql-query/.

W3C working group. 2010. Delivery context Ontology. Available at: http://www.w3.org/TR/dcontology/.

Zander, S. and B. Schandl. 2010. Context-driven RDF Data Replication on Mobile Devices. Semantic Web Journal—Special issue on Real-time and Ubiquotous Social Semantics, IOS Press.

CHAPTER 9

GeoBI Architecture Based on Free Software: Experience and Review

Elzbieta Malinowski

Introduction

Decision making is a process of establishing different alternatives and choosing one that better suits the specific situation at hand, according to users' knowledge and available data. For many years, different systems (with the name of Decision Support Systems, DSSs), were developed to assist users at different levels of management. Currently, these systems form a part of Business Intelligence (BI) solutions (Turban et al. 2010). BI requires current and historical data as well as adequate tools and techniques that allow users to perform different kinds of analyses to formulate hypotheses and to look for their confirmation or rejection in order to solve problems that companies or organizations are facing.

Data warehouses (DWs) play an important role in BI solutions providing storage alternatives that facilitate the expression of complex queries required for analysis purposes. Traditionally, DWs are implemented as relational databases and include integrated data supplied by different internal and external sources. Data from these sources can be integrated using extraction-transformation-loading (ETL) tools. Later, this data can be analyzed by taking into account a variety of tools, such as On-line Analytical Processing (OLAP) tools that many users consider as the first step for exploring DW

Department of Computer Science and Informatics, University of Costa Rica, Ciudad Universitaria "Rodrigo Facio", San Pedro, San José, Costa Rica.
Email: elzbieta.malinowski@ucr.ac.cr

data, since they provide a well-known spreadsheet-like environment extended by dynamic data manipulation and automatic aggregation.

On the other hand, the popularization of spatial data by internet providers, such as Google maps or Spatial Data Infrastructures (SDIs) developed on a national or global level opens different alternatives to include spatial data in various kinds of applications, including the ones used for decision making. However, traditionally, the manipulation and querying of spatial data rely on (often complex) software of Geographic Information Systems (GISs) that requires a geo-knowledge, e.g., knowledge about spatial reference systems, layers, storage formats, operations and/or functions necessary for spatial data analysis (Yeung and Hall 2007). This situation makes difficult the task of promoting decision-making processes based on spatial data since personnel responsible for making decisions, especially at lower administrative levels of municipalities or other administrative entities, do not always have the necessary geo-knowledge to explore data using GISs (Malinowski 2013). Furthermore, the available conventional data provided by many public institutions, e.g., census bureau or research centers, can be incorporated and joined with spatial data, thereby improving their diversity and augmenting the circle of users that can explore it for decision-making initiatives.

Spatial OLAP (SOLAP) is an option to deliver spatial and conventional data to the decision makers providing analysis capabilities without the necessity to dominate the geo-concepts required for manipulating spatial data in GIS software. SOLAP applications are more easily used by non-expert users than pure GIS applications (Rivest et al. 2005), even though in some occasions they have to be adjusted to satisfy user needs (Scotch et al. 2007). With a simple click, SOLAP provides aggregated or detailed data of interest in an analysis environment that includes tables, graphs, and maps (Rivest et al. 2005). In this way, non-experts users can analyze spatial and conventional data according to their needs which can be different from what the geo-specialists require, e.g., geographers, cartographers, and surveyors, among others. Nevertheless, the decisional process should rely on a wide variety of high quality data, otherwise, some aspects may be missing and incorrect decisions may be made that harm company/organization outcomes (Talhofer et al. 2011). ETL processes can be used for this purpose integrating, as well as transforming conventional and spatial data before loading it into spatial DWs (SDWs). Moving data to a SDW makes it available for analysis within existing organizational/enterprise data using different kinds of reporting and analytical tools (Badard et al. 2012).

SDWs and SOLAP are developed based on a relatively long and successful tradition in using conventional DWs and OLAP; however, SOLAP tools do not always meet users' expectations in providing similar functionalities as OLAP tools do. This situation is due to the fact that even though there are many works related to concepts of SDWs and SOLAP (e.g., Bédard et al. 2009; Glorio and Trujillo 2008; Gómez et al. 2009a; Malinowski and Zimányi 2008; Pourabbas 2003; Pedersen and Tryfona 2001; Rivest et al. 2005), there is no consensus among scientists and between research communities and practitioners about the features that these systems should have. Many proposed solutions are still used for research purposes and form part of scientific publications or prototype systems (e.g., Bimonte et al. 2006; Silva et al. 2006; 2010; Escribano et al. 2007). Other solutions convert from research projects to commercial solutions (e.g., Intelli3 2013) or respond to particular needs (e.g., Scotch and Parmanto 2005). On the other hand, different commercial companies provide SOLAP solutions (e.g., SAS 2013) and some of them are identifying their product as SOLAP, even though they are more GIS-like systems (Proulx et al. 2007). The situation worsens when free software solutions are required.

There is a widespread tendency, particularly in public institutions, to rely on free software. Although there are no costs related to the software acquisition, the development of a system based on free software may be a costly endeavor in terms of the required in-house expertise, limited functionalities, maintenance problems, as well as compatibility between different versions. Nevertheless, many public institutions take the associated risk and build their systems with free software. Among the existing possibilities of free software for developing GeoBI, spatially-extended DBMSs, e.g., PostgreSQL/PostGIS or MySQL can be used as a platform for implementing a SDW. Other tools, e.g., GeoKettle (Spatialytics 2013a) or Spatial Data Integrator (Talend 2013), extend the traditional ETL capabilities with spatial functions. Furthermore, implementers can create SOLAP cubes using the GeoMondrian server (Spatialytics 2013b). However, even though all these layers may deliver a required structure for spatial data exploration, a front-end layer that hides system complexity and allows an easy access and manipulation through a web browser is required. This layer is an important component because without it the SOLAP system could not be delivered to non-expert users. Currently, even though some front-end SOLAP software may be freely available, e.g., Location Intelligence (SpagoBI 2013a), the general practice is to make a programming effort to implement it (e.g., Bimonte et al. 2005; Scotch and Parmanto 2005; Silva, Times et al. 2008).

In this chapter, based on the experience in creating a GeoBI solution with a SDW and SOLAP, we present various stages of the development process that use available free software, while referring to different opportunities and challenges we faced during this process. Our goal is not to develop another SOLAP prototype; instead, we rely on existing solutions for building a SDW and SOLAP, delivering a Web-based GeoBI platform that is suitable for users without geo-knowledge. The development of our project allows us to assess the value of the research achievements related SDWs, spatial ETL, and SOLAP with the existing practice.

The first section briefly describes the basic concepts required to understand the remaining sections. The following section introduces our case study, which we use throughout this chapter to present our implementation of a SDW and SOLAP. Then, we synthesize the existing research related to SDWs, spatial ETL, and SOLAP, referring it to challenges found in practice and presenting the possible new research topics. Finally, we present conclusions.

Basic Concepts

A DW is commonly used to store historical data that can be analyzed using OLAP tools. Both systems rely on a multidimensional view of data that at the abstract level is called a **cube** and at the implementation level is often referred to as **star** and **snowflake** schemas. These kinds of schemas are organized into fact tables related to several dimension tables.

A **fact table** (*Sales* in Fig. 1) represents the focus of analysis (e.g., analysis of stores' sales) and typically includes attributes called **measures**. These are usually numeric values (e.g., *Quantity* and *Amount* in Fig. 1) that facilitate a quantitative evaluation of various aspects of interest. **Dimensions** (e.g., *Product, Time,* or *Store* in Fig. 1) are used to see the measures from different perspectives, e.g., according to geographic distribution of stores. Dimensions typically include attributes, some of which may form **hierarchies**. The difference between star and snowflake schemas resides in how hierarchies

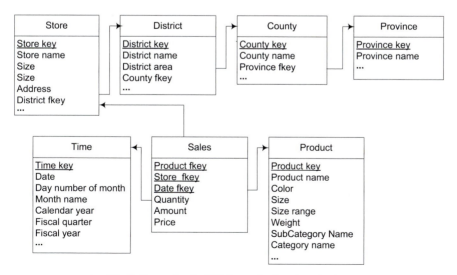

Fig. 1. Example of a DW for analyzing stores' sales.

are represented: the star schema uses a flat table (i.e., all attributes forming a hierarchy are included in the same table, e.g., *Product name, SubCategory Name, Category Name* in the *Product* table in Fig. 1), while the snowflake schema relies on representing a hierarchy as a normalized structure (e.g., tables *Store, District, County,* and *Province* in Fig. 1).

Hierarchies are important in analytical applications since they represent data at different levels of granularity (e.g., district, country, and province) allowing analyses at different levels of detail and exploiting OLAP systems to their full capabilities. When a hierarchy is traversed from finer to coarser levels, measures are automatically aggregated, e.g., moving in a hierarchy from a *Store* to a *District* level will give aggregated sales measures for different districts based on sales of their corresponding stores. This operation is called **roll-up**, while another opposite operation that provides more detailed data from aggregated values is called **drill-down**. The usual practice during the aggregation is to apply the sum operators. However, some measures, called **non-additive**, cannot use this operator, since the aggregation is meaningless, e.g., the *Price* measure in Fig. 1; other measures, called **semi-additive**, e.g., *Quantity* in the *Inventory* DW, can be aggregated in any dimension, but the *Time* dimensions, i.e., it is incorrect to add the quantity of existing products in inventory considering different periods of time.

As a consequence of the growing demand to incorporate spatial data into the decision-making process, SDW and SOLAP models based on multidimensional view of data have emerged (e.g., Bédard et al. 2009; Bimonte et al. 2010; Silva et al. 2010; Damiani and Spaccapietra 2006; Gómez et al. 2008; Jensen et al. 2004; Malinowski and Zimányi 2008; Pourabbas 2003). SDW and SOLAP strongly emphasize on spatial data represented in dimensions and measures, as well as on the necessity to include different spatial functions including the ones required for spatial measure aggregations.

Case study: Analysis of Cancer Incidence and Demographics

Through the years, the Central American Population Center (CCP in Spanish) at the University of Costa Rica has collected data about population, death and birth rates, incidence of different types of cancers, read/write literacy levels, professional occupation, and housing equipment, among others (CCP 2013). This data is provided by various public institutions; for example, the National Institute of Statistics and Census publishes census data from different time periods or several research centers at the University of Costa Rica deliver necessary data to expand social awareness about important issues that allow the country to improve its situation. Since available data uses different formats, CCP stores it in different systems.

Users interested in analyzing this data must access separate systems, which also limits the number of variables that can be requested. Furthermore, each aggregation or different combinations of variables requires another query to be issued.

On the other hand, CCP counts on publicly available spatial data that refers to a wide variety of topics, e.g., administrative distribution, health or economic regions, localization of beaches, bank agencies, schools, clinics, crop distribution, risk zones, national park distribution, among others. This spatial data is represented as a shape file (ESRI 1998) that uses hybrid architecture where spatial and related conventional data is stored separately. An example of a shape file for Costa Rican districts is given in Fig. 2. However, spatial data is not delivered to the CCP users due to the lack of the system that instead of displaying conventional data attached to the shape file allows the integration with other data related to the topics of interest of CCP users.

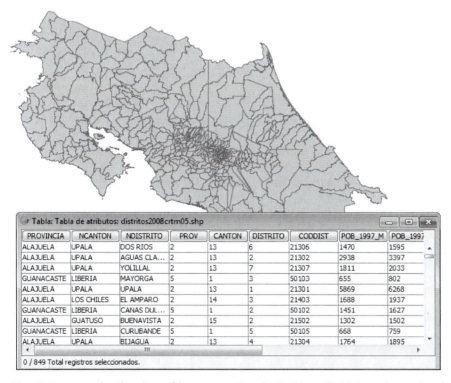

Tabla: Tabla de atributos: distritos2008crtm05.shp

PROVINCIA	NCANTON	NDISTRITO	PROV	CANTON	DISTRITO	CODDIST	POB_1997_M	POB_1997
ALAJUELA	UPALA	DOS RIOS	2	13	6	21306	1470	1595
ALAJUELA	UPALA	AGUAS CLA...	2	13	2	21302	2938	3397
ALAJUELA	UPALA	YOLILLAL	2	13	7	21307	1811	2033
GUANACASTE	LIBERIA	MAYORGA	5	1	3	50103	655	802
ALAJUELA	UPALA	UPALA	2	13	1	21301	5869	6268
ALAJUELA	LOS CHILES	EL AMPARO	2	14	3	21403	1688	1937
GUANACASTE	LIBERIA	CANAS DUL...	5	1	2	50102	1451	1627
ALAJUELA	GUATUSO	BUENAVISTA	2	15	2	21502	1302	1502
GUANACASTE	LIBERIA	CURUBANDE	5	1	5	50105	668	759
ALAJUELA	UPALA	BIJAGUA	2	13	4	21304	1764	1895

0 / 849 Total registros seleccionados.

Fig. 2. An example of a shape file representing Costa Rican districts and associated populations.

Color image of this figure appears in the color plate section at the end of the book.

CCP authorities have made a decision to develop a new system that not only integrates available spatial and conventional data of interest but also allows dynamic data manipulation and presentation in tables, graphs, and maps. Since CCP is a public institution that promotes development in Central American countries, three of the requirements were to ensure that (1) the new system is based on free software, (2) it does not require specialized knowledge to access data, and (3) users may be able to perform all manipulations and analysis through the Web. The necessity to use free software relies on the recently issued decree by the Government of Costa Rica that requires the use of free software, if available, in public institutions. Although using free software requires the expertise of developers to face many technical problems during implementation and maintenance, the CCP authorities decided to take this risk and the related costs of hiring highly-skilled developers. The requirement of relying on a system that does not require geo-knowledge is to encourage non-expert users to consider spatial data as "first-class citizens" in the decision-making processes without the necessity to overcome the (sometimes complex) problems of data integration and improvements in data quality. The last requirement of using the Web arises from considering geographically-spread users.

The available conventional and spatial data cover many years of Costa Rican history, as well as current data that allow decision-making users to better understand different aspects of regional evolution and development, and help make informed decisions towards progress; therefore, adequate structures and tools that facilitate storage of historical data and its analysis are required.

Data included in our case study for presentation in this chapter is currently limited to the period from 1981 to 2005 and refers to a number of cases resulting from different types of cancer, as well as to demographics data, e.g., populations, births, and deaths. Cancers are reported for each district according to cancer types (e.g., breast cancer, stomach cancer, leukemia) and gender, while demographics data does not consider cancer types. On the one hand, the data for cancers was retrieved from the existing database and it does not need any transformation. On the other hand, the demographics data is mainly delivered as spreadsheets in some cases with structures that demanded some programming effort. An example is given in Fig. 3 that shows the data related to population (Fig. 3a), births (Fig. 3b), and deaths (Fig. 3c) in Costa Rica for both male and female. The integration processes were unnecessarily more complex than expected since data, e.g., in Figs. 3a and 3b, is represented by a "nice" view where the first column mixes provinces, counties, and districts using bold, indentation, and empty rows for distinction of different granularity levels. Also, annual changes in population are represented in one file, while the data related to births uses one file for each year. The most complicated representation was given in the

a)

CUADRO 1

Población total por sexo, según provincia, cantón y distrito 2000-2011

Provincia, cantón y distrito	2000			2001		
	Total	Hombres	Mujeres	Total	Hombres	Mujeres
Costa Rica	3 928 966	1 996 415	1 932 551	4 005 587	2 035 124	1 970 463
San José	1 387 143	688 890	698 253	1 413 211	702 130	711 081
San José	319 024	157 106	161 918	322 550	159 092	163 458
Carmen	3 434	1 521	1 913	3 278	1 455	1 823
Merced	13 961	6 865	7 096	13 794	6 795	6 999
Hospital	24 946	13 200	11 746	24 636	13 101	11 535
Catedral	15 747	7 503	8 244	15 307	7 308	7 999
Zapote	21 339	10 229	11 110	21 308	10 220	11 088
San Francisco de Dos Rios	22 357	10 615	11 742	22 483	10 671	11 812
Uruca	27 996	13 994	14 002	30 272	15 176	16 096
Mata Redonda	9 567	4 408	5 159	9 378	4 318	5 060
Pavas	78 555	38 861	39 694	82 219	40 727	41 492

b)

Cuadro 1

Total de nacimientos por sexo, según provincia, cantón y distrito

2001

Provincia, cantón y distrito de residencia de la madre	Nacimientos		
	Total	Hombres	Mujeres
Costa Rica	76 400	39 214	37 186
San José	26 325	13 442	12 883
Cantón San José	6 602	3 310	3 292
Carmen	324	155	169
Merced	262	127	135
Hospital	542	252	290
Catedral	342	180	162
Zapote	347	176	171
San Francisco de Dos Rios	324	160	164
Uruca	968	484	484
Mata Redonda	110	64	46
Pavas	1 605	808	797

c)

Sexo

AREA # 10101 — 10101 Carmen

Categorias	Casos	%	Acumulado %
Masculino	92	57	57
Femenino	69	43	100
Total	161	100	100

AREA # 10102 — 10102 Merced

Categorias	Casos	%	Acumulado %
Masculino	43	52	52
Femenino	40	48	100
Total	83	100	100

AREA # 10103 — 10103 Hospital

Categorias	Casos	%	Acumulado %
Masculino	95	58	58
Femenino	69	42	100
Total	164	100	100

Fig. 3. Examples of spreadsheet data included in the project.

spreadsheet for deaths (Fig. 3c) since it is organized in sections according to each district and female/male distributions.

From the available shape files we have chosen several distributions existing in Costa Rica; however, for the purpose of this chapter we only present two spatial representations related to administrative distribution in district, counties, and provinces, as well as health areas. Due to the lack of mandatory nationwide standards and data governance initiatives in institutions responsible for generating spatial data, available shape files present many inconsistency problems that must be faced and corrected. The integration, cleaning, and transformation processes require the specialized knowledge in geo-informatics and the use of tools that facilitates this endeavor.

Development of a Spatial Data Warehouse and Spatial OLAP

The requirements expressed by users as described in the previous section, guided us to develop a SDW based on a multidimensional model that is able to represent changes in time and facilitates the SOLAP cube creation for dynamic data manipulation and aggregation. The development of a new system consists of several steps. Firstly, we refine requirement specification considering available data and user interests. Based on that, we created a SDW and developed the necessary ETL processes to clean, integrate, transform, and load data. Afterwards, the SOLAP cubes were implemented and a front-end application installed and configured for the manipulation of these cubes.

Creation of a Spatial Data Warehouse

The SDW implementation, similar to DWs, requires several phases: (1) requirements specification, (2) conceptual design, (3) logical design, and (4) physical design (Malinowski and Zimányi 2008).

Requirement specification

For requirements specification different methods exist (e.g., Golfarelli and Rizzi 2009; Guo 2006; Mazón et al. 2007; Malinowski and Zimányi 2008): (1) analysis/goal/user-driven which considers the business and/or user demands, (2) source/data/supply-driven which focus on the existing data provided by source systems, and (3) combination of both. Since our system is currently relatively small and users rely on the incremental development according to the acceptance of the system by outsiders, the CCP representative clearly specified foci of analysis that must be handled first and delivered the data needed for implementation as explained in the previous section.

Conceptual Design

The phase of conceptual design allows the representation of data requirements in a clear and concise manner that can be understood by the users. In practice, very few (S)DWs are represented at the conceptual level; instead, they are directly implemented as star or snowflake schemas mainly using object-relational databases with spatial extensions. This results from the fact that although there are several proposals for the conceptual SDW design (e.g., Silva 2010; Damiani and Spaccapietra 2006; Glorio and Trujillo 2009; Malinowski and Zimányi 2008; Pourabbas 2003), they are not known, except for a small group of researchers, and none of the proposals are well accepted. We represent our conceptual schema using a MultiDim model (Malinowski and Zimányi 2008) that allows multidimensional view of data considering spatial and conventional data types. The resulting schema is shown in Fig. 4.

The schema includes two foci of analysis (called fact relationships in the model) represented by the gray diamonds: *Cancer* and *Demographics*. These foci include measures, e.g., *Births*, *Deaths*, and *Population* in the *Demographics* fact relationship, which allow quantitative evaluation according to different

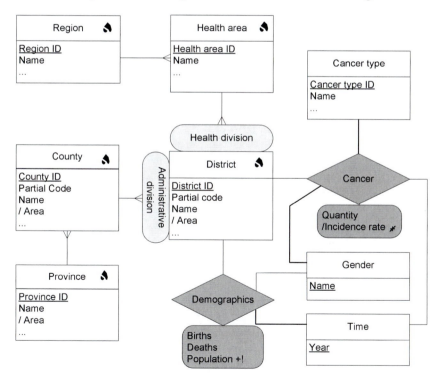

Fig. 4. Conceptual multidimensional schema of a spatial data warehouse.

dimensions. Dimensions are associated with fact relationships(s) and may be of one level, e.g., *Cancer type*, *Gender*, and *Time*, or may have several levels forming hierarchies. For example, the *District* dimension includes two hierarchies (with names placed in ovals): one called *Administrative division* composed by *District*, *County*, and *Province* levels, and another called *Health division* formed by *District*, *Health area*, and *Region* levels. To indicate that some levels are spatial, a pictogram next to the level name is placed, e.g., multi-polygons ◢ in Fig. 4, since many Costa Rican districts are formed by islands. The relationships between hierarchy levels indicate the many-to-one cardinalities, i.e., a district belong to one county and a county may be related to many districts. Other cardinalities can also be expressed, if needed. Notice that some levels include only one attribute (e.g., the *Gender* level). This representation on the conceptual level is important since it allows users to better understand what kind of analysis s/he can perform, e.g., considering population according to gender. As we will see later, during the implementation, this level does not need to be represented by a separate table.

By default, measures are considered additive (e.g., *Births* and *Deaths* in the *Demographics* fact relationship in Fig. 4). To indicate that measure is semi-additive (e.g., *Population*) or non-additive (e.g., *Incidence rate*), symbols +! and ✗ are used, respectively.

The MultiDim model allows sharing dimensions between different fact relationships as can be seen in the Fig. 4 for the *Time*, *Gender*, and *District* dimensions. Notice that the *Cancer* fact relationship in this figure includes an additional *Cancer type* dimension; as a consequence, the level of details (granularity) is smaller than in the *Demographics* fact relationship.

Furthermore, the conceptual level design requires the inclusion of an abstract specification of the required mapping between the source and SDW data. Table 1 includes some examples of these mappings that later on will be implemented during the ETL process. We only include those attributes from the sources that are used in the SDW schema.

Table 1. Some examples of abstract specification for the ETL processes.

Source file: District shape file	SDW dimension: District	Transformation
Source attribute	SDW Attribute	
CODDIST	DistrictID	Not required
DISTRITO	PartialCode	Not required
NDISTRITO	Name	Standardize the format of the names considering capitalization of only the first letter and make the necessary corrections for accent marks.
geom.	Geom	Union of geometries for the same district.

Logical Design

The logical design phase focuses on the transformation of the conceptual multidimensional schema into a logical schema and in the more detailed specification of the ETL processes. This schema is model dependent, but tool independent, i.e., it can be used for implementing in different DBMSs that rely on the same model, e.g., relational model. The mapping rules from the MultiDim model into the relational model consider differently the conventional and spatial levels (Malinowski and Zimányi 2008). For each conventional level, e.g., *Gender*, *Time*, and *Cancer type*, a relation is created that contains all attributes of the level and an additional attribute for a surrogate key. To avoid having relations with only one or two attributes that may affect system performance, we denormalize them, i.e., include the level attribute(s) in the relation representing a fact relationship. This kind of table is commonly known as **degenerate** or **fact dimensions** (Jensen et al. 2010; Luján-Mora et al. 2006; Malinowski and Zimányi 2008).

The mapping of a spatial level requires the addition of an attribute representing geometry. Furthermore, for representing relationships between hierarchy levels traditional mapping rules are applied that consider existing cardinalities. An example given in Table 2 shows a *District* level with an additional attribute representing geometry and two foreign keys for indicating its hierarchical relationship with the *County* and *Health area* levels. This mapping gives a snowflake schema where normalized tables are used for representing each level. However, this representation can be modified for performance reasons, using a star schema.

Finally, a table representing a fact relationship is mapped in the same way as *n*-ary relationships in the ER model, i.e., it includes foreign keys of all participating levels and attributes representing measures. However, not all measures can be stored in the SDW. Some of them, e.g., the *Incidence Rate* in the *Cancer* fact relationship in Fig. 4, that requires special calculations, should be defined during SOLAP cube creation, as we will see in the corresponding section.

Table 2. Logical representation of a District level.

Dimension District	District Key	District ID	Partial Code	Name	Area	Geom	County FK	Health area FK
Restrictions	PK	Unique	Not null, >0 and < 20	Not null	>0	Multipolygon	FK	FK
Data type	Integer	Integer	Integer	Character (20)	Decimal (10,2)	Geometry	Integer	Integer

Physical Design

Before implementing the SDW an appropriate DBMS should be selected. The selection process should consider many features that are also important for the DW implementation, e.g., different kinds of indices, parallel processing, fragmentation techniques, and materialized views management, among others (Jensen et al. 2010; Golfarelli and Rizzi 2009; Malinowski and Zimányi 2008). Furthermore, since our implementation takes spatial data into account, the existence of spatial features in a DBMS is fundamental. Currently, several DBMSs (e.g., Oracle, SQL Server, DB2, PostgreSQL, MySQL) provide spatial extensions with different numbers of spatial features related to data creation, manipulation, and analysis. However, since one of the requirements was to rely on free software, our options for choosing a DBMS were limited to PostgreSQL and MySQL. After thorough evaluation of these systems and comparing their functionalities and performance (Chen and Xie 2008), we have chosen PostgreSQL. It integrates with PostGIS extending the conventional capabilities to spatial features, e.g., spatial data types, spatial functions, spatial indexing, and a spatial query language.

PostGIS provides geography and geometry data types that allow spherical or planar representation of geometries, respectively. Currently, there are fewer functions that apply to geography types than geometry types, and the main difference in managing them is whether the calculations are performed over sphere or plane (PostgreSQL 2013). Since Costa Rica is contained in a small area, using geometry data type allows having more functions and they are also less expensive in the computation.

The spatial data type specified in the conceptual schema can be defined as the attribute with geometry type that includes a specific data type and a spatial reference identifier (SRID) as shown in Fig. 5 for the *District* relation. We use SRID 4326 that corresponds to the WGS84 (World Geodetic System 1984), accepted worldwide and required by law in Costa Rica since 2007. Different constraints can be specified, e.g., to check whether the geometry is valid (*st_isvalid* in Fig. 5). We included the last constraint due to the poor data quality even though it takes longer to insert the data. Currently, our ETL processes verify these aspects; however, in the future the system will

```
create table District
    (DistrictPK integer not null,
    --other conventional attributes
    Geom geometry (MULTIPOLYGON, 4326),
    constraint District_pkey primary key (DistrictPK),
    --other constraints
    constraint enforce_geom_valid check (st_isvalid(geom))
    );
```

Fig. 5. Definition of the *District* dimension in PostGIS.

be exposed to different kinds of users that could directly include spatial data in the SDW.

Implementing ETL Processes

Different options for developing spatial ETL processes that feed a SDW exist either as commercial tools (e.g., Data Interoperability from ESRI, Feature Manipulation Engine (FME) from Safe software, GeoMedia Fusion from Integraph) or free/open sources (e.g., GeoKettle from Spatialytics, Spatial Data Integrator from Talend). Since the requirement is to use free software and we were already familiar with GeoKettle, we chose this tool for our project.

GeoKettle (Spatialytics 2013a) operates under LGPL and is based on a Pentaho Data Integration (PDI), also known as Kettle (Pentaho 2013a), extending it with spatial features. It allows integration of different conventional and spatial data sources and includes connections to different DBMSs (e.g., Oracle spatial, PostgreSQL), GIS files (e.g., ESRI shape file), or geospatial web service (Badard and Dubé 2009). GeoKettle provides a graphical user interface allowing the specification of jobs and transformations. The jobs control the flow, i.e., execution order of the transformations during the execution process. The transformations are defined as a set of operations that need to be applied for the extracted data.

Figure 6 shows an example of the ETL process applied for cleaning a *District* shape file and loading the geometry into our SDW. Firstly, after loading the shape file, it establishes the required spatial reference system (SRS); then, it verifies and eliminates the invalid codes and unnecessary attributes (the components called *Verify district code* and *Select required data* in Fig. 6), and finally, it groups districts with the same name and performs a spatial union of all geometries conforming a district (the component called *Group districts* in Fig. 6). Some invalid geometries are corrected and the corresponding county foreign key is introduced. Transformed data is sent the corresponding table in our SDW.

Different ETL processes for cleaning and transforming shape files and for integrating conventional data as explained in the description of our study case were developed. Having the graphical interface with much functionality already implemented facilitates the development of ETL processes.

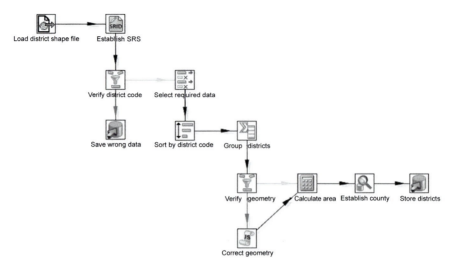

Fig. 6. Transformations applied to the geometries of the *District* shape file.

Color image of this figure appears in the color plate section at the end of the book.

Creation of SOLAP cubes

A typical architecture of DWs includes an OLAP server layer where the cubes are created and an OLAP front-end layer for cubes manipulation and analysis. The same architecture can be used for SDWs counting on spatial extensions for OLAP server and client.

The usual practice in implementing OLAP projects based on free software relies on a well-known Mondrian OLAP server developed by Pentaho (Pentaho 2013b). Mondrian is written in Java and requires an XML file to define a schema consisting of a cube with its dimensions and measures. This cube can be queried later on by applying the Multi-Dimensional Expressions (MDX) language. Schema and queries can be written manually or the Mondrian Workbench interface with visual components can be used instead.

Furthermore, a current stable version of Mondrian provides connections to any JDBC data source and includes an interface specification based on XML/A, i.e., SOAP, as well as olap4j. In addition, Mondrian represents memory-based architecture, i.e., reads data from the disk and copies it into the cache. Although it can put limits on Mondrian's performance, it provides an option of creating aggregated tables in the disk that can be used instead of bringing the base data and performing the calculations in the cache.

Mondrian was extended with spatial features (Spatialytics 2013b) allowing the definition of a vector geometry, as well as enriching MDX with spatial capabilities. Its first official version called GeoMondrian was released in September 2011 under the EPL and provides similar ways as Mondrian for schema creation and querying, i.e., it can be done manually or using the GeoMondrian Workbench tool. Currently, GeoMondrian supports only the PostgreSQL/PostGIS as a data source for geometry values.

Basic features

(Geo)Mondrian defines a cube by indicating, first, an underlying fact table with its measures and references to dimensions; then, for each dimension, it specifies its hierarchies with levels forming them; additional attributes may be associated with levels.

An example of a simple cube called *Demographics* is shown in Fig. 7. It refers to a fact table in the SDW called *DemographicsFactTable*. This cube contains a set of dimensions (Fig. 7 shows only details of the *CRDivision* dimension), each of which includes a foreign key referring to the name of the corresponding column in the fact table. The dimension may have several hierarchies; the first hierarchy does not include a name since it inherits the name from the dimension name. This is why we give this dimension a general name, *CRDivision*, instead of *District*, as was represented in the conceptual schema (see Fig. 4). Other hierarchies must have their names

```
<Schema>
    <Cube name="Demographics">
        <Table name="DemographicsFactTable"/>
        <Dimension name="CRDivision" foreignKey="DistrictFK">
            <Hierarchy hasAll="false" primaryKey=" DistrictID">
                <Table name="District"/>
                <Level name="Province" column="ProvinceName" type="String" />
                    <Property name="ProvinceGeom" column="ProvinceGeom" type="Geometry" />
                    <Property name="ProvinceArea" column="ProvinceArea" type="Numeric" />
                    <! –Other level attributes defined as properties -->
                <Level name="County" column="CountyName" type="String" />
                    <Property name="CountyGeom" column="CountyGeom" type="Geometry" />
                    <Property name="CountyPartialCode" column="CountyPartialCode" type="Integer" />
                    <! --Other level attributes defined as properties -->
                <Level name="District" column="DistrictName" type="String" />
                    <Property name="DistrictGeom" column="DistrictGeom" type="Geometry" />
                    <! --Other level attributes defined as properties -->
            </Hierarchy>
            <Hierarchy name = "HealthDivision" hasAll="false" primaryKey="DistrictID">
                <!-- Definition of levels -->
            </Hierarchy>
        </Dimension>
        <!-- Definitions of other dimensions -->
        <Measure name="Births" column="Births" aggregator="sum" datatype="integer" format string "##,###"/>
        <Measure name="Deaths" column="Deaths" aggregator="sum" datatype="integer" format string "##,###"/>
        <Measure name="Population" column="Population" aggregator="avg" datatype="numeric" format string
            "#,###,###,###.00" />
        <CalculatedMember="Growth" dimension "measure" >
            <Formula> [Measures].[Births] - [Measures].[Deaths] </Formula>
        </CalculatedMember>
    </Cube>
</Schema>
```

Fig. 7. A simple example for defining a GeoMondrian cube.

assigned, e.g., the *HealthDivision* hierarchy. The hierarchy name is built from dimension and hierarchy names, separated by dot, e.g., *CRDivision. HealthDivision*. The additional attributes of a level can be included as a level property indicating its name, the corresponding attribute name in the SDW table, and its type. Using GeoMondrian, the type of attributes is extended by geometry as can be seen in Fig. 7 for defining spatial levels for provinces, counties, and districts. This property can be displayed in a cube browser as shown in Fig. 8 for a part of geometry of the San José province.

Furthermore, two additional measures (i.e., *Births* and *Deaths*) with the *sum* aggregate operator are defined in Fig. 7; the third measure called *Population* is semi-additive, i.e., it cannot be summed in the hierarchies of the *Time* dimension. In our example, this dimension does not include the hierarchy since the data is collected on an annual basis. According to user requirements, if the *Population* for all periods is required, the *avg* function could be applied as specified in Fig. 7. Notice also that, in our example, in order to aggregate the *Time* dimension members according to all periods, in the definition of the hierarchy for the *Time* dimension the value true for the *has All* parameter must be established. The additional measure with label *CalculatedMember* will be explained in the next sub-section.

To simplify the definition, Fig. 7 considers denormalized hierarchies of the *CRDivision*, i.e., having all attributes in one table. If several tables are used representing a snowflake schema, e.g., *District*, *County*, and *Province* tables, before defining levels the corresponding join operations must be specified for each hierarchy as can be seen in Fig. 9 for the administrative division.

Test Query uses Mondrian OLAP

Distritos	geom
−Todas las provincias	
+1 San José	MULTIPOLYGON (((-83.430889 9.369758, -83.431326 9.369958, -83.431498 9.37008 -83.43187 9.370448, -83.432043 9.370566, -83.432161 9.370675, -83.432605 9.3712(-83.432804 9.371752, -83.432766 9.372475, -83.432701 9.372647, -83.432699 9.373: -83.432915 9.374446, -83.433113 9.374575, -83.43316 9.37452, -83.433389 9.37424! -83.433872 9.373889, -83.434045 9.373835, -83.434711 9.373493, -83.434857 9.373: -83.43503 9.373322, -83.435276 9.37316, -83.435449 9.373107, -83.435886 9.37305- -83.436341 9.372983, -83.43676 9.37302, -83.437898 9.373132, -83.438526 9.37323: -83.439226 9.373407, -83.439698 9.373616, -83.440179 9.37416, -83.44027 9.37432: -83.440342 9.374594, -83.44034 9.375263, -83.440394 9.375525, -83.440474 9.3763: -83.440636 9.377063, -83.440789 9.377515, -83.441151 9.378294, -83.441577 9.379(-83.442012 9.379444, -83.442358 9.379671, -83.442539 9.379717, -83.442731 9.379: -83.443331 9.379746, -83.444015 9.379603, -83.444124 9.379613, -83.44508 9.3794! -83.448485 9.379336, -83.449077 9.379337, -83.449677 9.379339, -83.450442 9.379: -83.450733 9.37945, -83.451233 9.379606, -83.452216 9.379843, -83.452797 9.3800: -83.45297 9.380162, -83.453807 9.380625, -83.455889 9.381381, -83.456379 9.3816!

Fig. 8. A part of geometry for defining a San José province in Costa Rica (GeoMondrian).

```
<Schema>
    <Cube name="Demographics">
        <Table name="DemographicsFactTable"/>
        <Dimension name="CRDivision" foreignKey="DistrictFK">
            <Hierarchy hasAll="false" primaryKey="DistrictID" primaryKeyTable = "District>
                <Join leftKey="CantonFK" rightAlias="Canton" rightKey = "CantonPK">
                    <Table name="District"/>
                    <Join leftKey="ProvinceFK"  rightKey = "ProvincePK">
                        <Table name="County"/>
                        <Table name="Province"/>
                    </Join>
                </Join>
                <!—Level declarations -->
            </Hierarchy>
            <!—Other declarations follow -->
```

Fig. 9. Definition of a hierarchy based on normalized tables.

More advanced features

(Geo)Mondrian also includes other more advanced features, e.g., shared dimensions, calculated members, virtual cubes, parent-child hierarchies, and table aggregations. Some of them were required in our project.

Shared dimensions. A cube definition can be simplified by reusing already existing definitions for those dimensions, whether they are conventional or spatial, that are shared among different cubes. In our example, we shared the *CRDivision* dimension as defined in Fig. 7 between the *Demographics* and *Cancer* cubes. As can be seen in Fig. 10, the *CRDivision* dimension with its hierarchies is defined only once in the schema outside of any other cube definitions. Then, each cube that requires the inclusion of this dimension uses the label *DimensionUsage* (instead of the label *Dimension*) indicating as a source the dimension globally defined for this schema.

```
<Schema>
    <Dimension name="CRDivision" >
        <!-- Dimension definition as presented in Fig. 7. -->
    </Dimension>
    <Cube name="Demographics">
        <Table name="DemographicsFactTable"/>
        <DimensionUsage name="CRDivision" source="CRDivision" foreignKey="DistrictFK"/>
        <!-- Definition of other specific for the cube dimensions and measures -->
    </Cube>
    <Cube name="Cancer">
        <Table name="CancerFactTable"/>
        <DimensionUsage name="CRDivision" source="CRDivision" foreignKey="DistrictFK"/>
        <!-- Definition of other specific for the cube dimensions and measures -->
    </Cube>
</Schema>
```

Fig. 10. A shared *District* dimension between the *Demographics* and *Cancer* cubes.

Calculated measures. In many occasions, users require calculations of the measures based on already existing measures in a SDW. Different methods can be applied for that: (1) store a calculated measure as a part of a SDW, (2) create a calculated member as a part of (Geo)Mondrian XML schema, or (3) express a new member in the MDX formula (as we will see later). For example, we want to include a simple calculated measure indicating population growth as: Growth = Birth-Deaths.

We did not apply the first method to avoid re-structuring a SDW every time a user requires a new calculated measure. Instead, we use the second and third (as explained latter) methods. The inclusion of a calculated measure in the XML schema is straightforward; it requires inserting the MDX formula between opening and closing labels of *CalculatedMember*, as shown in Fig. 7, for the *Growth* measure. (Geo)Mondrian automatically creates a special dimension called *Measures*, to which calculated measures belong. However, to the best of our knowledge, even though GeoMondrian has defined different spatial functions, it does not allow using them for calculated members in the schema.

Virtual cubes. The idea behind the virtual cubes is to offer users the possibility to analyze measures from different cubes while hiding the implementation details, e.g., measures represented at different levels of granularities. Furthermore, this kind of cube can help create calculated measures that use in their formula measures belonging to different cubes. To create a virtual cube, all cubes forming it must share some dimension(s). An example is given in Fig. 11 that shows a virtual cube *Cancer_Demographics* created from *Demographic* and *Cancer* cubes; it includes two shared dimensions, i.e., *CRDivision* and *Time*, as well as a private dimension of *Cancer type* from the *Cancer* cube; additionally, the *Population* measure from the *Demographics* cube and *Quantity* from the *Cancer* cube is taken into account. Notice that shared dimensions do not require the specification of a cube name, since both cubes contain them.

```
<VirtualCube name="Cancer_Demographics">
    <VirtualCubeDimension cubeName="Cancer" name="CancerType"/>
    <VirtualCubeDimension name="CRDivision"/>
    <VirtualCubeDimension name="Time"/>
    <VirtualCubeMeasure cubeName="Demographics" name="[Measures].[Population]"/>
    <VirtualCubeMeasure cubeName="Cancer" name="[Measures].[Quantity]"/>
    <CalculatedMember name="IncidenceRate" dimension="Measures">
        <Formula>
            <!—revision for null or cero values for population to avoid execution errors-->
            [Measures].[Quantity]/(Aggregate({[CancerType].[All CancerType]},[Measures].[Population]))*100000
        </Formula>
    </CalculatedMember>
</VirtualCube>
```

Fig. 11. Virtual cube with its shared and private dimensions and a calculated measure.

The virtual cube in Fig. 11 includes a calculated measure called *IncidenceRate*. According to user requirements, this measure should be obtained as a division of the number of cancer cases and the corresponding district population multiplied by 100,000, to obtain the number of cases for every 100,000 inhabitants. Notice that both measures, *Population* from the *Demographics* cube and *Quantity* from the *Cancer* cube are represented at a different granularity level. Therefore, if the formula is expressed as (*Quantity*/*Population*)*100000 and the user requests this measure for a specific type of cancer, e.g., leukemia, and for particular district, e.g., Carmen, (Geo)Mondrian will look for *Quantity* and *Population* according to these two filters: leukemia and Carmen. Since the population for a specific type of cancer and district does not exist, the calculation fails. Considering this situation and in order to ensure adequate calculation, it is necessary to obtain a population figure at the district level, i.e., ignoring the *Cancer type* dimension, as shown in Fig. 11. This is achieved by using the aggregate function and representing the *Population* measure at the level called *All*, where all members of the *Cancer type* dimensions are considered together.

Querying a cube data

To obtain SDW data according to the cube schema expressed by the XML schema, a query must be executed. (Geo)Mondrian uses a MDX for writing these queries. MDX was first introduced by Microsoft in 1997 and then adopted by a wide range of OLAP software providers. Despite being similar to the traditional SQL, MDX is not an SQL extension; it is a special query language with many analytical functions. It not only expresses selections, calculations, and some metadata definitions against an OLAP cube, but also provides some capabilities for specifying how query results should be represented (Smith et al. 2009). MDX for OLAP cubes takes a similar role as SQL for relational databases.

A simple query in MDX uses the *select-from-where* form indicating what should be displayed on rows and columns. For example, the following query accesses the *Cancer* cube (the *from* clause) to retrieve (the *select* clause) (1) the quantity of cancer cases that should be displayed on columns and (2) members of highest level of the *CRDivision* hierarchy, i.e., provinces, that should be displayed on rows. Additionally the restriction for considering only the year 2005 is added (the *where* clause):

```
select    {[Measures].[Quantity]} on columns,
          {[CRDivision].[Province].Members} on rows
from      [Cancer]
where     [Time].[2005]
```

This query can be filtered for a selection of a subset of members satisfying some condition, e.g., to select only those provinces whose number of cancer cases is higher than 500:

```
select    {[Measures].[Quantity]} on columns,
          filter({[CRDivision].[Province].Members}, [Measures].[Quantity] > 500) on rows
from      [Cancer]
where     [Time].[2005]
```

Using GeoMondrian, a filter with spatial functions and topological relationships can be applied as shown in the following examples. In the first case, only provinces with areas larger than 7,000 km^2 are selected; in the second example, only the provinces that do not have any topological relationship (disjoint) with the Alajuela province are chosen. Other topological relationships, such as *ST_Intersects*, *ST_Within*, *ST_Contains* can also be used. Notice that in order to calculate the area size, the *ST_Transform* function is used to convert geometries expressed in the SRS using longitude/latitude (WGS with the number of 43264) to the format that allows metric calculations (UTM zone 17N with the number of 32617). Furthermore, the execution time for these queries can become high if large data set is used and the limited amount of memory is available since GeoMondrian, similar to Mondrian, represents memory-based architecture, i.e., reads data from the disk and copies it into the cache. Although Mondrian provides an option of creating aggregated tables in the disk to speed-up queries, this feature is still not available for spatial data.

```
select    {[Measures].[Quantity]} on columns,
          filter({[CRDivision].[Province].Members}}, ST_Area(ST_ Transform
          ([CRDivision].[Province].CurrentMember.Properties("geom"),4326,
          32617))/1E6 > 7000) on rows
from      [Cancer]
where     [Time].[2005]

select    {[Measures].[Quantity]} on columns,
          filter({[CRDivision].[Province].Members},ST_Disjoint([CRDivision].[Province].
          CurrentMember.Properties("geom"),
          [CRDivision].[Province].[Alajuela].Properties("geom") )) on rows
from      [Cancer]
where     [Time].[2005]
```

As was stated in the previous section, MDX can also be used to create calculated members on the fly. For example, to create a calculated measure *Growth* for the *Demographic* cube and display it together with the *Population* measure, the following MDX may be used:

```
with        member [Measures].[Growth] as '[Measures].[Births] - [Measures].[Deaths]'
select      {[Measures].[Growth], [Measures].[Population]}    on columns,
            {[CRDivision].[Province].Members}   on rows
from        [Demographics]
where       [Time].[2005]
```

Furthermore, calculated members can use the spatial functions, such as area or distance, as shown in the following specification for calculating the distance of each province from the Alajuela province:

```
with        member [Measures].[Distance] as
            'ST_Distance(ST_Transform([CRDivision].[Province].CurrentMember.Properties
            ("geom"),4326,32617),
            ST_Transform([CRDivision].[Province].[Alajuela].Properties("geom"),4326,
            32617))'
select      {[Measures].[Population],[Measures].[Distance] } on columns,
            {[CRDivision].[Province].Members} on rows
from        [Demographics]
where       [Time].[2005]
```

Additionally, using calculated members coordinates can also be retrieved:

```
with        member [Measures].[ProvinceCoordinates] as '([CRDivision].[Province].
            CurrentMember.Properties("geom"))'
select      {[Measures].[Population], [Measures].[ ProvinceCoordinates]} on columns,
            {[CRDivision].[Province].Members} on rows
from        [Demographics]
where       [Time].[2005]
```

The solution of using calculated measures "on the fly" has a drawback, since the calculation must be included in every query where this measure is required. Instead, having a calculated measure as a part of the XML schema allows users to access it as other measures of the cube.

SOLAP cube visualization

MDX is not a full report-formatting language and requires some APIs to present results and to hide its complexity from non-expert users. Currently, JPivot is provided together with Mondrian serving as an OLAP client. Other APIs can also be used, e.g., STPivot, JPalo, JasperSoft, or Saiku. These kinds of client tools provide a graphical interface for selecting data based on hierarchies and measures of interest and include different OLAP functionalities, e.g., drill-down, roll-up, pivot, delivering the data in a table and graph representations. An example of the tabular and graphic displays for the MDX query related to the *Cancer* cube is shown in Fig. 12.[1] Users familiar with MDX may use the editor to modify the query for displaying data (the low part Fig. 12) or choose components (the left part of Fig. 12) that automatically create a new MDX query and present results on the screen. In this way non-expert users can perform different operations without being aware of the implementation details.

However, even though there is a variety of free software for conventional OLAP data visualization and manipulation, the situation is different when the spatial component is added. When our project initiated, the available free solutions allow the typical OLAP operations, e.g., drill-down, roll-up, pivot, to be performed over conventional data, lacking synchronization with and manipulation over maps. Looking for some solutions to start with, the software called GeoOLAP from Camptocamp (Camptocamp 2012) was available to provide some basic SOLAP operations in managing spatial dimensions and conventional measures and offered three environments for displaying results: tables, graphs, and maps.

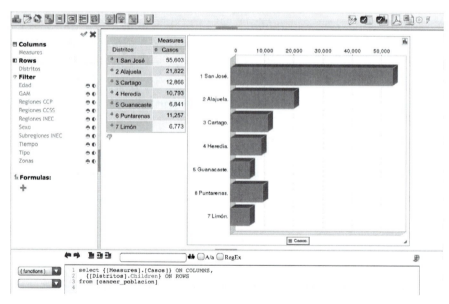

Fig. 12. OLAP client (STPivot from StrateBI) representing results and MDX query.

Color image of this figure appears in the color plate section at the end of the book.

---

[1] The names in the screenshot do not correspond to cube definition in this chapter, since we translate them from Spanish in order to make XML and MDX codes more understandable.

An example is given in Fig. 13 showing the distribution of population in 2005 according to gender. Maps allow two types of displays: simultaneous (as shown in Fig. 13) or separate: (1) the traditional coloring method indicating different ranges for population, independently for female and male distribution and/or (2) the bars (or pies) for each province indicating the female/male distributions that are placed over each province. Furthermore, in the lower part of the display, the table with aggregated data is also included and optionally two different graphs for comparing the female and male distributions among seven Costa Rican provinces (other displays are also available) can be requested.

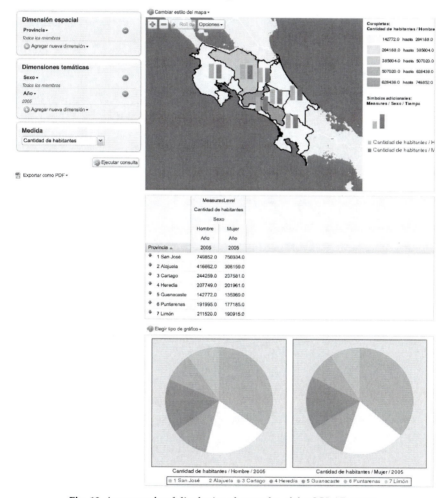

Fig. 13. An example of displaying the results of the SOLAP query.

Color image of this figure appears in the color plate section at the end of the book.

Furthermore, the operation drill-down and roll-up can be performed over a table or over the map, synchronizing results over all three environments, i.e., table, graphs, and map. In a similar way, the slice-and-dice operation can be applied to constraint the data analysis to a subset of provinces. An example in Fig. 14 results from applying the drill-down operation over the Alajuela province.

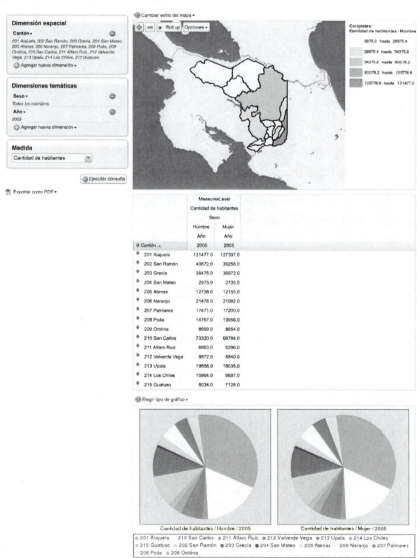

Fig. 14. Applying the drill-down operation over the province of Alajuela.

Color image of this figure appears in the color plate section at the end of the book.

An interesting feature provided by GeoOLAP is the possibility of combining two spatial hierarchies, as shown in Fig. 15. In this example, to make the display simpler we chose only the total population of the Alajuela province and combined two spatial hierarchies: administrative and health divisions. Since this province belongs to four different Health regions, the map in the Fig. 15 includes this division.

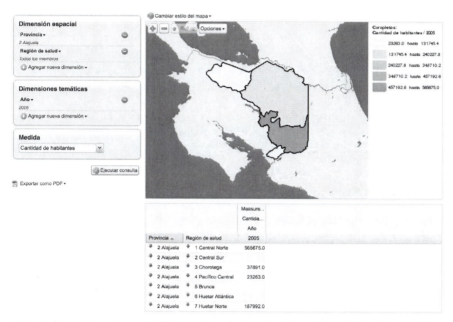

Fig. 15. Combining two spatial hierarchies of administrative and health divisions for the Alajuela province.

Color image of this figure appears in the color plate section at the end of the book.

Challenges in Research and Practice

Research related to spatial data warehouses (SDWs) and Spatial OLAP (SOLAP) is not new. Research works started emerging at the end of the 1990s (Han et al. 1998), and their number increased in the following years. Different aspects are considered, among others:

- Spatial dimensions and measures (e.g., Baltzer 2011; Fidalgo et al. 2004; Gómez et al. 2009b; Han et al. 1998; Malinowski and Zimányi 2008; Pourabbas 2003; Rivest et al. 2001; Silva, Times et al. 2008).
- Different types of spatial hierarchies (e.g., Baltzer 2011; Jensen et al. 2004; Malinowski and Zimányi 2008).

- Design process and methods (e.g., Fidalgo et al. 2004; Glorio and Trujillo 2008; Luján-Mora et al. 2006; Malinowski and Zimányi 2008).
- Conceptual models (e.g., Bimonte et al. 2010; Aguila et al. 2011; Silva et al. 2010; Damiani and Spaccapietra 2006; Jensen et al. 2004; Malinowski and Zimányi 2008; Pourabbas and Rafanelli 2002).
- Spatial multidimensional querying (e.g., Baltzer 2011; Bimonte et al. 2010; Camossi et al. 2008; Escribano et al. 2007; Ferri et al. 2000; Gómez et al. 2008; Pourabbas and Rafanelli 2002; Silva 2010).
- Optimization of spatial queries and aggregations (e.g., Baltzer 2011; Costa et al. 2010; Silva, Times et al. 2008; Glorio and Trujillo 2009; Gómez et al. 2009a,b; Han et al. 1998; Lopes et al. 2012; Papadias et al. 2001; Pedersen and Tryfona 2001; Silva, Manhães and Gitahy 2008; Shekhar and Chawla 2003).
- Integration between GISs and data warehouse or OLAP environments (e.g., Escribano et al. 2007; Ferri et al. 2000; Kouba et al. 2000; Miksovsky and Kouba 2001; Pourabbas 2003).

Relatively recent surveys about research related to SDWs and SOLAP can be found in several works, e.g., Gómez et al. (2009b), Viswanathan and Schneider (2011), Bimonte et al. (2010). Many of the proposed concepts are implemented in prototypes to ensure the feasibility of the suggested solutions; however, they rarely reach the world of practitioners.

On the other hand, the extension of OLAP systems to include spatial data is also a concern for software companies. Many commercial companies involved in BI and companies dedicated to spatial data manipulation focus on creating partnerships to join their solutions in order to enhance multidimensional reporting and analysis with interaction through a map-based interface, e.g., ESRI established several partnerships with SAP (ESRI 2013a), IBM Cognos (ESRI 2013b), or SAS (SAS 2013). On the other hand, different outlooks exist in relation to spatial solutions for OLAP relying on free software. There is a large number of free software related to GIS (OSGIS 2013). It can be used with different purposes, among others: ready-to-go applications, e.g., Kosmo, GeoKettle, and QuantumGIS, extensions to existing software, e.g., PostGIS, libraries, frameworks, APIs in C++ or Java that can be used in developing other applications, e.g., JTS, JUMP, CGAL, GEOS, GeOxygeme, or servers for enabling spatial data over the internet, e.g., MapServer, GeoServer. However, although some of this software is used in developing SOLAP prototypes, as far as we know, there is a very small number of publically available SOLAP software that can be used in a "ready-to-go" manner.

To facilitate the analysis of research achievements and current solutions for implementation of a SDW and SOLAP based on free software, we group our observations according to different areas, as explained in the previous sections, even though these areas clearly overlap.

Design and Implementation of SDWs

SDWs relay on the concept of spatial dimensions and measures. Three types of spatial dimensions based on the spatial references of the hierarchy members were proposed (Stefanovic et al. 2000; Rivest et al. 2001): non-spatial (the usual conventional hierarchy), spatial-to-non-spatial (a level has a spatial representation that rolls up to a non-spatial representation), and fully spatial (all hierarchy levels are spatial). Furthermore, spatial measures were classified as the collection of pointers to spatial objects (Stefanovic et al. 2000; Rivest et al. 2001), as spatial objects themselves (i.e., their geometries), or calculated using spatial operators, e.g., area or length (Rivest et al. 2001). Many authors rely on these definitions (e.g., Baltzer 2011; Glorio and Trujillo 2008; Gómez et al. 2009a; Malinowski and Zimányi 2008; Times et al. 2008) or propose some modifications. For example, Fidalgo et al. (2004) introduced a GeoDWFrame only with spatial dimensions that were created, in a rather complex manner, from spatial measures. Later on, this model was extended to include spatial measures (Silva, Times et al. 2008). Bimonte et al. (2005; 2006) and Damiani and Spaccapietra (2006) looked for symmetry in defining spatial dimensions and measures, allowing the latter to include spatial measures represented at different levels of granularity, i.e., forming hierarchies. Shekhar and Chawla (2003), on basis of the classification used for non-spatial data, presented a classification and examples of various types of spatial measures. Jensen et al. (2004) extended their model by allowing partial containment relationships for spatial members of lower hierarchy level in relation to its parent level, while Malinowski and Zimányi (2008) referred to different types of spatial hierarchies.

To the best of our knowledge, very few conceptual multidimensional models with spatial support have been proposed (e.g., Aguila et al. 2011; Bimonte et al. 2005; Damiani and Spaccapietra 2006; Glorio and Trujillo 2008; Jensen et al. 2004; Malinowski and Zimányi 2008; Pourabbas and Rafanelli 2002; Times et al. 2008). These models, especially those that included a graphical interface, could help implementers to better represent user requirements and implement underlying structures more adequately, since they could indicate which levels or hierarchies are shared, evidence the aspect of measure additivity, or distinguish different kinds of spatial hierarchies. However, even though some of the proposed solutions may represent the multidimensional elements adequately, a systematic analysis is still missing to determine whether all these elements are necessary in real-life applications or whether some are omitted. In particular, not only typical scenarios of the DWs representing administrative divisions as spatial dimensions (as the ones that we use in this chapter), but also other kinds of scenarios should be considered. For example, is the multidimensional model adequate for implementing SDWs and SOLAP that consider spatial data related to zones of flooding, fire, or earthquake risks, localization of fire stations, clinics, schools, crop, and vegetation distributions, among others? Could the development of SOLAP solutions with these kinds of data extended with other conventional data help the users at the different levels of administration to improve the planning of local development or risk prevention? Could professional from other fields, e.g., agronomy, biology, medicine, hydrology, benefit from SOLAP solutions for exploring spatial and conventional data in a multidimensional manner before other kinds of analysis are applied?

Furthermore, there are other aspects that make the acceptance of conceptual modeling difficult, before the SDW and SOLAP implementations take place: (1) lack of tradition in GIS and DW communities to use conceptual models, relying instead on either shape files (for GIS) or star or snowflake logical schemas (for DWs), (2) a variety of research proposals with their own specification and presentation, (3) lack of a well-known and accepted model for representing DW or SDW structures, and (4) the proposed models seldom include rules that are required for transformations from conceptual to logical or physical levels in order to implement a SDW.

Regarding implementation, there is a clear tendency of converging GIS and DBMS technologies by having a spatial extension in DBMSs and promoting geo-databases (that in fact are spatial DBMSs) in the GIS community (ESRI 2012). As indicated for conventional DW (Costa et al. 2006; Golfarelli and Rizzi 2009; Jensen et al. 2010), through an adequate physical design (e.g., aggregated tables, indexes) query performance may be improved. This aspect should also be investigated in the context of SDWs. Furthermore, the implementation process should consider whether normalized (snowflake) or denormalized (star) tables are a better choice. For conventional data, there are several recommendations (e.g., Jensen et al. 2010; Martyn 2004) and, fortunately, Costa et al. (2010), based on several experiments with spatial dimensions represented as star and snowflake schemas, also give advice that may be useful before choosing either option. In addition, star and snowflake schemas for conventional DW were extended (Jensen et al. 2010), thus enabling the implementation of other features, e.g., many-to-many relationships between facts and dimensions, degenerate dimensions, and different kinds of hierarchies, e.g., parent-child, non-covering, non-strict. Whether these features are required, useful, and can be implemented for SDWs, it is still an open research topic. Nevertheless, based on our experience, there are still several aspects that may influence negatively the possibility to consider spatial DBMSs in the SDW implementation:

- Spatial DBMSs are relatively a new technology, thus, they are not well known outside the small circle of implementers. There are still not enough computer science specialists willing to accept the challenge and learn about spatial data, its manipulation, and problems.

- Most of the projects related to spatial data require the knowledge of geo-specialists and they are usually more familiar with GIS products with their traditional data storage than spatial DBMSs.
- The common supposition that spatial DBMSs are "heavy", i.e., map retrieval is slower, prevails among geo-specialists and to our knowledge there are no publications that contradict this assumption.
- Web map services that do not require DBMSs are currently very popular.
- There is no evaluation (i.e., benchmarking) for spatial DBMSs or a guide for choosing one according to specific needs, e.g., economic factor, performance, storage, ease of use, compliance with standards, among others.

Spatial ETL

The practical importance of the ETL processes is high, since this time-consuming process is responsible for data integration and customization. Several works refer to conceptual modeling of ETL processes for conventional data (Akkaoui et al. 2011; Albrecht and Nauman 2008; Silva et al. 2012; Trujillo and Luján-Mora 2003; Vassiliadis et al. 2005; 2009). These proposals, sometimes complex in nature, intend to provide formal foundations and vendor-independent models for the design of ETL processes. To the best of our knowledge, there is not a formal proposal for spatial ETL processes. Some works refer to specific problems, e.g., integrating spatial data related to water quality (Wang et al. 2010) or provide data for health resource and service administration (Cohen and Baitty 2005). As a consequence, spatial ETL processes are usually implemented using specific tools according to the problem at hand.

ETL processes not only integrate data and transform it according to user analysis requirements, but they also are important in improving spatial data quality. If data collectors are not aware of data quality and systems implementers are not familiar with the methods to improve quality, the last element of the chain, i.e., data customers, obtain incorrect or deficient data that can be harmful for the company/organization, whether this data is operational or decisional (Talhofer et al. 2011). However, traditionally the digitalization and analysis of spatial data relied on geo-specialists who were aware that GISs would not help them improve spatial data quality since these systems work under the assumption that data is perfect and do not provide the capabilities to establish data quality control (Delavar and Devillers 2010). In contrast, currently, different users with a profile that is usually unrelated to GISs are in charge of either collecting data from existing repositories or creating it from aerial photographs and satellite images (Mäkelä 2006). As a consequence, many users and producers of spatial data are not trained in geo-field and may not be aware of the existence and number of problems that this kind of data may have (Delavar and Devillers 2010; Boin and Hunter 2005; Wang and Strong 1996).

On the other hand, even though the research related to spatial data quality has a 30-year long tradition, it introduces concepts and techniques that are complex and difficult to understand by typical users and there is lack of approaches for solving real problems (Devillers et al. 2010). Furthermore, spatial data quality can also be seen as a relative concept depending on user expectations (Olsen 2003; ISO19113 2002), e.g., a detailed representation of different spatial objects may be required for analyzing hazard risks, but may not be crucial to represent population for different districts for decision-making users. Several works refer to spatial data quality in SDWs and SOLAP proposing different mechanisms (1) to improve the prevention of inappropriate uses of spatial data (Gervais et al. 2010), (2) to facilitate the reuse of spatial data in a distributed and heterogeneous environment (Sboui et al. 2010), (3) to establish an iterative process during the phases of spatial cube development for identifying possible risks in using spatial data by decision-making users (Levesque et al. 2007), or (4) to include tolerant integrity constraints for merging spatial objects with different representations (Bejaoui et al. 2009).

Different international standards and specifications (e.g., ISO19113.2002, ISO19114.2003, ISO19138.2006, OGC 2011) were established to tackle the problem of spatial data quality. These standards can serve as a basis for the development of national standards. The standardization must not only refer to spatial object representations, but also used measures, town names, and zip codes, among others (Badard et al. 2012). However, this task demands the leadership and coordination of various government authorities with long-term financial support, which is not always readily available.

The shortage of national standards, gazetteers, domain-specific dictionaries, and metadata that help data providers and consumers to count on better spatial data quality, decreases users' trust in using data for informed decisions. Changing this situation is not an easy task and, meanwhile, different techniques and tools may be applied, such as spatial ETL tools. However, although our implementation of the ETL processes, based on a graphical interface provided by GeoKettle, decreases the programming effort, the cleaning process in some occasions requires the Java code or complex SQL statements. In addition, each implementer must rely on her/his own knowledge about spatial data cleaning methods, even though some patterns may be common for different applications. Therefore, there is a need to consider already existing research related to ETL processes applied for conventional data and extend them with spatial features, taking into consideration real-world applications and analyzing whether there are some common steps for cleaning patterns. This could help to expand ETL tools by including a (complex) component in charge of (at least some) cleaning processes, e.g., spatial aggregation of geometries for objects with the same identifier, and validation of geometries, among others. Furthermore, when using the same shape file for different DBMSs, we found that they differ in criteria to check the validity of a geometry. There are already solutions for transforming invalid geometry into a valid one that depends on chosen DBMSs and they could also be included as a component of the spatial ETL tool. By doing so, users are not forced to have knowledge about buffering, inverting the specification order for coordinates, or self-intersection, among other aspects.

Spatial OLAP server layer

Multidimensional modeling concepts applied to spatial data have been used in various spatial OLAP proposals and solutions. Different publications have considered the integration between GISs and DW or OLAP environments. Pourabbas (2003) and Ferri et al. (2000) referred to common key elements of spatial and multidimensional databases: time and space. Based on a formal model they achieved the integration by applying mapping between the hierarchical structures of the OLAP and GIS environments. The concept of this kind of mapping was also exploited by other authors (Escribano et al. 2007; Kouba et al. 2000). Different implementations with server and client layers, e.g., GeoWOlap (Bimonte et al. 2006), Piet (Escribano et al. 2007), Map4Decision (Intelli3 2013), SAS Web OLAP Viewer (SAS 2013), SOVAT (Scotch and Parmanto 2005), GOAL (Kouba et al. 2000), GeoMDCube (Silva, Manhães and Gitahy 2008), include a user interface that hides the fact that two separate systems compose the SOLAP architecture: OLAP system (e.g., Microsoft SQL Server Analysis Services, Mondrian, GOLAPE, or SAS Enterprise BI Server) with GIS solutions (e.g., ESRI ArcGIS or some customized applications that facilitate map display based on Java MapXtreme or others). The above mentioned SOLAP systems are not always available to a wide spectrum of users since if they are commercial, e.g., SAS Web OLAP Viewer (SAS 2013) or Map4Decision (Intelli3 2013), they need some investment that may be difficult to justify for inexperienced users with little or no knowledge about OLAP. On the other hand, some of the solutions are prototypes that (1) are not publicly available (to the best of our knowledge), e.g., GeoCube (Bimonte et al. 2005), GeoWOlap (Bimonte et al. 2006), Piet (Escribano et al. 2007; Gómez et al. 2011), GOLAPE (Silva et al. 2006), GeoOlap (Soares et al. 2007), (2) are customized (no general-purpose solutions), e.g., SOVAT developed for public health decision support systems (Scotch and Parmanto 2005), or (3) may be discontinued, e.g., SOLAPLayer (Spatialytics 2013c).

The prototypes and customized solutions may confirm the veracity of the concepts and the feasibility to implement the proposed solutions; however, since they are not publicly available for use, they cannot be considered as an option for implementing the SOLAP system. Furthermore, another aspect undertaken by several authors (e.g., Badard et al. 2012; Fidalgo et al. 2003; 2010; Dubé et al. 2009) is the need to have a SOLAP server separate from the client application that allows cube expression in XML format. Additionally, Dubé et al. (2009) refer to different SOLAP characteristics: (1) representation of data and metadata, (2) usage of existing standards, (3) independence of particular implementation in OLAP and GIS products, (4) incorporation of calculated measures and members expressed in the schema and within a query, and (5) separation of contents and presentation. Therefore, instead of integrating two systems in the user's interface, another approach can be taken by providing the management of spatial and conventional data at the lower, SOLAP server layer, along with the help of a spatial database layer (Baltzer 2011). Some examples are GeoMondrian (Spatialytics 2013b) or GOLAPE (Silva et al. 2006) servers that directly interface with the Mondrian OLAP server and extend its capabilities to handle spatial data. These SOLAP servers, in turn, use a spatially enabled DBMS, e.g., PostGIS, as data storage.

GeoMondrian (Spatialytics 2013b) and GOLAPE (Silva et al. 2006) by extending Mondrian with spatial features provide XML for defining multidimensional structures. However, to the best of our knowledge only GeoMondrian is available for free. It is relatively new and still lacks some features. For example, although it allows users to define geometries for each spatial level in the XML schema, this extension does not apply to the measures, i.e., spatial measures and their aggregation functions cannot be included in the XML schema. The only possibility to create spatial measures is by specifying calculated measures "on the fly" using the *with member* clause, as we saw in the previous examples. GeoMondrian also limits the spatial aggregation function to spatial union missing other functions, e.g., intersection or difference that could be considered during roll-up operations (Malinowski and Zimányi 2008). This situation represents a clear gap between the research community and practitioners. On the one hand, the research community proposes models with spatial measures (e.g., Bimonte et al. 2010; Malinowski and Zimányi 2008; Rivest et al. 2005; Stefanovic et al. 2000) and even a prototype with a sophisticated classification of spatial measures and aggregation operations was developed (Silva, Times et al. 2008). Other authors (e.g., Damiani and Spccapietra 2006) take yet a stronger position indicating that multidimensional spatial models must include a spatial measure and the presence of spatial dimensions is optional. On the other hand, implemented SOLAP systems rely on spatial dimensions and do not allow the inclusion of spatial measures with corresponding spatial aggregation functions (e.g., Spatialytics 2013b; Intelli3 2013; SAS 2013; Scotch and Parmanto 2005).

Furthermore, since Mondrian has very few functions for conventional measure aggregations and analysis, e.g., sum, count, min, max, distinct-count, lacking a more advanced one, e.g., median, standard deviation, GeoMondrian inherits this problem. This situation can be improved using PostgreSQL/PostGIS and defining measures in terms of the SQL expression that is executed in PostgreSQL and the result is sent back to (Geo)Mondrian. These solutions may be complex and require a good (not basic) knowledge of (Geo)Mondrian and PostgreSQL.

The situation aggravates when new programmers are required to implement (S)OLAP cubes using (Geo)Mondrian. Although there is documentation with basic features, many problems require advanced knowledge and it is necessary to look for them among different support communities. Proposed solutions are not always intuitive or easy to understand and depend on the implementers' luck to find a response for the particular question at hand.

Another challenging issue may be the changes between different versions. Currently, GeoMondrian is based on Mondrian 3.5, however, there are new features in the beta version of Mondrian 4.0 including several changes in the schema definition and getting closer to Microsoft SQL Server Analysis Services features. When this new version is accepted by the users, it will put more pressure on GeoMondrian developers to include modifications.

Therefore, there is a need to join the efforts of the research community and apply in practice many concepts that are already known in academic works and prototypes. This could help not only to extend SOLAP server functionalities, but also focus on other still unresolved aspects, e.g., improving performance of SOLAP queries considering pre-aggregations (e.g., Pedersen and Tryfona 2001) as is done transparently for OLAP servers, including multiple representations of spatial objects forming hierarchies, among other issues. Furthermore, the question whether the integrated approach of GIS and OLAP or spatial extensions of conventional OLAP servers is more convenient in terms of performance and complexity should be addressed.

Spatial OLAP front-end layer

Different front-end solutions for SOLAP were developed using free software mainly based on the Mondrian server as we have already mentioned in the previous section. For example, Bimonte et al. (2005) used JPivot to provide JSP pages with OLAP functionality and MapXtreme Java to support map visualization and interaction. On the other hand, Silva, Times et al. (2008) based their SOLAP front tool on the Java Plugin Framework (JPF), OpenJUMP, and JRubik technologies. These SOLAP clients can be also integrated with Google Earth providing 2D (Silva, Manhães and Gitahy 2008) or 3D (Di Martino et al. 2009) view of spatial data. These systems differ in the presentation of multidimensional elements and also in functionalities since the authors' goal is to demonstrate feasibility of implementation of the concepts proposed by them.

There are various OLAP front-end solutions that use Mondrian as an OLAP server providing basic OLAP functionalities, e.g., JPivot, STPivot, JPalo. However, implementers of SOLAP systems based on a GeoMondrian server have a difficult time finding free SOLAP front-end software. The SOLAPLayers (Spatialytics 2013c) tool was first announced to be released together with GeoMondrian; however, it is no longer supported and a new GeoBIExt will be developed to replace it. To the best of our knowledge, this extension is still unavailable. On the other hand, some initiatives are currently emerging with the aim to join their effort in order to develop fully-functional SOLAP software. For example, the SpagoBI initiative (SpagoBI 2010) proposed to create a group of developers to take advantage of existing solutions, e.g., GeoOLAP from Camptocamp or GeoMondrian from Spatialytics. Currently, SpagoBI includes a component called Location Intelligence (SpagoBI 2013a) that combines GIS visualization and spatial analysis with BI features related to conventional data. However, this extension is relatively new and still requires manual configuration using a new document composition internal engine (SpagoBI 2013b) in order to be able to synchronize tables, graphs, and maps.

Another solution is Saiku[2] (Saiku 2013) which allows the selection of cubes, dimensions, measures, and typical OLAP operations with the connection to Mondrian, SSAS, SAP BW and Oracle Hyperion. Although Saiku does not include spatial support natively, Lamas et al. (2013) propose the extension based on a map server developed to deliver maps. However, this extension again is not freely available and required LeafLet, an open-source JavaScript library for mobile-friendly interactive maps, in order to integrate OLAP cubes and maps. Furthermore, a new tendency has also emerged that relies on Web services as providers for SOLAP environments (Barros and Fidalgo 2010).

We believe that the front-end tools as a free option are currently the weakest component considering different layers forming a SOLAP system. The lack of this layer introduces an additional GIS-like level complexity

---

[2]Previously called Pentaho Analysis Tool and currently in development to replace Pantaho JPivot.

for non-expert users where the knowledge of MDX with spatial extension is required to query spatial data. GeoOLAP from Campotocamp used in our project still need other features to improve its interface, e.g., the pivot operation is not allowed, it is hard to analyze historical data covering several periods of time, the layout is not always user-friendly when graphs and tables are included in addition to maps, and some modifications were applied to correct the wrong display, among others. According to Rivest et al. (2005) different features should be included in SOLAP tools in order to fully explore their potential. They present them using their product that was originally commercialized under the name JMap and currently as Map4Decision (Intelli3 2013). We categorized these features into various groups: (1) data visualization (e.g., tabular, graph, and map displays, synchronization between different environments during SOLAP operations, modification of graphical semiology, among others), (2) data exploration (e.g., drill-down, roll-up, and slice-and-dice operations, manipulation of temporal dimension with a timeline, addition of calculated members), and (3) data structures (e.g., support for a complete geometric primitives, support for several spatial and non-spatial dimensions, support for the storage of geometries that changes over time). On the other hand, Viswanathan and Schneider (2011) present a list of 25 requirements that SOLAP systems should fulfill. These requirements refer to conventional and spatial DW and OLAP and by mixing both environments, i.e., DW and OLAP, it is not clear which system is responsible for satisfying them. Furthermore, these requirements are compilations from different works (e.g., Pedersen et al. 2001; Vassiliadis and Sellis 1999; Rivest et al. 2005) and represent a complex list that even conventional OLAP tools cannot satisfy. Therefore, the specification of requirements that the SOLAP front-end layer should fulfill is still an open research topic. Relying on implementers' intuition and knowledge does not always give the desired results as demonstrated in the SOVAT system (Scotch et al. 2007), where adjustments and modifications were required to be implemented in order to better satisfy user needs.

Conclusions

Current progress in managing spatial data as a part of traditional storage of object-relational databases opens up the possibility to implement SDWs using a multidimensional view of data expressed as star or snowflake schemas. The implementation should consider user requirements, as well as available and possibly acquire data. In addition, different phases of development that include conceptual, logical, and physical design of schemas should be conducted. These phases should be augmented by the ETL processes that allow integrating data from different sources and improving their quality. As a next step, SOLAP cubes could be implemented at the server level allowing their analysis, visualization, and manipulation at the front-end level. These SOLAP system development phases may look easy to achieve. However, based on our experience in the development of a GeoBI solution with a SDW and SOLAP using free software, there is still many challenges that must be overcome.

Research related to SDWs and SOLAP has more than 15 years of tradition and many interesting and valuable proposals were developed showing the possible directions that these systems could take. The solutions cover a wide range of topics, e.g., SDW or SOLAP design concepts and methods, queries, performance, indexing, and implementation, to mention a few. However, many times these solutions are presented as research papers or prototypes without the possibility to be used by the wider public. Furthermore, the separation of content and presentation is not always clear (Dubé et al. 2009) and the front-end layer is in charge to hide the complexity of the system and the fact that two separate underlying systems are used, i.e., OLAP and GIS. In addition, concepts developed by researchers are seldom included in free SOLAP products, e.g., spatial measures, different types of spatial hierarchies, and spatial aggregation functions, among others. In particular, it is difficult to satisfy the requirement of having freely-available SOLAP server and front-end software that rely on an open architecture. To the best of our knowledge, the GeoMondrian server (Spatialytics 2013b) is currently an option for including spatial dimensions in the SOLAP schema. However, there are still limitations that this software must overcome in order to allow users to exploit OLAP features to their full capabilities. The situation gets more complicated when searching for free SOLAP front-end solutions based on GeoMondrian server. There is a very limited number of options and they require good technical skills in order to make them functional.

We believe, as stated by Devillers et al. (2010) and Goodchild (2008), that even though a number of research projects led to methods and tools used by the nonacademic community, a large body of scientific knowledge is still only in the hands of researchers and embedded in scientific publications, like the one related to SDWs and SOLAP. Combining research efforts, using open architectures, and improving existing solutions based on research proposals will not only benefit a research community but also deliver solutions to non-expert users that may help them make informed decision based on spatial and conventional data analysis. In this chapter, we mentioned several issues that the research community and practitioners should face. Furthermore, some additional evaluations of SOLAP solutions—such as the one made for conventional BI (Golfarelli 2009; Thomsen and Pedersen 2008)—could help better evaluate the current situation. In addition, the research should consider that SDW/SOLAP solutions are not replacing existing GIS tools; on the contrary, not only do both environments have relatively distinct work division, but they also play highly complementary roles (Bédard et al. 2009; Rivest et al. 2001).

Another aspect is the need to develop a geomatics vision by joining knowledge from different geo-fields and informatics, as already done in several universities around the world, e.g., Laval University in Quebec, Canada, California State University, USA, and Politécnico de Milano, Italy. In this way, professionals who have sufficient informatics background may be ready to take advantage of technological progress in providing solutions for spatial data management and analysis.

Acknowledgment

I would like to express my gratitude to Leonardo Pandolfi for the effort of implementing the system that serves as a basis for the examples used in this chapter.

References

Aguila, P., R. Fidalgo and A. Mota. 2011. Towards a More Straightforward and More Expressive metamodel for SDW Modeling. Proc. of the ACM 14th Int. Workshop on Data Warehousing and OLAP. 31–36.

Akkaoui, Z., E. Zimányi, J. Mazón and J. Trujillo. 2011. A Model-Driven Framework for ETL Process Development. Proc. of the ACM 14th Int. Workshop on Data Warehousing and OLAP. 45–52.

Albrecht, A. and F. Nauman. 2008. Managing ETL Process. Proc. of the Int. Workshop on New Trends in Information Integration. 12–15.

Baltzer, O. 2011. Computational Methods for Spatial OLAP. Ph.D. Thesis. Dalhousie University, Nova Scotia.

Badard, T. and E. Dubé. 2009. Enabling Geospatial Business Intelligence. Technology Innovation Management Review. Retrieved April 2013 from http://timreview.ca/node/289.

Badard, T., M. Kadillak, G. Percivall, S. Tamage, C. Reed, M. Sandersen, R. Singh, J. Sharma and L. Vaillancourt. 2012. Geospatial Business Intelligence. OGC White paper. Retrieved April 2013 from http://www.opengeospatial.org/pressroom/papers.

Barros, D. and R. Fidalgo. 2010. An Architecture and a Metamodel for Processing Analytic and Geographic Multilevel Queries. Proc. of the Int. Conf. on Computational Science and Its Applications.

Bejaou, L., Y. Bédard, F. Pinet and M. Schneider. 2009. Qualified Topological Relations between Spatial Objects with the Possibly Vague Shape. Int. Journal of GIS. 23(7): 877–921.

Bédard, Y., T. Merret and J. Han. 2009. Fundamentals of Spatial Data warehousing for Geographic Knowledge Discovery. pp. 53–73. In: J. Miller and J. Han (eds.). Geographic Data Mining and Knowledge Discovery. CRC Press.

Bimonte, S., A. Tchounikine and M. Miquel. 2005. Towards a spatial multidimensional model. Proc. of the 8th ACM international workshop on Data warehousing and OLAP. 39–46.

Bimonte, S., A. Tchounikine, M. Miquel and F. Pinet. 2010. When spatial analysis meets OLAP: multidimensional model and operators. Int. Journal of Data Warehousing and Mining (IJDWM). 6(4): 33–60.

Bimonte, S., P. Wehrle, A. Tchounikine and M. Miquel. 2006. GeWOlap: A Web Based Spatial OLAP Proposal. Proc. of the 2nd Int. Workshop on Semantic-based Geographical Information Systems. 1596–1605.

Boin, A.T. and G.J. Hunter. 2005. Do Spatial Data Consumers Really Understand Data Quality Information? Proc. of the 7th Int. Symposium on Spatial Accuracy Assessment in Natural Resources and Environmental Sciences. 215–224.

Camossi, E., M. Bertolotto and E. Bertino. 2008. Querying MultigranularSpatio-temporal Objects. Proc. Int. Conf. on Database and Expert Systems Applications. 390–403.

Camptocamp. 2012. GeoOLAPProject. Retrieved April 2012 from https://github.com/camptocamp/GeoBI/zipball/master.

CCP. 2013. Centro Centroamericano de Población. Tasas demográficas. Retrived April 2013 from http://ccp.ucr.ac.cr/tasas_demograficas/tasas.html#.

Chen, R. and J. Xie. 2008. Open sources databases and their spatial extensions. pp. 105–130. In: G. Brent Hall and M. Leahy (eds.). Open Sources Approaches in Spatial Data Handling.

Cohen, T. and J. Baitty. 2005. Health Resources and Service Administration Geospatial Data Warehouse: Integrating Spatial and Tabular Extract, Transform, and Load Processes. ESRI International Users Conference.

Costa, M., A. Gomes and C. Souza. 2006. Towards a Logical Multidimensional Model for Spatial Data Warehousing and OLAP. Proc. of the ACM 9th Int. Workshop on Data Warehousing and OLAP. 83–90.

Costa, R., T. Lopes, V. Times, R. Rodrigues and C. de Aguilar. 2010. How does the Spatial Data Redundancy Affect Query performance in Geographic Data Warehouses. Journal of Information and Data Management. 1(3): 519–534.

Damiani, M. and S. Spaccapietra. 2006. Spatial Data Warehouse Modeling. pp. 1–27. In: J. Darmont and O. Boussaïd (eds.). Processing and Managing Complex Data for Decision Support. Idea Group Publishing.

Delavar, M. and R. Devillers. 2010. Spatial Data Quality: From Process to Decisions. Transaction in GIS. 14(4): 379–386.

Devillers, R., Y. Bédard, A. Stein and N. Chrisman. 2010. Thirty Years of Research on Spatial Data Quality: Achievements, Failures, and Opportunities. Transactions in GIS. 14(4): 387–400.

Di Martino, S., S. Bimonte, M. Berttolotto and F. Ferrucci. 2009. Integrating Google Earth within OLAP Tools for Multidimensional Exploration and Analysis of Spatial Data. Proc. of the Int. Conf. on Enterprise Information Science. 940–951.

Dubé, E., T. Badard and Y. Bédard. 2009. XML Enconding and Web Services for Spatial OLAP Data Cube Exchange: an SOA Approach. Journal of Computing and Information Technology. 4: 347–358.

Escribano, A., L. Gomez, B. Kuijpers and A. Vaisman. 2007. Piet: a GIS-OLAP implementation, Proc. of the ACM 10th Int. Workshop on Data Warehousing and OLAP. 73–80.

ESRI. 1998. Shapefile Technical Description. White Paper. Retrieved January 2013 from http://www.esri.com/library/whitepapers/pdfs/shapefile.pdf.

ESRI. 2012. Geotababase: overview. Retrieved January 2013 from http://www.esri.com/software/arcgis/geodatabase/index.html.

ESRI. 2013a. EsriCoorporate Alliances: SAP. Retrived April 2013 from http://www.esri.com/partners/alliances/sap.

ESRI. 2013b. ESRI Maps for IBM Cognos. Retrieved April 2013 from http://www.esri.com/software/esri-maps-for-ibm-cognos.

Ferri, F., E. Pourabbas, M. Rafanelli and F. Ricci. 2000. Extending Geographic Databases for a Query Language to Support Queries Involving Statistical Data. Proc. of the 12th Int. Conf. on Scientific and Statistical Database Management. 220–230.

Fidalgo, R.N., J. da Silva, V. Times, F. Souza and A. Salgado. 2010. Revisiting Providing Multidimensional and Geographical Integration Based on a GDW and Metamodel. Journal of Information and Data Management. 1(1): 107–110.

Fidalgo, R.N., V. Times, J. da Silva and F. Souza. 2004. GeoDWFrame: A Framework for Guiding the Design of Geographical Dimensional Schemas. Proc. of the 6th Int. Conf. on Data Warehousing and Knowledge Discovery. 26–37.

Fidalgo, R.N., J. da Silva, V. Times, F. Souza and R. Barros. 2003. GMLA: a XML Schema for Integration and Exchange of Multidimensional Geographical Data. Proc. V SimpósioBrasileiro de Geoinformática.

Gervais, M., Y. Bédard, M.A. Levesque, E. Bernier and R. Devillers. 2010. Data Quality Issues and Geographic Knowledge Discovery. pp. 99–116. In: H. Miller and J. Han (eds.). Geographic Data Mining and Knowledge Discovery. Taylor & Francis.

Glorio, O. and J. Trujillo. 2008. An MDA Approach for the Development of Spatial Data Warehouses. Proc. of the 10th Int. Conf. on Data Warehousing and Knowledge Discovery. 23–32.

Glorio, O. and J. Trujillo. 2009. Designing Data Warehouses for Geographic OLAP Querying by Using MDA. Proc. of the Int. Conf. on Computational Science and Its Applications. 505–519.

Golfarelli, M. and S. Rizzi. 2009. Data Warehouse Design: Modern Principles and Methodologies. McGrow-Hill Osborne Media.

Golfarelli, M. 2009. Open Source BI Platforms a Functional and Architectural Comparison. Proc. of the 11th Int. Conf. on Data Warehousing and Knowledge Discovery. 287–297.

Gómez, L., S. Haesevoets, B. Kuijpers and A. Vaisman. 2009a. Spatial Aggregation: Data Model and Implementation. Information Systems. 34(6): 551–576.

Gómez, L., B. Kuijpers, B. Moelans and A. Vaisman. 2009b. A Survey of Spatio-Temporal Data Warehousing. Int. Journal of Data Warehousing and Mining. 5(3): 28–55.

Gómez, L., A. Vaisman and S. Zich. 2008. Piet-QL: a query language for GIS-OLAP integration. Int. Symposium on Advances in GIS. 27.

Gómez, L., B. Kuijpers and A. Vaisman. 2011. A Data Model and Query Language for Spatio-Temporal Decision Support. Geoinformatica. 15: 455–496.

Goodchild, M.F. 2008. Epilogue: Putting research into practice. pp. 345–56. In: A. Stein, W. Shi and W. Bijker (eds.). Quality Aspects in Spatial Data Mining, CRC Press.

Guo Y., S. Tang, Y. Tong and D. Yang. 2006. Triple-Driven Data Modeling Methodology in Data Warehousing: a Case Study. Proc. of the ACM 9th Int. Workshop on Data Warehousing and OLAP. 59–66.

Han, J., N. Stefanovic and K. Koperski. 1998. Selective materialization: An efficient method for spatial data cube construction. Research and Development in Knowledge Discovery and Data Mining. 1394: 144–158.

Intelli3. 2013. Map4Decision. Retrieved April 2013 from http://www.intelli3.com/.

ISO 19113. 2002. Geographic information – Quality principles.

ISO 19114. 2003. Geographic Information – Quality Evaluation Procedures.

ISO 19138. 2006. Geographic Information – Data Quality Measures.

Jensen, C., A. Kligys, T. Pedersen and I. Timko. 2004. Multidimensional Data Modeling for Location-based Services. VLDB. 13(1): 1–21.

Jensen, C., T. Pedersen and Ch. Thomsen. 2010. Multidimensional Databases and Data Warehousing. Morgan and Claypool Publishers.

Kouba, Z., K. Matousek and P. Miksovsky. 2000. On Data Warehouse and GIS Integration. Proc. of the 11th Int. Conf. on Database and Expert Systems Applications. 604–613.

Lamas, A., F. Sotelo, M. Borobio and J. Varela. 2013. Creación de un Módulo Espacial OLAP para Saiku. VII Jornadas de SIG Libre. University of Girona. Retrived April 2013 from http://www.sigte.udg.edu/jornadassiglibre/uploads/articulos_13/a26.pdf.

Levesque, M., Y. Bédard, M. Gervais and R. Devillers. 2007. Towards Managing the Risks of Data Misuse for Spatial Datacubes. Proc. of the 5th Int. Symposium on Spatial Data Quality.

Lopes T., C. de Aguilar, V. Times and R. Rodrigues. 2012. The SB-index and the HSB-index: Efficient Indices for Spatial Data Warehouses. Geoinformatica. 16(1): 165–205.

Luján-Mora, S., J. Trujillo and I. Song. 2006. A UML Profile for Multidimensional Modeling in Data Warehouses. Data & Knowledge Engineering. 59(3): 725–769.

Malinowski, E. and E. Zimányi. 2008. Advanced Data Warehouse Design From Conventional to Spatial and Temporal Applications. Springer.

Malinowski, E. 2013. Decision-making Processes Based on Knowledge Gained from Spatial Data. In: Y. Al-Bastaki and A. Shajera (eds.). Building a Competitive Public Sector with Knowledge Management Strategy. IGI Global. To be published. http://www.igi-global.com/book/building-competitive-public-sector-knowledge/75831.

Martyn, T. 2004. Reconsidering Multi-dimensional Schemas. SIGMOD Record. 33(1): 83–88.

Mäkelä, J.M. 2006. The Impact of Spatial Data Quality on Company's Decision Making. The International Archives of the Photogrammetry, Remote Sensing and Spatial Information Sciences. 34.

Mazón, J., J. Trujillo and J. Lechtengörger. 2007. Reconciling Requirement-driven Data Warehouses with Data Sources Via Multidimensional Normal Forms. Data Knowledge & Engineering. 63: 725–251.

Miksovsky, P. and Z. Kouba. 2001. GOLAP—Geographical On-Line Analytical Processing. Proc. of the Int. Conf. on Database and Expert Systems. 201–205.

OGC. 2011. OpenGIS Implementation Standard for Geographic Information—Simple Feature Access—Part1: Common Architecture, Geospatial Business Intelligence (GeoBI). White paper. Retrieved April 2013 from http://www.opengeospatial.org/standards/sfa.

Olsen, J. 2003. Data Quality: The Accuracy Dimension. Morgan Kaufmann Publishers.

OSGIS. 2013. Open source GIS. Retrieved April 2013 from http://opensourcegis.org/.

Papadias, D., P. Kalnis, J. Zhang and Y. Tao. 2001. Efficient OLAP Operations in Spatial Data Warehouses. Proc. of the Int. Symposium on Advances in Spatial and Temporal Databases. 443–459.

Pedersen, T., C. Jensen and C. Dyreson. 2001. A Foundation for Capturing and Querying Complex Multidimensional Data. Information Systems. 26(5): 383–423.

Pedersen, T. and N. Tryfona. 2001. Pre-aggregation in Spatial Data Warehouses. Proc. of the 7th Int. Symposium on Advances in Spatial and Temporal Databases. 460–478.

Pentaho. 2013a. Kettle project. Retrieved April 2013 from http://kettle.pentaho.com/.

Pentaho. 2013b. Analysis Services: Mondrian Project. Retrieved April 2013 from http://mondrian.pentaho.org/.

PostgreSQL. 2013. PostGIS documentation. Retrieved April 2013 from http://postgis.net/docs/manual-2.0/.

Pourabbas, E. 2003. Cooperation with Geographic Databases. pp. 393–432. In: M. Rafanelli (ed.). Multidimensional Databases: Problems and Solutions. Idea Group.

Pourabbas, E. and M. Rafanelli. 2002. A Pictorial Query Language for Querying Geographic Databases using Positional and OLAP Operators. SIGMOD Record. 31(2): 22–27.

Proulx, M.-J., S. Rivest and Y. Bédard. 2007. Évaluation des produits commerciaux offrant des capacités combinées d'analyse OLAP et de cartographie. Industrial Research Chair in Geospatial Databases for Decision Support, Departement of Geomatics Sciences. University of Laval, Quebec, Canada.

Rivest, S., Y. Bédard and P. Marchand. 2001. Towards better support for spatial decision making: Defining the characteristics of spatial online analytical processing (SOLAP). Geomatica. 55(4): 539–555.

Rivest, S., Y. Bédard, M. Proulx M. Nadeau, F. Hubert and J. Pastor. 2005. SOLAP technology: Merging Business Intelligence with Geospatial Technology for Interactive Spatio-temporal Exploration and Analysis of Data. ISPRS Journal of Photogrammetry and Remote Sensing. 60(1): 17–33.

Saiku. 2013. SaikuAnalystics. Retrieved April 2013 from http://analytical-labs.com/index.html.

SAS. 2013. SAS Web OLAP Viewer for Java. Retrieved April 2013 from http://www.sas.com/technologies/bi/query_reporting/webolapviewer/.

Sboui, T., M. Salehi and Y. Bédard. 2010. A Systematic Approach for Managing the Risk Related to Semantic Interoperability between Geospatial Datacubes. Int. Journal of Agricultural and Environmental Information Systems. 1(2): 20–41.

Scotch, M. and B. Parmanto. 2005. SOVAT: Spatial OLAP Visualization and Analysis Tool. Proc. of the 38th Annual Hawaii International Conference onSystem Sciences. 142–144.

Scotch, M., B. Parmanto and V. Monaco. 2007. Usability Evaluation of the Spatial OLAP Visualization and Analysis Tool (SOVAT). Journal of Usability Studies. 2(2): 76–95.

Shekhar, S. and S. Chawla. 2003. Spatial Databases: A Tour. Prentice Hall.

Silva, R., R. Manhäes and P. Gitahy. 2008. WebGeoOlap: Visualização de Dados Multidimensionais e Geográficos para Ambiente Web. Revista de Sistemas de Informação da FSMA. 1: 9–16.

Silva, J., A. de Oliveira, R. Fidalgo, A. Salgado and V. Times. 2010. Modeling and Querying Geographical Data Warehouses. Information Systems. 35(5): 592–614.

Silva, J., V. Times, A. Salgado, C. Souza, R. Fidalgo and A. Salgado. 2008. A Set of Aggregation Functions for Spatial Measures. Proc. of the ACM 11th Int. Workshop on Data Warehousing and OLAP. 25–32.

Silva, J., V. Times and A. Salgado. 2006. An Open Source and Web Based Framework for Geographic and Multidimensional Processing. Proc. of the ACM Symposium on Applied Computing. 63–67.

Silva, J., V. Times and M. Kwakye. 2012. A Framework for ETL Systems Development. Journal of Information and Data Management. 3(3): 300–315.

Smith, B., R. Clay and H. Consulting. 2009. Microsoft SQL Server 2008 MDX Step by Step. Microsoft Press.

Soares, R., S. Montenegro, G. Colonese, R. de Carvalho and A. Kiyoshi. 2007. GeoOlap: An Integrated Approach for Decision Support. Proc. of the Int. Conf. On Research and Practice Issues of Enterprise Information Systems. 359–369.

Spatialytics. 2013a. GeoKettle. Retrieved April 2013 from http://www.spatialytics.org/projects/geokettle/.

Spatialytics. 2013b. GeoMondrian. Retrieved April 2013 from http://www.spatialytics.org/projects/geomondrian/.

Spatialytics. 2013c. SOLAPLayers. Retrieved April 2013 from http://sourceforge.net/projects/spatialytics/.

SpagoBI. 2010. GeoBI Initiative: Open Source Location Intelligence. Retrieved April 2013 from http://www.spagoworld.org/spw-resources/Resources/Presentations/GeoBI@fOSSa2010.pdf.

SpagoBI. 2013a. Location Intelligence. Retrieved April 2013 from http://www.spagoworld.org/xwiki/bin/view/SpagoBI/Geo.

SpagoBI. 2013b. Document Composition Internal Engine. Retrieved April 2013 from https://wiki.spagobi.org/xwiki/bin/view/spagobi_server/Document+Composite.

Stefanovic, N., J. Han and K. Koperski. 2000. Object-based selective materialization for efficient implementation of spatial data cubes. IEEE Trans. Knowl. Data Eng. 12(6): 938–958.

Talend. 2013. Spatial Data Integrator Retrieved April 2013 from http://sourceforge.net/projects/sdispatialetl/.

Talhofer, V., A. Hofmann, S. Hošková – Mayerová and P. Kubícek. 2011. Spatial Analyses and Spatial Data Quality. Proc. of the 14th Agile Int. Conf. on Geographic Information Science.

Times, V., R. Fidalgo, R. Fonseca, J. da Silva and A. Oliveira. 2008. A Metamodel for the Specification of Geographical Data. Annals of Information Systems: New Trends in Data Warehousing and Data Analysis. 1–22.

Thomsen, Ch. and T. Pedersen. 2008. A Survey of Open Source Tools for Business Intelligence. Technical report. Retrieved April 2013 from http://vbn.aau.dk/ws/files/14833824/DBTR-23.pdf.

Trujillo, J. and S. Luján-Mora. 2003. A UML Based Approach for Modeling ETL Processes in Data Warehouses. Proc. of the 22nd Int. Conf. on Conceptual Modeling. 307–320.

Turban E., R.E. Sharda and D. Delen. 2010. Decision Support and Business Intelligence Systems (9th Edition), Prentice Hall.

Vassiliadis, P. and T. Sellis. 1999. A Survey of Logical Models for OLAP Databases. SIGMOD Record. 28(4): 64–69.

Vassiliadis, P., A. Simitsis and E. Baikous. 2009. A taxonomy of ETL activities. Proc. of the ACM 2nd Int. Workshop on Data Warehousing and OLAP. 25–32.

Vassiliadis, P., A. Simitsis, P. Georgantas, M. Terrovitis and S. Skiadopoulos. 2005. A Generic and Customizable Framework for the Design of ETL Scenarios. Information Systems. 30(7): 492–525.

Viswanathan, G. and M. Schneider. 2011. On the Requirements for User-Centric Spatial Data Warehousing and SOLAP. DASFAA Workshops. 144–155.

Wang, R. and D. Strong. 1996. Beyond Accuracy: What Data Quality Means to Data Consumers. Journal of Management Information Systems. 12(4): 5–34

Wang, B., Ch. Li, X. Fu, M. Li, D. Wang, H. Du and Y. Xing. 2010. Design of ETL Process on Spatio-temporal Data and Study of Quality Control. Proc. Conf. on Computer and Computing Technologies in Agriculture. 487–494.

Yeung, A. and G. Hall. 2007. Spatial Database Systems: Design, Implementation, and Project Management. Springer.

CHAPTER 10

Semantic Similarity based on Weighted Ontology

Elaheh Pourabbas

Introduction

Semantic similarity has been the focus of interest in linguistics (Jiang and Conrath 1997; Tanga and Zheng 2006), biomedical studies (Wang et al. 2007; Guzzi et al. 2012), and artificial intelligence (Resnik 1999; Gabrilovich and Markovitch 2007). Essentially, they have been conceived to compare concepts, to facilitate searching through ontologies and to improve matching and aligning ontologies (Ehric 2007; Jean-Marya et al. 2009).

In the context of Geographic Information Science (*GIScience*), similarity methods have been used to measure the degree of semantic interoperability between Geographic Information Systems (GIS) or geographic data. With the rapid development of Internet technology and the emergence of the Semantic Geospatial Web, the computation of semantic similarity studies play a core role in understanding and handling semantic heterogeneity and, hence, in enabling interoperability between services and data repositories on the Web (Sheth 1999; Egenhofer 2002). The role of such methods in dealing with approximate query answering and reasoning as the basis for semantic retrieval and integration is growing in importance (Uitermark et al. 1999; Lutz and Klien 2006; Baglioni et al. 2009; Janowicz et al. 2011).

In GISs, *ontology* as a kind of information science research methodology was mainly introduced to solve the sharing, integration, as well as the interoperability (Uitermark et al. 1999; Buccella et al. 2009; Buccella et

National Research Council, Institute of Systems Analysis and Computer Science "Antonio Ruberti", Viale Manzoni 30, 00185 Rome, Italy.
Email: elaheh.pourabbas@iasi.cnr.it

al. 2011; Sánchez et al. 2010; 2012). An Ontology is an explicit and formal specification of sharing information (Gruber 1993). Generally, ontologies in GISs constitute the hierarchy structures, which can be derived from appropriate classification of data in geographical information domain. Some proposals in the GIS domain use a given *reference ontology* in the semantic similarity measuring process, for instance in (Rodriguez and Egenhofer 2003; 2004; Formica and Pourabbas 2009). In particular, in (Formica and Pourabbas 2009) the authors propose a semantic similarity method, called *GSim*, which is conceived by combining the similarity of geographical concepts organized as a *Part-of* taxonomy and the feature (or attribute) similarity of geographical classes. To this end, the information content approach of Lin (1998) has been extended and integrated with a method for attribute similarity, which is based on the maximum *weighted matching problem* in bipartite graphs (Kuhn 1955). The information content of a concept c is computed according to the following expression: $-logw(c)$, where $w(c)$ is the weight of the concept. The GSim method, as large majority of proposals in the literature, indicates the use of weights of concepts (or geographical classes) derived from WordNet (Miller 1995) frequencies. WordNet (a lexical ontology for the English language) provides, for a given concept (noun), the natural language definition, hypernyms, hyponyms, synonyms, etc., and also a measure of the *frequency* of the concept, by using noun frequencies from the Brown Corpus of American English (Francis and Kucera 1979). However, weights are often not available for all possible concepts (as discussed in the section devoted to weight assignment).

In this chapter, we focus on the problem of weight assignment to concepts. We propose and formalize a *probability-based* approach as an alternative to the *frequency-based* measures provided by WordNet. This approach is characterized by three notions, called *uniform probabilistic approach*, *uniform probabilistic weighted arc* and *non-uniform probabilistic weighted arc*. They essentially differ on the basis of nodes, arcs and depth of the reference ontology. Successively, we describe in detail the selected methods, Lin (1998), Dice (Maarek et al. 1991), Matching-Distance Similarity Measure—for short *MDSM* (Rodriguez and Egenhofer 2003; 2004), and the GSim method as representative similarity methods in the literature. As anticipated, Lin method is an information content-based approach, Dice and MDSM methods are based on the feature-based approach, and the GSim method is a combined approach, which is conceived to capture both the concept similarity within the hierarchy, and the attribute similarity (or tuple similarity) of geographic concepts or classes. In the section of experimental analysis, these methods are investigated in detail and the results of the contrast of the selected methods are illustrated. An experiment about two different weight assignment approaches is provided and their impact on the selected methods is illustrated.

The chapter is structured as follows. In the next section the state of the art of semantic similarity methods is given. In the third section, the basic notions of geographic data and knowledge base are presented. In the fourth section, we discuss the problem of weight assignment to ontology and introduce the probability-based approaches. In the fifth section, the selected similarity methods are described. In the sixth section, the experimental analysis is provided. Finally, the seventh section contains conclusions and future works.

State of the Art

Studies on semantic similarity have been stemmed in the cognitive sciences and particularly in psychology (Lakoff 1988; Medin et al. 1990; 1993; Schwering 2008). They mainly refer to similarity between individuals, classes, complex (pictorial) scenes, and processes. However, semantic similarity as reasoning support in information retrieval and approximate query answering has been extensively discussed in information sciences. These studies have been mainly focused on similarity of concepts and relationships. Concerning the relationships, different types have been considered as follows: *associative* (e.g., cause-effect), *equivalence* (or synonymy), *hierarchical* (e.g., *Is-A* or hyperonym-hyponym, *Part-of* or meronym/holonym, etc.). The hierarchical relationship, particularly, has attracted a lot of interest in the research community since it has been considered very suitable for mapping the human cognitive view of classification (i.e., taxonomy).

In the literature, the traditional approach to evaluate semantic similarity in taxonomy is the so-called *edge/node-counting* approach (Lee et al. 1993; Rada et al. 1989; Wu and Palmer 1994), which corresponds to the semantic distance approach. In such a method, the taxonomy is treated as an undirected graph and the semantic distance is equal to the minimal path length between concepts. The greater the distance between two concepts, the less similar they are. As mentioned above, the semantic distance between concepts can be measured by using edge counting or node counting approaches. In the former, the distance is given by the number of edges connecting the concepts, whereas in the latter it is the number of nodes along the shortest path that connects concepts, including the end nodes representing the concepts. In Fig. 1, which represents an ontology of water system, the semantic distance between *River* and *Aqueduct* is 4 using edge counting, while it is 5 using node counting. Given an ontology formed by a set of nodes and a root node, E_1 and E_2 represent two ontology elements of which we will calculate the similarity.

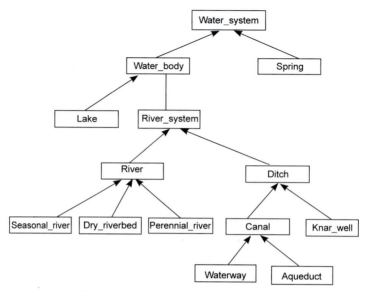

Fig. 1. Example of a water system ontology.

The Wu and Palmer (1994) approach is based on the distance between each of nodes E_1 and E_2 and their closest common ancestor E_3 as well as the distance between E_3 and the root node as follows:

$$2D_{node}(E_3, root)/(D_{node}(E_1, E_3) + D_{node}(E_2, E_3) + 2D_{node}(E_3, root))$$

where $D_{node}(E_i, E_j)$ for $i, j = 1,2,3$ is the distance, based on node counting, between E_i, E_j. For instance, in Fig. 1 we consider as elements E_1 and E_2 two nodes, *River* and *Aqueduct*, respectively. Their closest common ancestor is the node *River_system*, the distance of *River_system* from the root node *Water_system*, i.e., D_{node}(*River_system, Water_system*), is equal to 3.

Similarly, D_{node}(*River, River_system*) = 2, and D_{node}(*Aqueduct, River_system*) = 4. Thus, the Wu and Palmer similarity measure of *River* and *Aqueduct* is equal to 0.5.

This semantic distance has been reformulated by Resnik in (Resnik 1999) according to edge counting. The principle of this approach is based on the distance (D_1 and D_2) which separates nodes E_1 and E_2 from the root node and the distance (D_3) which separates the closest common ancestor of E_1 and E_2 from the root node. It is given by: $2D_3/(D_1 + D_2)$, where D_i represents the depth of node E_i (i.e., the distance of the node form the root in a taxonomy). In the case of the example above, the Resnik similarity measure is equal to $\frac{2*2}{3+5} = 0.5$.

The similarity measures of both Resink, Wu and Palmer approaches score between 1 and 0. The score 1 occurs when the two concepts are the same, while the score 0 is possible when the closest common concept is the root node.

A different measure based on semantic distance and depths was proposed by Leacock and Chodorow (1998). According to this approach, the similarity between two concepts E_1 and E_2 is given by the number of nodes along the shortest path between them, divided by double the maximum depth (from the lowest node to the root) in the taxonomy in which E_1 and E_2 occur as follows: $-\log (D_{node}(E_1, E_2)/(2*D))$, where D is the maximum depth of the taxonomy. Hence, the number of nodes between two siblings, i.e., two nodes with the same parent node, is three. For instance, the Leacock and Chodorow measure of similarity between *River* and *Aqueduct* is $-\log (5/(2*6)) = 0.38$.

The main limitation of node/edge counting approaches is related to the underlying assumption for defining a taxonomy, according to which the distances of adjacent nodes in each level are equivalent. In fact, according to these approaches for instance the distances of nodes *Seasonal_river*, *Dry_riverbed* and *Perennial_river* from the root node *Water_system* coincide and are equal to 4 in the case of edge counting and 5 in the case of node counting. Similarly, their distances from the node *River_system* are 2 and 3 in edge and node counting approaches, respectively.

In the 1990s, a different approach, namely *information content* (or node-based) approach has been introduced (Resnik 1995; 1999), which has been successively refined by Lin (1998). Essentially, it relies on the association of probabilities with the nodes of the taxonomy. The similarity between concepts is measured by the ratio between the amount of information shared by the concepts and the sum of the amounts of information of concepts. This approach is recalled in the Similarity Methods section. With respect to other existing proposals, the Lin approach shows a higher correlation with human judgment as has been discussed in Jiang and Conrath (1997).

However, *feature-based* (or tuple) similarity models are the most prominent and one of the key approach is the *Dice's* function (Maarek et al. 1991; Rasmussen 1992; Castano et al. 1998). It provides the coefficient of correlation between feature vectors, and it is given by the ratio between the number of features that are common to two vectors and the sum of the numbers of the features of each vector. Although, Dice's function is the most commonly used approach, it does not allow tuple similarity to be computed by explicitly considering the similarity degrees of components. However, the similarity degrees of components through the notion of information content similarity can be addressed and this approach will be recalled in the Similarity Methods section.

The research in the field of GIScience has been aligned with the advent of studies on similarity in the literature. Similarity has a long tradition in GIScience, and has been mainly discussed from the *spatial* point of view. In fact, spatial similarity has been defined as matching and ranking according to a certain context (function, goal), scale (coarse or finer level), and technology (for searching, retrieving, and recognizing data). The research on spatial similarity ranges from *data retrieval* (Papadias and Delis 1997; Samet 2004), *problem solving* (Yeh and Shi 2003), *conflation* (Cobb et al. 1998), to *interoperability* (Harvey 1999), etc.

The importance of semantic similarity of geographical concepts in GIScience has been highly emphasized by tasks of sharing and interoperation of geographical information (Buccella et al. 2011). These tasks are addressed to support the sharing and reuse of formally represented knowledge among GISs and are supported through a common ontology in which shared knowledge is represented. Typically, ontologies can be regarded as general tools of information representation on a specific subject. In fact, there are several examples of using ontologies in the spatial data domain. In the Spatial Data Transfer Standard (SDTS),[1] for instance, an ontology has been used to describe the underlying conceptual model and the detailed specifications for the content, structure, and format of spatial data, their features and associated attributes. In SDTS, concepts are commonly used on topographic quadrangle maps and hydrographic charts (USGS 2012). The GeoNames[2] ontology makes it possible to add geospatial semantic information to the Web as well. GeoNames is a community-driven database which contains place names and contains 8 million entries. It provides services to describe the relation between toponyms or place names. A further example is LinkedGeoData[3] that has been emerged to enhance OpenStreetMap[4] semantics. LinkedGeoData (LGD) has taken the entire OpenStreetMap data set and linked it to a formal ontology, which is a simple and shallow tree structure, representing keys and values (Ballatore et al. 2012). Its semantic content is limited to *Is-A* relationships between tags and respective values.

Principally, semantic similarity measures are necessary to quantify similarities of concepts belonging to the same or different ontologies (Gruber 1993; Visser et al. 2002), and are applied to a wide range of applications including the integration of geographic information from different sources, data mining, semantically enabled gazetteer services, semantics based geographic information search and retrieval as well as in conjunction with any Web search engine for enhancing the performance of retrieval (Jones et al. 2003).

---

[1] http://mcmcweb.er.usgs.gov/sdts/
[2] http://www.geonames.org/
[3] http://linkedgeodata.org/
[4] http://www.openstreetmap.org/

Semantic similarity measures are also widely used to perform query approximation in GISs. In fact, they provide approximate answers to the user queries which are formulated in terms of geographic data that have no match or are missing in the database. For instance, if the user asks the list of waterways (i.e., navigable transportation canals used for carrying ships and boats) in a certain region of USA, and the concept waterway is not provided in the database, then canal, which is more abstract concept than waterway in the hierarchy (see Fig. 1), is proposed. Similarly, suppose only a database of river_system is provided and the user erroneously asks the list of ditches. The approximate answer to this query can be obtained by applying a similarity method, which is addressed in detecting the most similar concept to ditch in the database that is river.

The proposal of Rodriguez and Egenhofer (2003; 2004) was one of the first similarity measures introduced for the geospatial domain, in which the set of spatial entity classes and their relations are described as an ontology. The authors define a computational method for assessing semantic similarity among geospatial entities by using the Tversky's feature-based model (Tversky 1977). Tversky's model is a set-theoretic measure expressing the similarity between concepts *a* and *b* defined by two description sets *A* and *B*, respectively, as a function of their common and distinct features as follows:

$$S_{Tversky} = \frac{f(A \cap B)}{f(A \cap B) + \alpha f(A - B) + \beta f(B - A)}$$

where the similarity measure $S_{Tversky}$ in the above ratio model lies between 0 and 1 and $\alpha, \beta \geq 0$ are parameters to be defined appropriately. If $\alpha = \beta = 1$, the ratio model reduces to $f(A \cap B)/f(A \cup B)$.

The method proposed by the authors (2003; 2004), called *Matching-Distance Similarity Measure* (MDSM), is a linear combination of weighted shared features and weighted distinct features (see the Similarity Methods section for more details). Their method determines the similarity by using a matching process over synonym sets, semantic neighborhoods, and distinguishing features (such as parts, functions, and attributes) of a source ontology. In order to determine features' relevance, two approaches, namely the variability and commonality have been introduced in Rodriguez and Egenhofer (2004). These proposals have inspired works on the context dependency of similarity (Keβler et al. 2007; Janowicz 2008) and other approaches like the proposals in (Pirrò and Seco 2008) and (Formica and Pourabbas 2009) as well. In particular, the latter proposal is an ontology— centered and information content-based method that is conceived to capture both the concept similarity within the hierarchy, and the attribute similarity (or tuple similarity) of geographic classes. It will be recalled in the Similarity Methods section as a representative of combined approaches.

The model of Tversky has also been used in the context of ontologies (Pinto and Martins 2004) in (Kavouras et al. 2005) to determine semantic similarity of geographic categories (or entities). In particular, the problem of cross-mapping of geographic ontologies has been addressed and a systematic methodology for comparing categories has been presented. Essentially, this approach is addressed to establish semantic similarity among categories from three geographic ontologies (i.e., CORINE LC,[5] MEGRIN[6] and WordNet) by using information derived from categories' definitions. More specifically, the focus is to extract semantic properties-relations and values for measuring the similarity of categories. For instance, the category "lake" has been defined as "a body of water surrounded by land" in MEGRIN, while the same category in WordNet is defined as "a body of (usually fresh) water surrounded by land". In both cases, the semantic properties and relations that can be identified are as follows: *Hypernym* with value "body", *Material* with value "water" and "water (usually fresh)", respectively, and *Surrounded-by* with value "land". Obviously, the above mentioned ontologies equivalently define the category "lake". Similarly, "Ditch" in WordNet is defined as "any small natural waterway", while in MEGRIN it is defined as "A canal for irrigation and drainage". These categories share only *Hypernym* with value "waterway" and "canal" (in Fig. 1, *Canal* is a generalized concept of *Waterway*), while "irrigation and drainage" can be identified as *Purpose* in MEGRIN, and "small", "Natural" as *Size* and *Nature*, respectively in WordNet. Thus their similarity measure, according to Tversky's model (see above for $\alpha = \beta = 1$) is equal to 0.25.

In Schwering and Raubal (2005), the authors propose a method based on spatial relations between different geospatial concepts. To capture the semantics of geo-objects with spatial relations the authors use a reduced group of spatial relations extracted from the set of spatial relations in natural language formalized by Shariff et al. (1998). In order to investigate similarities, they consider hydrological geo-objects within a large-scale topographic map for defining spatial relations. As a case study, they consider OS MasterMap ontology and a shared vocabulary that contains terms

---

[5] CORINE LC is a land cover categorization schema intended to provide consistent localized geographical information on the land cover of the member states of the European Community, by using satellite data. The CORIINE Land Cover has three hierarchies of categories. The upper level consists of 5 categories, the middle level of 15, and the lowest one of 44 categories. European Environmental Agency: CORRINE: Land Cover Methodology and Nomenclature. http://www.eea.europa.eu/publications/COR0-part1, http://www.eea.europa.eu/publications/COR0-part2

[6] GDDD-Geogrpahical Data Description Directory, MEGRIN's GDDD contains information on the digital geographic information available from Europe's National Mapping Agencies. (NMAs). Layers names, feature type names and feature attribute types names correspond to the nomenclature used in the DIGEST Feature and Attribute Coding Catalogue (FACC). http://www.eurogeographics.org/

describing properties and relations between concepts such as "flooding area is next to river" in which "next" is a spatial relation and "flooding area" and "river" are concepts or terms. In Buccella et al. (2009), a survey of approaches based on ontology for geographic information integration has been given.

In the rest of this chapter, we focus on Lin, Dice, MDSM, and GSim approaches, which are selected as representative semantic similarity methods defined in the literature.

Basic Concepts

In this section, we refer to a simple object-oriented geographic data model, which is characterized by the notion of *geographic class* informally defined below (Pourabbas 2003; Formica and Pourabbas 2009; Formica et al. 2012; Formica et al. 2013). Essentially, a *geographic class* describes a set of geographic objects having the same set of *attributes* (or *properties*). It is specified by a *name* and a class *expression*. The class expression contains a *tuple* of typed attributes, and one or more *geometric* types from {*point, polyline, polygon*}. Each attribute is associated with an atomic type (e.g., *integer, string, boolean*, etc.). A set of geometric types can be associated with a geographic class depending on the scale or because in multiple-representation database the geometric type may change. For instance, *county* can be conceived as a *polygon* or as a *point*. With each geometric type a set of (unary or binary) operations involving the topological relationships among geographic objects is associated. These relationships have been extensively discussed in the literature (for more details see, Egenhofer and Franzosa 1991; 1995) and their definition go beyond the scope of this work.

In order to formalize the definition of geographic class, we assume that countable sets $\mathcal{N}$, $\mathcal{A}$ of class names and attribute names, are given, respectively. Let $\mathcal{T}$ and $\mathcal{G}$ be the sets of atomic and geometric types defined, respectively, as follows:

$\mathcal{T} = \{string, boolean, integer, real\}$
$\mathcal{G} = \{point, polyline, polygon\}$

Definition 1: A *geographic class* (*class* for short) is defined by a name and an expression as follows:

$n = < \{a_1: t_1, \dots, a_k: t_k\}, G >, k \geq 1$

where n is the *name* of the class from $\mathcal{N}$, the a_i's are attribute names from $\mathcal{A}$, the t_i's are types from $\mathcal{T}$, and $G \subseteq \mathcal{G}$ is the set of geometric types of the class n.

Example 1. The following geographic class of name *Municipality*:

Municipality =
< {*identity*: *String*; *councilHead*: *String*, *countryCode*: *Integer*, *inhabitant*: *String*, *area*: *Integer*}, {*point*, *polygon*} > is defined by a set of alphanumeric attributes (i.e., identity, councilHead, countryCode, inhabitant, area) and a set of geometric types (i.e., point, polygon).

A *geographic knowledge base* is essentially a set of geographic classes where each class is identified by a name, and every name is associated with a class expression (i.e., there are no dangling class names). It is characterized by an ontology, where the geographic classes can be related by different types of relationships, e.g., *Is-A* or *Part-of*. In particular, according to these mentioned relations, an ontology of a geographic knowledge base can be organized as a *Is-A* hierarchy or as a *partition* hierarchy as well (Pourabbas 2003; Pourabbas and Rafanelli 2003). The former indicates the well-known *is-a* relationship, and the latter captures the *is-in* relationship or *inclusion* property.

In the literature, basically, three different kinds of semantics for inclusion have been identified: *class*, *meronymic*, and *spatial* (Storey 1993; Winston et al. 1987). Class inclusion[7] is the standard subtype/supertype relationship which has been widely discussed in the database literature, and it is indicated by *Is-A* (Codd 1979; Tsichritzis and Lochovsky 1982; Teorey et al. 1986). Concerning meronymic (*Part-whole*, or *Part-of*) inclusion, many studies have been carried out. One of the various semantics of meronymic relationships discussed in Storey (1993) is the *place-area*. It concerns parts which are similar to whole, and they cannot be separated (for instance, the reception area is part of an office). Finally, the semantics of spatial inclusion differs from place-area in that it represents objects that are surrounded by others but they are not part of them (as for instance, car *is-in* city).

We consider the place-area semantics for the inclusion relationship that is more suitable for capturing the meaning of inclusion in the geographic context and, in particular, of geographic classes organized as partition hierarchies (Storey 1993). Furthermore, we can apply the information content approach of Lin, which is conceived for *Is-A* hierarchies, to partition hierarchies by using the place-area semantics. Henceforth, for the sake of simplicity, we will use terms *Is-A* and *Part-of* hierarchies to mean these kinds of hierarchies.

---

[7] As highlighted in (Storey 1983), class inclusion is easily confused with member-collection relationships (as well as other meronymic relationships) because both involve membership of individuals in a larger set. Meronymic relationships are determined on the basis of characteristics that are extrinsic to the individual members themselves. Class inclusion is determined by similarity to other members based on an intrinsic characteristic.

Note that in the sequel, the term *concept* will be used to denote a class, an attribute, or a type name. Before introducing the notion of *geographic knowledge base*, the notion of *partial order on partitions* is given below.

Definition 2: Let E be a subset of the plane, and P a partition of E, i.e., $\bigcup P = E$, and $\forall p, p' \in P, p \cap p' = \varnothing$. Let $\mathcal{P}(E)$ be a set of partitions of E, and P, $P' \in \mathcal{P}(E)$. The partial order, $\sqsubseteq$, on $\mathcal{P}(E)$, indicated as $(\mathcal{P}(E), \sqsubseteq)$, is given as follows:

$P \sqsubseteq P'$ iff $\forall p \in P, p' \in P', p \cap p' = \{p, \varnothing\}$

Example 2: Let E be a subset of the plane called *Country*, and let $\mathcal{P}(Country)$ = {*Region, State, Department, Province, County, Municipality*} be a set of partitions. The partial order $\sqsubseteq$ on the above partitions is the following, as shown in Fig. 2:

Department $\sqsubseteq$ Country
Municipality $\sqsubseteq$ Province
Province $\sqsubseteq$ Region
Region $\sqsubseteq$ Country
County $\sqsubseteq$ State
State $\sqsubseteq$ Country

Definition 3: Let E be a subset of the plane and $(\mathcal{P}(E), \sqsubseteq)$ be a partial order on the set of partitions of E. A *geographicknowledge base* $K_E = (C, A, Cls)$ consists of finite sets $C \subset \mathcal{N}, A \subset \mathcal{A}$ and a finite set Cls of geographic classes such that, for each $n \in C$, n is the name of precisely one geographic class in Cls, and Cls also contains a class expression of name E. Thus $C = \mathcal{P}(E)$, and $(\mathcal{P}(E), \sqsubseteq)$ is referred to as *Part-of* hierarchy.

Example 3: Let us consider the partitions of the class *Country* of Example 2. A geographic knowledge, called *GeoKB*, can be defined by the geographic classes shown in Table 1. In Fig. 2, these geographic classes are organized

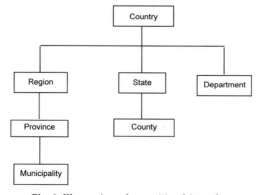

Fig. 2. Illustration of a partition hierarchy.

Table 1. Example of geographic classes.

$Country =< \{name: String, countryCode: Integer, president: String, flag: String\}, \{polygon\} >$
$Region =< \{label: String, healthCenter: String, president: String\}, \{polygon\} >$
$State =< \{tag: String, countryCode: Integer, governor: String\}, \{polygon\} >$
$Department =< \{tag: String, countryCode: Integer, governor: String\}, \{polygon\} >$
$Province =< \{name: String, prefecture: String, localEductionOffice: String\}, \{polygon\} >$
$Municipality =$ $< \{identity: String; councilHead: String, countryCode: Integer, inhabitant: String, area: Integer\}, \{point, polygon\} >$
$County =< \{countyID: Integer, population: Integer, surface: Integer\}, \{polygon\} >$

as a partition hierarchy. In this figure, *Country* is partitioned according to different administrative organizations, i.e., *Region, State, Department,* which correspond to the territorial subdivision of Italy, USA, and France, respectively. Similarly, *Region* and *State* are partitioned again into *Province* and *County,* respectively, and are related to *Region* and *State* by a partial order relationship. Analogously, *Province* into *Municipality.*

Definition 4: Given a geographic knowledge base $K_E = (C, A, Cls)$, let $S_i \subset N \cup A$, $i = 1, ..., n$, be sets of synonyms according to a lexical database for the English language. Then, $SynSet_K$ is a (possibly empty) *set of synonyms* for the geographic knowledge base K_E if:

$SynSet_K = \{S_1, ..., S_n\}$, $n \geq 0$, and $S_i \cap (C \cup A) \neq \emptyset$ for $i = 1, ..., n$.

Example 4: The set of synonyms for the geographic knowledge *GeoKB*, shown in Table 1 (i.e., $SynSet_{GeoKB}$), defined according to WordNet, are (for the sake of illustration we consider the following set):

$SynSet_{GeoKB} =$
$\{\{name, label, identity, mark\}, \{surface, area\},$
$\{inhabitant, indweller, denizen, dweller, population\}\}$

Weight Assignment Methods

In this section, we investigate the problem of assigning weights to a reference ontology (or H) of geographic classes. We address a simplified notion of hierarchy, H, consisting of a set of concepts organized according to essentially *Is-A* or *Part-of* relations, here referred to as R.

Definition 5: Let R be a relation. H is a *taxonomy* defined by the pair $H =< C, H >$, where C is a set of concepts and H is the set of pairs of concepts of C that are in R relation as follows:

$H = \{(c_i, c_j) \in C \times C \mid R(c_i, c_j)\}$

Note that, given two concepts, $c_i, c_j \in C$, the *least upper bound* of $c_i, c_j, lub(c_i, c_j)$, is always uniquely defined in C (we assume the hierarchy is a lattice).

It represents the least abstract concept of the hierarchy that is more general concept with respect to both c_i and c_j in the case of *Is-A* relation, or contains both c_i and c_j in the case of *Part-of* relation.

In Fig. 1, an example of an *Is-A* hierarchy is shown, where for instance the *lub* of the concepts *Seasonal_river* and *Aqueduct* is *River_system*. Whereas, Fig. 2 represents a *Part-of* hierarchy, in which as an example *Country* is the *lub* of concepts *Municipality* and *Department*.

Definition 6: A *Weighted Reference Ontology* ($\mathcal{H}_w$) is a pair:

$$\mathcal{H}_w =< \mathcal{H}, w >$$

where w is a function defined on C, such that given a concept $c \in C$, $w(c)$ is a rational number in the interval [0,1].

In this chapter, we focus on *Part-of* hierarchies, however the selected approaches mentioned in the Introduction section can also be applied to *Is-A* hierarchies (Beeri 1990). Below the notions of *weighted* ontology according to different weighting approaches are introduced.

Frequency-based approach

In the large majority of papers proposed in the literature, the assignment of weights to the concepts of a reference hierarchy (or a taxonomy) is often performed by using WordNet (see for instance; Kim and Candan 2006; Li et al. 2003; Resnik 1995; Lin 1998; Budanitsky and Hirst 2006; Patwardhan and Pedersen 2006). WordNet provides a measure of the *frequency* of the concept. The *frequencies* of concepts are estimated using noun frequencies from large text corpora, as for instance the Brown Corpus of American English (Francis and Kucera 1979). Then, the *SemCor* project (Fellbaum et al. 1997) made a step forward by linking subsections of Brown Corpus to senses in the WordNet lexicon. According to *SemCor*, the total number of observed instances of nouns is 88312.

In Fig. 3, for instance, WordNet has been used to assign weights to the geographic classes. According to frequency-based approach, the *weight* of the concept c, indicated as, $w_f(c)$ is defined as follows:

$$w_f(c) = \frac{freq(c)}{M}$$

where $freq(c)$ is the frequency of the concept c, and M is the total number of observed instances of nouns in the corpus. In Table 2, frequencies of geographic classes according to WordNet 2.0 are illustrated. Note that, in WordNet there are concepts for which the frequency is not given, such as for instance *Department*, and *Municipality*. For this reason, in Table 2, the frequencies of such concepts are assumed to be equal to 1. Another limitation of WordNet is related to multi-word terms, for which the frequency are not

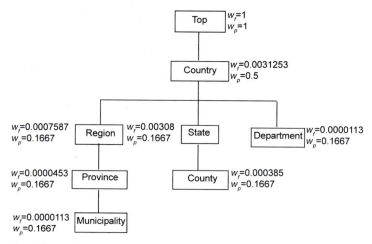

Fig. 3. *Part-of* hierarchy weighted by frequencies and uniform probabilistic distribution.

Table 2. Descriptions and frequencies of geographic classes according to WordNet.

Description	Freq
Country—the territory occupied by a nation	276
Region—the extended spatial location of something;	67
State—the territory occupied by one of the constituent administrative districts of a nation	272
County—an area created by territorial division for the purpose of local government	34
Department—the territorial and administrative division of some countries	1
Province—the territory occupied by one of the constituent administrative districts of a nation	4
Municipality—.........	1

given. For instance, "Seasonal_river" or "Knar_well", shown in Fig. 1, do not exist. Figure 3 illustrates the weighted *Part-of* hierarchy of the *GeoKB* example, where we assume it has a unique *Top* node (i.e., the most abstract Concept) such that for each c, *Part-of(Top; c)* holds, and $w_f(Top) = 1$.

Some proposals in the literature tackle the problem of computing weights of ontology without relying on large text corpora. For instance, the proposal in (Fang et al. 2005) makes a joint use of an ontology and a typical Natural Language Processing method, based on *term frequency* and *inverse document frequency* (*tf-idf*). In order to measure the similarity between terms and elements of the ontology, the authors propose an approach based on five fixed relevance levels corresponding to five constants: *direct(1.0)*, *strong(0.7)*, *normal(0.4)*, *weak(0.2)*, *irrelevant(0.0)*. In this chapter, we focus on approaches in which the weights and the similarity between concepts may take any value between 0 and 1.

Probability-based approach

The probabilistic approach is based on a uniform probabilistic distribution along the reference hierarchy. The root of the hierarchy, referred to as *Top* has weight equal to 1 (i.e., $w_p(Top)=1$), and weights are assigned to the concepts of the hierarchy according to a top-down approach, as follows.

Definition 7: Given a concept c, let c be a part (or meronym) of c', $w_p(c)$ is equal to the probability of c' divided by number of parts, for short P, of c' as follows:

$$w_p(c) = \frac{w_p(c')}{|P(c')|} \tag{1}$$

This approach is based on the discrete uniform distribution notion in probability theory and statistics (Balakrishnan and Nevzorov 2005), according to which a finite number of values are equally likely to be observed, i.e., each one of n values has equal probability $1/n$. Accordingly, the rationale behind the definition above is that all parts (or meronyms) of a holonym are equally probable.

Example 5: Assume the root of the hierarchy shown in Fig. 3 has two parts. Thus, the weight associated with *Country* is 0.5 (according to Eq. (1), $w_p(Country) = \frac{w_p(Top)}{2}$). Now, let us consider the concept *Region*. The associated w_p is 0.1667 because *Country* has three parts (i.e., *Region*, *State*, *Department*).

In Fig. 3, the uniform probabilistic-based weights associated with the geographic classes are indicated. As we can observe nodes (except the root node and the *Country* node) have only one part. It means that the weight associated with whole and its parts coincide (e.g., (*Region, Province*), (*Province, Municipality*), (*State, County*)). As we will see in the next section, the similarity of the above mentioned concepts (e.g., *Region, Province*) with a given concept (e.g., *State*) coincide as well. Thus, similar cases will not be distinguished by using the uniform probabilistic approach.

In order to distinguish such cases, we propose a different probabilistic approach to assign weights as described below. In this approach, called *uniform probabilistic weighted arc*, the weight of nodes is calculated on the basis of the weights of its predecessor *node* and *arc*. The weights assigned according to this approach are shown by w_{pa}. Starting from the Top node (we assume $w_{pa}(Top) = 1$), we distinguish different paths along its successor nodes. Thus, given a path defined by k nodes, each arc along this path is labeled with a weight equal to $1/(k-1)$. We proceed to label arcs in the remaining paths. If an arc is labeled only by one weight, then this will be assigned to the arc. Otherwise, in the case of an arc labeled by more than one weight, the minimum weight among the labeled ones will be assigned to the arc. In the following the definition of *uniform probabilistic weighted arc* is given

Definition 8: Given a concept c_i, let c_i be a part (or meronym) of c_{i-1}. The weight of c_i is given by the following formula:

$$w_{pa}(c_i) = w_{pa}(c_{i-1}) * w_{a(i-1,i)} \qquad (2)$$

where, $w_{a(i-1,i)}$ indicates the weight of arc connecting the concepts c_{i-1}, c_i.

Example 6: Let us consider Fig. 4. We identify three paths, which are

Path1: <*Top, Country, Department*>
Path2: <*Top, Country, State, County*>
Path3: <*Top, Country, Region, Province, Municipality*>

We observe that the number of arcs for paths is 2, 3, and 4, respectively. Hence, the uniform weights of arcs are $1/2$, $1/3$, and $1/4$ along each path, as well. In this approach, weights of arcs along a given path are equal. In other words, the weight of each arc $w_{a(i-1,i)}$ is constant and it is equal to $1/(k-1)$, where k is the number of nodes of the belonging path. In Fig. 4, the arc connecting *Top* and *Country* nodes are labeled with three weights, 0.5, 0.33, and 0.25. Thus, minimum value is assigned as weight to this arc.

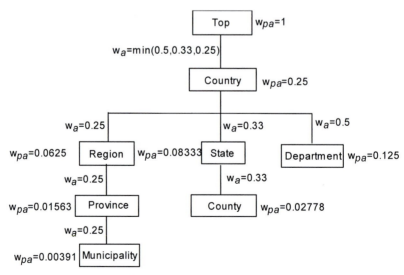

Fig. 4. Hierarchy weighted by uniform probabilistic weighted arc approach.

Following the discrete uniform probability distribution mentioned above, the weights of arcs along a path are equally probable and depend on the number of belonging nodes. Accordingly, the weights of arcs of longer paths are less than the weights of arcs of shorter paths. The benefits of this approach consist in weighting differently both the sibling nodes (e.g., *Region, State, Department*) and nodes belonging to a path (e.g., *Region, Province, Municipality*) that otherwise are not distinguished by uniform probability distribution. In fact, according to the uniform probabilistic approach the weights of *Region, State, Department* coincide and are equal to 0.1667 (see Fig. 3), whereas conforming to the uniform probabilistic weighted arc approach they are distinct and are equal to 0.0625, 0.08333, and 0.125 respectively (see Fig. 4).

In accordance with the general theory of the standard argumentation of information theory (Shannon 1948) a concept weight will be used to determine its information content. The information content of a concept c is defined as follows:

$$ic(c) = - \log w(c)$$

where w is the weight associated with the concept c in the reference hierarchy. The basic intuition behind the use of the negative log likelihood is that the more probable a concept is of appearing then the less information it conveys, in other words, specialized concepts are more informative than more general ones. Thus, as the weight of a concept increases the informativeness decreases, hence, the more general a concept the lower its information content.

Correspondingly, in Fig. 4, we observe that the weight of nodes at the lower level of a path is less than the ones at the higher level. In fact, the nodes at lower level represent more specified parts of concepts with respect to the concepts at higher level, e.g., *Municipality* is more specified (or detailed) part of *Region* with respect to *Province*. Thus, the probability of a meronym is less than the probability of its holonym. Accordingly, the information content of a *low* probable concept is higher than the information content of a *high* probable concept. For instance, the information content of Municipality $ic(Municipality) = -\log(0.00391) = 7.9986$ is higher than $ics(Province) = - \log(0.01563) = 5.9995$. Similarly, the information content of *Province* is higher than $ic(Region) = -\log(0.0625) = 4.0$.

We can observe in Fig. 4, as the probability-based weight increases in the hierarchy in bottom-up direction, the information content decreases. In other words, the information content increases in an opposite direction, i.e., top-down.

With respect to this approach, a different method consists in taking into account the weight of *predecessor arcs* of a given arc, which is called *non-*

uniform probabilistic weighted arc. Accordingly, in formula (2), the weight of arc is $w_{a(i-1,i)} = (\frac{1}{(k-1)})^{(i-1)}$, and weights are denoted by $w_{\overline{pa}}$.

For instance, in Fig. 5, consider Path3: <*Top, Country, Region, Province, Municipality* >. In this case, $k=5$. The weight of arc connecting *Region* (node 3) and *Province* (node 4) (i.e., $w_{a(Region,Province)} = w_{a(3,4)} = (\frac{1}{(5-1)})^{(4-1)}$) is equal to $(1/4)^3 = 0.016$. Similarly, the weights of remaining concepts are computed.

In this approach, with respect to the previous uniform probabilistic weighted arc approach, the weights associated with arcs along a path are different and the weight decreases as the length of a path increases.

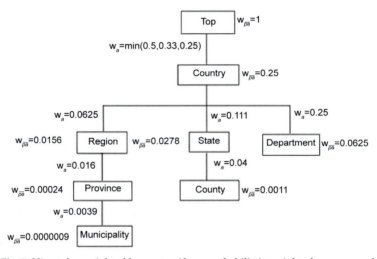

Fig. 5. Hierarchy weighted by non-uniform probabilistic weighted arc approach.

Similarity methods

We focus on three types of similarity methods, namely information content-based, feature based and combined approaches.

Information content-based method

As anticipated, the information based approach to measuring semantic similarity is based on the proposal of Resnik (1995), which has been successively refined by Lin (1998). Resnik views noun synsets as a class of words, where the class is made up of all words in a synset as well as, directly or indirectly, words in all subordinate synsets. Thus, conceptual similarity is considered in terms of class similarity.

According to the Resnik's approach, the similarity between two hierarchically organized classes (or concepts), Sim_R, is given by the maximum information content shared by the classes. Given two classes c_i and c_j, the similarity between them is defined by the information content of the concept that is the least upper bound (*lub*) of c_i, c_j in the hierarchy, and is given as follows:

$$Sim_R = -\log w(lub(c_i, c_j))$$

In the Resnik's approach, $w(c)$ is calculated by estimating the probability of occurrence of the class in a large text corpora.

For instance, let us consider weights w_f of geographic classes shown in Fig. 3. Let us consider the geographic classes *Region* and *County*. Since their *lub* is *Country*, the following holds:

$$Sim_R = -\log w_f(Country) = 8.3218$$

Successively, concept similarity has been refined by Lin (1998) as the maximum information content shared by the concepts (Sim_R) divided by the information contents of the comparing concepts as follows:

$$Sim_{Lin}(c_i, c_j) = \frac{2 \log w(lub(c_i, c_j))}{\log w(c_i) + \log w(c_j)}$$

For instance, the Sim_{Lin} of the classes mentioned in the previous example is:

$$Sim_{Lin}(Region, County) = \frac{2 \log w_f(Country)}{\log w_f(Region) + \log w_f(County)}$$

$$= \frac{2 * 8.3218}{10.3642 + 11.3429} = 0.76674$$

Observation on weights

As anticipated in the Probability-based approach subsection of the previous section, depending on the reference ontology, sometimes the weight associated with whole (holonym) and its parts (meronyms) coincide. For instance, referring to the weighted ontology shown in Fig. 3, we observe that the weight associated with *Region*, *Province*, and *Municipality* is equal to 0.1667. Thus, the similarity of pairs (*Region, State*), (*Province, State*) coincide too. In fact, $Sim_{Lin}(Region, State) = Sim_{Lin}(Province, State) = 0.3869$. Thus, uniform probabilistic approach is not appropriate for weighting concepts of reference ontologies similar to the one shown in Fig. 3.

However, according to the uniform probabilistic weighted arc approach, we observe that the similarity of the pairs mentioned above is different. In fact, they are: $Sim_{Lin}(Region, State) = 0.5274$, $Sim_{Lin}(Province, State) = 0.4173$.

Following the approach of non-uniform probabilistic weighted arc, we obtain:

$Sim_{Lin}(Region, State) = 0.3581$, $Sim_{Lin}(Province, State) = 0.233$.

Feature-based method

Feature-based methods have been conceived to measure the similarity between concepts as a linear combination of the measures of their common and distinctive features or attributes (Tversky 1977). Common features tend to increase the similarity, and conversely, non-common features tend to decrease the similarity of two concepts. One of the well-known feature-based approaches is Dice's function (Maarek et al. 1991; Castano et al. 1998). According to such approach, given two concepts, say c_i, and c_j, each described by a set of features, say $F(c_i)$, $F(c_j)$, respectively, their similarity is defined as follows:

$$Dice(c_i, c_j) = \frac{2 \mid B(c_i, c_j) \mid}{\mid F(c_i) \mid + \mid F(c_j) \mid}$$

where $B(c_i, c_j) = \{(x, y) \mid x \in F(c_i), y \in F(c_j), x, y \in D \sqsubseteq Aff\}$, Aff is the set of sets of attributes showing affinity that is computed similar to the maximum weighted matching problem in bipartite graphs (Kuhn 1955), and for any set X, $\mid X \mid$ indicates the cardinality of X.

For instance, let us consider *Municipality*, and *County* in Table 1. The pairs of attributes showing affinity are (*population, inhabitant*), (*area, surface*). Thus:

$$Dice(Municipality, County) = \frac{2 * 2}{5 + 3} = 0.5$$

A different approach based on features is the proposal of Rodriguez and Egenhofer (2004), called *Matching-Distance Similarity Measure* (MDSM for short). This method is a weighted sum of the similarity measures for *parts, functions* and *attributes*, which are the distinguishing features of spatial entity classes. Given two classes, say s and t, their similarity according to MDSM, indicated as $MDSM(s, t)$ is given by the following formula:

$$MDSM(s, t) = w_{att} * S_{att}(s, t) + w_{func} * S_{func}(s, t) + w_{part} * S_{part}(s, t) \tag{3}$$

where w_{att}, w_{func}, w_{part}, are the weights of similarity vales for attributes, functions and parts, respectively, such that $w_{att} + w_{func} + w_{part} = 1$ (defined below), and $S_q(s, t)$, q standing for attributes (*att*), functions (*func*), and parts (*part*) is defined as follows:

$S_q(s, t)$

$$= \frac{\mid S \cap T \mid}{\mid S \cap T \mid + \alpha(s, t) \mid S - T \mid + (1 - \alpha(s, t)) \mid T - S \mid} \tag{4}$$

In formula (4),which is borrowed from the Tversky's model (see $S_{Tversky}$ discussed in the State of the Art section) S, and T are description sets of the classes s, t, respectively, $S \cap T$, $S - T$ $(T - S)$ are the set-theory intersection and difference of the sets S, T, respectively, and $0 \le \alpha(s, t) \le 1$ is a function defining the "relative importance" of the non-common attributes of the classes. This function is essentially defined in terms of the distance among the classes s, and t the class representing their least upper bound in the hierarchy. Such a distance is given by the number of arcs along the shortest path between classes, and is defined as follows:

$$\alpha(s, t) = \begin{cases} \dfrac{d(s, lub)}{d(s, t)} & d(s, lub) \le d(t, lub) \\[2ex] 1 - \dfrac{d(s, lub)}{d(s, t)} & d(s, lub) > d(t, lub) \end{cases}$$

where $d(s, t) = d(s, lub) + d(t, lub)$.

For instance, in Fig. 2, α (*Municipality, County*) = 0.4 because the *lub* of these concepts is *Country*.

In the MDSM approach, the weights w_{att}, w_{func}, w_{part} indicate the relevance of attributes, functions and hierarchical parts, respectively. They are defined on the basis of two different approaches, namely *variability* and *commonality*. These approaches are recalled in the following, where the term "feature" in accordance with (Rodriguez and Egenhofer 2004) stands for attributes, functions or parts.

According to the variability approach, the relevance of a feature is related to the feature's informativeness, therefore the feature's relevance decreases if it is shared by most of geographic classes in the knowledge base. Whereas, according to the commonality approach, high frequency of a feature corresponds to high relevance. Given a geographic knowledge base, assume q is a type of feature. Then, let P_q^v and P_q^c (where v stands for *variability* and c for *commonality*, respectively) be defined as follows:

$$P_q^v = 1 - \sum_{i=1}^{m} \frac{Oi}{nm} \tag{5}$$

$$P_q^c = \sum_{i=1}^{m} \frac{Oi}{nm} = 1 - P_q^v \tag{6}$$

where O_i is the number of occurrences of a feature in the geographic knowledge base, n is the number of geographic classes and m is the number of distinct features in the geographic knowledge base. Since in this chapter, we deal with the evaluation of the relevance of both the hierarchical (parts) and attribute components of a geographical knowledge base, we focus on determining the weights w_{att}, w_{part}.

According to the variability approach, the weights are determined as follows:

$$w^v_{att} = \frac{p^v_{att}}{p^v_{att} + p^v_{part}}$$

$$w^v_{part} = \frac{p^v_{part}}{p^v_{att} + p^v_{part}}$$

Whereas, according to the commonality approach, they are:

$$w^c_{att} = \frac{p^c_{att}}{p^c_{att} + p^c_{part}}$$

$$w^c_{part} = \frac{p^c_{part}}{p^c_{att} + p^c_{part}}$$

Example 7: In our running example, the number of geographic classes is $n = 7$, in the case of parts $m = 7$, and in the case of attributes $m = 13$. Note that the number of occurrences of attributes and parts are 25, and 9, respectively. Thus, we have the following weights:

$w^v_{att} = 0.470$, $w^v_{part} = 0.530$
$w^c_{att} = 0.599$, $w^c_{part} = 0.401$

Let us consider *Municipality*, and *County* shown in Table 1. We assume that S and T are the sets of attributes of *Municipality* and *County*, respectively. Since, similar to Dice's function, the pairs of synonym attributes between them are (*population, inhabitant*), (*area, surface*), we have:

$|S \cap T| = 2$

And the set difference are:

$S-T = \{identity,\ councilHead, countryCode\}$
$T-S = \{country\ ID\}$

Thus,

$$S_{att}(Municipality, County) = \frac{|S \cap T|}{|S \cap T| + 0.4|S - T| + 0.6|T-S|}$$

$$= 0.526$$

For parts we have:

$S_{part}(Municipality, County) = 0$

Finally, the measures calculated according to the variability and commonality approaches, $MDSM^v$ and $MDSM^c$, respectively, are:

$MDSM^v$(*Municipality*,
County)$=w_{att}^v * S_{att}$(*Municipality*, *County*)$+w_{part}^v *$
S_{part}(*Municipality*, *County*)

$$=0.470*0.526+0.530*0=0.247$$

$MDSM^c$(*Municipality*,
County)$=w_{att}^c * S_{att}$(*Municipality*, *County*)$+w_{part}^c *$
S_{part}(*Municipality*, *County*)

$$=0.599*0.526+0.401*0=0.315$$

Combined approach

Most of similarity methods proposed in the literature are based on the hierarchical structure of concepts, in contrast with the work of psychologists who typically focus on concept features (or attributes) (Rodriguez and Egenhofer 2004). Recently, there has been a growing effort to capture both the concept similarity within the reference ontology, and the similarity of properties of geographic classes. For instance, the proposal in (Formica and Pourabbas 2009), called *GSim*, is an approach that combines both such similarities. In particular, it focus on geographic partition hierarchies, and the attribute similarity (here referred to as *tuple* similarity) of geographic classes. In this mentioned paper, the concepts of the reference ontology (or hierarchy) have been weighted by using *WordNet* taxonomy concepts.

The *GSim* method has been defined by revisiting, extending, and integrating two approaches, which are the *information content (ics)* approach as defined in Lin (1998), and a method for tuple similarity which is based on the *maximum weighted matching* problem in bipartite graphs (Kuhn 1955). The first approach regards the formal definition of information content approach, which is composed of three points (see below, Definition 9), i.e., point (i) that states the *ics* of two concept names is equal to 1 if they coincide or are synonyms, otherwise in all the other cases (point (iii)), their similarity is equal to zero. In point (ii), the information content similarity of hierarchically related concepts, as discussed in the previous section, is formally defined.

Definition 9: Given a geographic knowledge base $K_E = (C, A, Cls)$, a weighted reference hierarchy $\mathcal{H}_w$ and a $SynSet_K = \{S_1,..., S_n\}$, $n \geq 0$ for K_E. Let us consider the sets T of atomic types, as defined previously. Given $m_1, m_2 \in C \cup A \cup T$, the *information content similarity* of m_1, m_2, indicated as $ics(m_1, m_2)$, is defined as follows:

(i) If $m_1 = m_2$ or $m_1, m_2 \in S_k \in SynSet_{\mathcal{JC}}$, for some $1 \leq k \leq n$

$$ics(m_1, m_2) = 1$$

(ii) If $m_1, m_2 \in C \cup A$, and $m_1, m_2 \notin S_k \in SynSet_{\mathcal{JC}}$, for any k

$$ics(m_1, m_2) = \frac{2 \log w(lub(m_1, m_2))}{\log w(m_1) + \log w(m_2)}$$

where $lub(m_1, m_2)$ is the least upper bound of both m_1, m_2 providing maximum information shared by m_1, m_2;

(iii) Otherwise $ics(m_1, m_2) = 0$.

Example 8: Let us consider the weighted hierarchy shown in Fig. 3 (see, w_f), and the geographic classes *Municipality* and *County*. Their least upper bound is the geographic class *Country*, and their maximum shared information is given by the following:

$$ics(Municipality, County)$$
$$= \frac{2 \log w_f(Country)}{\log w_f(Municipality) + \log w_f(County)}$$

$$= \frac{2 * 8.32180}{16.43032 + 11.34286} = 0.5993$$

The tuple similarity approach is based on the definition of *candidate sets of pairs*, which is recalled below. The rationale behind such definition is, given two class names m_1, m_2, it allows to identify the sets of pairs of attributes, each pair formed by one attribute from m_1 and the other from m_2, which have to be considered in order to maximize the sum of the *ics* of the pairs.

Definition 10: Let $m_1 =< \{a_1: t_1, ..., a_h: t_h\}, G_1 >, h \geq 1$ and $m_2 =< \{b_1: t_1, ..., b_k: t_k\}, G_2 >, k \geq 1$ be the names of two classes of a geographic knowledge base where $h \leq k$, and $A(m_1) = \bigcup_i (a_i), A(m_2) = \bigcup_j (b_j)$. The set of *candidate sets of pairs* of m_1, m_2, indicated as $P(m_1, m_2)$, is defined by all possible sets of h pairs of attribute names, as follows:

$$P(m_1, m_2)=$$
$$\left\{\{< r_1, s_1 >, ..., < r_h, s_h >\} \mid r_q \in A(m_1), \ s_q \in A(m_2), for \ q = 1, ..., h \atop r_q \neq r_v, s_q \neq s_l, for \ all \ v, l \neq q \right\}$$

Intuitively, given two geographic classes, namely m_1, m_2, $P(m_1, m_2)$ consists of all the sets of pairs of attributes (each pair formed by attributes not belonging to the same class), such that there are no two pairs of attributes sharing an element. For instance, assume that $A(m_1)$ and $A(m_2)$ represent a set of boys and a set of girls, respectively. A candidate set of pairs defines a possible set of marriages (when polygamy is not allowed) (Galil 1986).

Definition 11: Let m_1, m_2 be the names of two geographic classes as in the previous definition, such that $h \leq k$. The *tuple similarity* of m_1, m_2, indicated as, $tSim(m_1, m_2)$ is defined as follows:

$$tSim(m_1, m_2) = \frac{1}{k}\left(\underset{B \in P(m_1, m_2)}{max} \Sigma_{(a,b) \in B}\ ics(a, b) * ics(\tau(m_1, a), \tau(m_2, b)) \right)\ (7)$$

where *ics* is the information content similarity, $P(m_1, m_2)$ is the set of candidate sets of pairs of m_1, m_2, and $\tau(m_1, a)$, $\tau(m_2, b)$ are the types associated with a, b in the classes m_1, m_2, respectively.

Example 9: Let us consider the geographic classes *Municipality* ($k = 5$) and *County* ($h = 3$). A possible set of pairs that maximizes the formula (7) is:

$$\{< countryCode, countyID >, < inhabitant, population >,$$
$$< area, surface >\}$$

According to Definition 9, we have: *ics*(*inhabitant, population*) = *ics*(*area, surface*)=1

and the same holds for related types, i.e., *ics*(*string, string*) = *ics*(*integer, integer*)=1, whereas *ics*(*countryCode, countyID*) = 0.

Thus $tSim(Municipality, County) = \frac{2}{5} = 0.4$.

Note that two possible sets that maximizes the sum above each contains the pair (*identity, countyID*) and (*councilHead, countyID*) in place of (*countryCode, countyID*).

In fact *ics* (*identity, countyID*) = 0, *ics*(*councilHead, countyID*) = 0.

An important parameter in the similarity of geographic classes is related to the similarity of their geometric components, as defined in the following:

Definition 12: Let G_i and G_j be sets of geometric types. The similarity of geometric types, indicated as $\lambda Sim(G_i, G_j)$, is defined as below:

$\lambda Sim(G_i, G_j) = 1$ if $G_i \cap G_i \neq \emptyset$

$\lambda Sim(G_i, G_j) = 0$ otherwise

For instance, in the case of geographic classes in Example 9, the sets of geometric types of *Municipality* and *County* are {*point, polygon*} and {*polygon*}, respectively. Thus, $\lambda Sim(Municipality, County) = 1$.

Definition 13: Let $m_1 = < \{a_1: t_1, ..., a_k: t_h\}, G_1 >, h \geq 1$ and $m_2 = < \{b_1: t_1, ..., b_k: t_k\}, G_2 >, k \geq 1$ be the names of two classes of the geographic knowledge base $\mathcal{K}_E = (C, A, Cls)$, $\mathcal{H}_w$ be a weighted *Part-of* hierarchy, and a $SynSet_{\mathcal{K}} = \{S_1, ...,$

$S_n\}$, $n \geq 0$ for $\mathcal{K}_E$. Assume $h \leq k$. The *geographic class similarity, GSim(m_1, m_2)*, is given as follows:

$$GSim(m_1, m_2) = \lambda Sim(G_1, G_2)$$

$$* \left(w_{part} * ics(m_1, m_2) + w_{att} * tSim(m_1, m_2) \right)$$

where, λSim is the geometric type similarity of class names m_1, m_2 ics is the information content similarity, $tSim$ is the tuple similarity, and w_{part}, w_{att} are weights, which express the relevance to be given to the hierarchy (ics) and attributes ($tSim$), respectively, such that $w_{part} + w_{att} = 1$.

In the above combined approach, the weights w_{part}, w_{att} are defined in line with both the variability and commonality approaches of MDSM recalled in the previous subsection. Accordingly, we obtain two measures for *GSim*, namely $GSim^v$, $GSim^c$. Note that, in the definition above if $w_{att} = 0$, then *GSim* is given by the shared information content between m_1, and m_2, as derived from the partition hierarchy of the geographic classes. On the contrary, if $w_{part} = 0$, then *GSim* corresponds to the tuple similarity.

In the following, an example of such measures is given.

Example 10: Let us consider *ics(Municipality, County)*, and *tSim(Municipality, County)* given in Example 8, and Example 9 and weights (i.e., w_{part}, w_{att}) of Example 7. The measures of *GSim* for *Municipality* and *County*, according to the variability and commonality approaches, are given below:

$GSim^v(Municipality, County) = 1 * (0.530 * 0.5993 + 0.470 * 0.4)$
$\qquad\qquad = 0.5056$
$GSim^c(Municipality, County) = 1 * (0.401 * 0.5993 + 0.599 * 0.4)$
$\qquad\qquad = 0.4799$

Experimental analysis

The experimental analysis is focused on the approaches shown in the previous sections, i.e., *Lin, Dice, MDSM* and *GSim*. For the last two approaches, the variability and commonality measures are given. Note that, we refer to the weighted *Part-of* hierarchy shown in Fig. 3, and in particular, we use the weights computed according to the frequency-based approach indicated by w_f.

In order to evaluate the approaches mentioned above, we refer to the "right" values, that are established according to *Human Judgment (HJ)*. In the literature, studies on human judgments was performed by Rubenstein and Goodenough (Rubenstein and Goodenough 1965) in 1965, and repeated on a subset consisting of 30 pairs, in 1991 by Miller and Charles (Miller and Charles 1991). These experiments were not addressed to evaluate similarity measures; they were essentially concerned to study the relationship between similarity of context and similarity of meaning (synonyms). In fact, only a part of these experiments was dedicated to ask humans to judge the similarity between words, and this later became the basis for similarity measures evaluation.

In general, a qualitative assessment of similarity measures is achieved in terms of calculating *correlation* between two sets of rating, i.e., estimated similarity measures and human judgment. The correlation reflects the noisiness in the linear relationship between human judgment and estimated similarity measures and essentially means that higher scores on human judgment tend to be paired with higher scores on estimated values and analogously for lower scores. It varies in the interval [−1,1]. The formal definition of the correlation (r) between human judgment (HJ) and estimated similarity values (v) is given as follows:

$$r = \frac{1}{N-1} \sum_{i=1}^{N} \left(\frac{HJ_i - \overline{HJ}}{\sigma_{HJ}} \right) \left(\frac{v_i - \bar{v}}{\sigma_v} \right)$$

where N is the total number of pairs of words (or concepts), $\overline{HJ}$, $\bar{v}$ are the means of HJ, and v, respectively, and σ_{HJ}, σ_v are the standard deviation as follows:

$$\sigma_{HJ} = \sqrt{\frac{1}{N} \sum_{i=1}^{N} (HJ_i - \overline{HJ})^2}$$

$$\sigma_v = \sqrt{\frac{1}{N} \sum_{i=1}^{N} (v_i - \bar{v})^2}$$

In the following analysis, we refer to the set of human judgment rates indicated in our previous work (Formica and Pourabbas 2009). This set of values has been established by asking 24 students to assess the similarity among all pairs of geographic classes (shown in Fig. 2). They were asked to assign a similarity score, on a scale of 0 to 1, to each 42 pairs of classes. In Table 3, second column, for each pair the average of values of human judgment (HJ) is shown.

In the same table the similarity values obtained according to *Lin*, *Dice*, *MDSM^v*, *MDSM^c*, *tSim*, *GSim^v*, and *GSim^c* are illustrated. Note that as anticipated, *MDSM* and *GSim* are based on the variability and commonality measures. Hence, in this experiment, the weights are w^v_{part}=0.474 (w^v_{att}=0.526), and w^c_{part}=0.558 (w^c_{att}=0.442). Note that these weights are computed by applying Eq. (5) and Eq. (6) and the number of occurrences of attributes and parts are 25, and 17, respectively.

Table 3. Similarity results and correlations of the selected methods with HJ.

Pairs of classes	HJ	Lin	Dice	MDSM^v	MDSM^c	tSim	GSim^v	GSim^c
(Municipality, Province)	0.9200	0.9352	0.2500	0.1800	0.1367	0.2000	0.5485	0.6102
(Municipality, Region)	0.7500	0.7736	0.2500	0.1800	0.1367	0.2000	0.4719	0.5201
(Municipality, Country)	0.7300	0.6724	0.4444	0.2700	0.2050	0.4000	0.5291	0.5520
(Municipality, State)	0.6500	0.6718	0.5000	0.3086	0.2343	0.4000	0.5289	0.5517
(Municipality, County)	0.6300	0.5993	0.5000	0.2842	0.2158	0.4000	0.4945	0.5112
(Municipality, Department)	0.7100	0.5065	0.6667	0.3812	0.2894	0.6000	0.5557	0.5478
(Province, Municipality)	0.9100	0.9352	0.2500	0.5680	0.6720	0.2000	0.5485	0.6102
(Province, Region)	0.8900	0.8360	0.3333	0.1800	0.1367	0.3333	0.5716	0.6138
(Province, Country)	0.7600	0.7315	0.2857	0.1350	0.1025	0.2500	0.4782	0.5187
(Province, State)	0.6900	0.7308	0.3333	0.1800	0.1367	0.3333	0.5218	0.5551
(Province, County)	0.6000	0.6458	0.0000	0.0000	0.0000	0.0000	0.3061	0.3603
(Province, Department)	0.5700	0.5393	0.2857	0.1473	0.1118	0.2500	0.3871	0.4114
(Region, Municipality)	0.7100	0.7736	0.2500	0.5680	0.6720	0.2000	0.4719	0.5201
(Region, Province)	0.8500	0.8360	0.3333	0.6400	0.7267	0.3333	0.5716	0.6138
(Region, Country)	0.9200	0.8907	0.5714	0.2700	0.2050	0.5000	0.6852	0.7180
(Region, State)	0.8200	0.8897	0.3333	0.1800	0.1367	0.3333	0.5970	0.6438
(Region, County)	0.6600	0.7667	0.0000	0.0000	0.0000	0.0000	0.3634	0.4278
(Region, Department)	0.7100	0.6212	0.2857	0.1543	0.1171	0.2500	0.4259	0.4571
(Country, Municipality)	0.7200	0.6724	0.4444	0.6760	0.7540	0.4000	0.5291	0.5520
(Country, Province)	0.6900	0.7315	0.2857	0.6400	0.7267	0.2500	0.4782	0.5187
(Country, Region)	0.9200	0.8907	0.5714	0.8200	0.8633	0.5000	0.6852	0.7180
(Country, State)	0.8700	0.9987	0.5714	0.8200	0.8633	0.5000	0.7364	0.7783
(Country, County)	0.6300	0.8464	0.0000	0.4600	0.5900	0.0000	0.4012	0.4723
(Country, Department)	0.7900	0.6724	0.2500	0.5950	0.6925	0.2500	0.4502	0.4857
(State, Municipality)	0.6500	0.6718	0.5000	0.2400	0.1822	0.4000	0.5289	0.5517
(State, Province)	0.6900	0.7308	0.3333	0.1800	0.1367	0.3333	0.5218	0.5551
(State, Region)	0.8200	0.8897	0.3333	0.1800	0.1367	0.3333	0.5970	0.6438
(State, Country)	0.9200	0.9987	0.5714	0.2700	0.2050	0.5000	0.7364	0.7783
(State, County)	0.7300	0.8476	0.0000	0.4600	0.5900	0.0000	0.4018	0.4730
(State, Department)	0.7300	0.6718	0.2857	0.1543	0.1171	0.2500	0.4500	0.4854
(County, Municipality)	0.6300	0.5993	0.5000	0.2571	0.1952	0.4000	0.4945	0.5112
(County, Province)	0.6000	0.6458	0.0000	0.0000	0.0000	0.0000	0.3061	0.3603
(County, Region)	0.6600	0.7667	0.0000	0.0000	0.0000	0.0000	0.3634	0.4278
(County, Country)	0.6700	0.8464	0.0000	0.0000	0.0000	0.0000	0.4012	0.4723
(County, State)	0.7600	0.8476	0.0000	0.0000	0.0000	0.0000	0.4018	0.4730
(County, Department)	0.7200	0.5993	0.5714	0.2945	0.2236	0.5000	0.5471	0.5554
(Department, Municipality)	0.7100	0.5065	0.6667	0.3411	0.2589	0.6000	0.5557	0.5478

Table 3. contd....

Table 3. contd.

Pairs of classes	HJ	Lin	Dice	MDSM^v	MDSM^c	tSim	GSim^v	GSim^c
(Department, Province)	0.5700	0.5393	0.2857	0.1620	0.1230	0.2500	0.3871	0.4114
(Department, Region)	0.7100	0.6212	0.2857	0.1543	0.1171	0.2500	0.4259	0.4571
(Department, Country)	0.8200	0.6724	0.2500	0.1350	0.1025	0.2500	0.4502	0.4857
(Department, State)	0.7300	0.6718	0.2857	0.1543	0.1171	0.2500	0.4500	0.4854
(Department, County)	0.7200	0.5993	0.5714	0.3240	0.2460	0.5000	0.5471	0.5554
Correlation	1.0000	0.7127	0.2972	0.3774	0.3419	0.3318	0.7364	0.8037

We observe that the highest correlation of similarity measure with *HJ* is achieved by *GSim^c* (0.8037), and the correlation of *GSim^v* (0.7364) with *HJ* is slightly higher than the correlation of *Lin* (0.7127). We also observe a significant difference between similarity scores obtained by the hierarchical-based approach of *Lin* and the feature-based approaches of *Dice* and *MDSM*. The scores obtained by the combined approach *GSim*, in most cases are greater than the values computed according to *Dice* and *MDSM*, and they are less than the measures calculated according to the information content approach by *Lin*. The reason is twofold. Firstly, similar to *Dice* and *MDSM*, the *GSim* approach considers the structure of classes, and the heterogeneity of the attributes, in some cases, significantly impacts on the similarity values that are considerably less than the ones obtained according to *Lin*. Secondly, the *GSim* method captures the informativeness of geographic classes organized as a hierarchy, while *Dice* and *MDSM* are mainly based on the common and non-common characteristics of classes. In Table 3, some scores by *Dice* and *MDSM* are equal to zero, indicating that the pairs of considered classes have no common features and they are completely distinct. Similar results we obtain by *tSim* component of *GSim*, which is addressed to capture the similarity of classes' structures.

Analysis of parameters

In this section, we analyze the variability and commonality parameters in the combined *GSim* method. These parameters change depending on the variation of the number of occurrences of attributes and parts in formulas (5), and (6). To this end, we perform two experiments. In the first experiment, we change attributes occurrences, and fix the occurrences of parts, while in the second one we change parts' occurrences and fix the occurrences of attributes.

We start with the first experiment, for which we fix the number of occurrences of parts, say 19, and change occurrences of attributes. Intuitively, as the number of occurrences of attributes increases, consequently, the number of common attributes between classes increases as well. This impacts significantly on the commonality parameters w^c_{att}, and w^c_{part} as are shown in Table 4, and presented graphically in Figs. 6 and 7. In Fig. 6, we observe that, as w^c_{att} increases, w^c_{part} decreases (note that $w^c_{att} + w^c_{part}=1$). We also observe that the correlation of $GSim^c$ (GC^c for short) increases with the increase of w^c_{att} and varies from 0.72 to 0.83.

Table 4. Correlation of *GSim* and parameters by varying attributes' occurrences.

Occ. of Att.	w^v_{part}	w^v_{att}	w^c_{part}	w^c_{att}	*GC*^{v}	*GC*^{c}
15	0.423	0.577	0.702	0.298	0.3335	0.7191
17	0.430	0.570	0.675	0.325	0.3902	0.7248
19	0.436	0.564	0.650	0.350	0.5708	0.7751
21	0.443	0.557	0.627	0.373	0.6067	0.7772
23	0.450	0.550	0.605	0.395	0.6503	0.7903
25	0.458	0.542	0.585	0.415	0.7214	0.8198
27	0.465	0.535	0.567	0.433	0.7748	0.8310
29	0.473	0.527	0.549	0.451	0.7997	0.8340
31	0.481	0.519	0.532	0.468	0.8148	0.8342
33	0.490	0.510	0.517	0.483	0.8242	0.8343
35	0.499	0.501	0.502	0.498	0.8332	0.8345

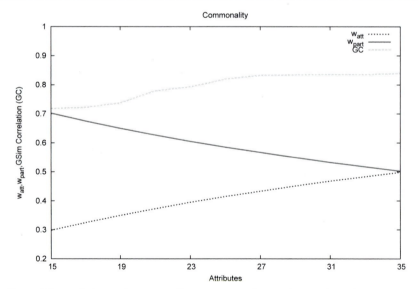

Fig. 6. GC^c correlation and commonality parameters by varying attributes' occurrences.

Color image of this figure appears in the color plate section at the end of the book.

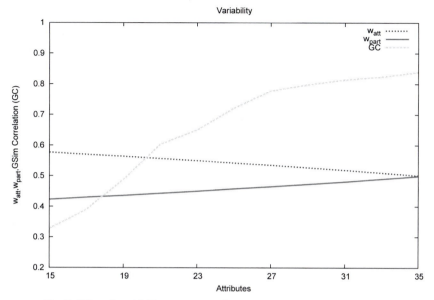

Fig. 7. GC^v and variability parameters by varying attributes' occurrences.
Color image of this figure appears in the color plate section at the end of the book.

Regarding the variability parameters, shown in Fig.7, since $w^v_{att} + w^v_{part} = 1$, as w^v_{part} increases, w^v_{att} decreases. In the case of this last parameter, we note that it decreases because w^c_{att} increases. We observe that the *correlation of GSim^v* (GC^v for short) increases as well, and varies from 0.33 to 0.83. In Table 4, for attributes' occurrences equal to 15, $GC^c=0.72$ (Fig. 6), and $GC^v=0.33$ (Fig. 7). Whereas for attributes' occurrences equal to 35, both commonality and variability parameters tend to coincide and are approximately equal to 0.50. Consequently, the difference between GC^c and GC^v is very low and both are approximately equal to 0.83. In fact, the difference between GC^c and GC^v is high for low occurrences of attributes and it is low for high occurrences.

The impact of variation of attributes' occurrences on GC^v is high with respect to GC^c. Since commonality of attributes increases, their variability decreases, this leads to increase the variability of parts. In fact, GC^v increases with the increase of w^v_{part}. In this case, commonality highlights attributes, whereas variability highlights parts.

Concerning the second experiment, we fix the number of attributes' occurrences, say 25, and change the number of occurrences of parts. The results are shown in Table 5, and presented graphically in Figs. 8, and 9. In Fig. 8, we observe that as w^c_{part} increases, w^c_{att} decreases. We also observe that GC^c increases with the increase of w^c_{part} and varies from 0.61 to 0.85.

Table 5. Correlation of and parameters by varying parts' occurrences.

Occ. of Att.	w^v_{part}	w^v_{att}	w^c_{part}	w^c_{att}	GC^v	GC^c
7	0.542	0.458	0.342	0.658	0.7928	0.6056
9	0.530	0.470	0.401	0.599	0.7839	0.6652
11	0.517	0.483	0.450	0.550	0.7737	0.7137
13	0.503	0.497	0.491	0.509	0.7621	0.7517
15	0.489	0.511	0.527	0.473	0.7499	0.7816
17	0.474	0.526	0.558	0.442	0.7364	0.8037
19	0.458	0.542	0.585	0.415	0.7214	0.8198
21	0.441	0.559	0.609	0.391	0.7050	0.8312
23	0.423	0.578	0.631	0.369	0.6873	0.8392
25	0.403	0.597	0.650	0.350	0.6672	0.8440
27	0.382	0.618	0.667	0.333	0.6460	0.8469
29	0.360	0.640	0.683	0.317	0.6237	0.8482
31	0.336	0.664	0.697	0.317	0.5996	0.8483

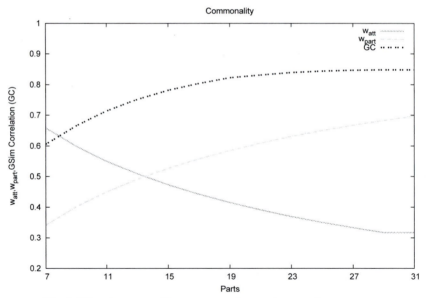

Fig. 8. GC^c and commonality parameters by varying parts' occurrences.
Color image of this figure appears in the color plate section at the end of the book.

The increase of w^c_{part} has a large effect on w^v_{part}. In fact, it decreases, and consequently, w^v_{att} increases as are shown in Fig. 9. Form this graph, we observe that GC^v decreases with the decrease of w^v_{part}, as well, and varies from 0.79 to 0.60. We observe, as the commonality of parts increases the

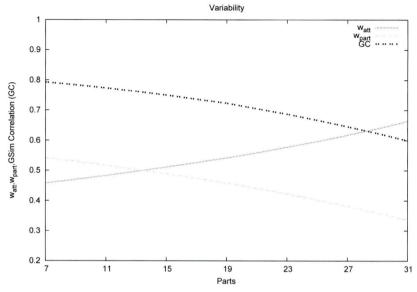

Fig. 9. GC^v and variability parameters by varying parts' occurrences

Color image of this figure appears in the color plate section at the end of the book.

variability of parts decreases. Consequently, the variability of attributes increases, and the related commonality decreases. In this case, commonality and variability highlight parts.

Comparing methods

We apply the experiments shown in the previous subsection to *Lin, Dice* and *MDSM* similarity methods as well in order to contrast the *GSim* method against them. To this end, first we compare their correlation achieved by changing the occurrences of attributes, which are shown in Table 6 and Fig. 10. We observe that the correlation of $GSim^c$ with human judgment is always higher than the correlation achieved by applying other methods. In particular, the correlations obtained by applying $GSim^v$, and $GSim^c$ are better than $MDSM^v$, $MDSM^c$, respectively, and in most cases the ones obtained by *Dice* is worse than the others. The reason is that in *Dice,* an exact feature (or attribute) matching approach is adopted, which essentially relies on the cardinalities of the intersection and/or union of the comparing features. In the case of lower occurrences of attributes (i.e., 15, 17 in Table 6), which represent lower number of shared features, the correlations by *Dice* are even less than 0 (i.e., –0.1539, –0.1021). It is interesting to note that, in the case of higher number of occurrences of attributes (i.e., from 27 to 31, see Table 6), the correlations achieved by *Dice* increase and are higher than

Table 6. Results about similarity methods by changing occurrences of attributes.

Att.	Correlation					
	Lin	*Dice*	*MDSM^v*	*MDSM^c*	*GSim^v*	*GSim^c*
15	0.7127	−0.1539	0.1495	0.2290	0.3335	0.7191
17	0.7127	−0.1021	0.1836	0.2361	0.3902	0.7248
19	0.7127	0.0332	0.2570	0.2640	0.5708	0.7751
21	0.7127	0.1337	0.2984	0.2865	0.6067	0.7772
23	0.7127	0.1855	0.3351	0.3103	0.6503	0.7903
25	0.7127	0.2972	0.3774	0.3419	0.7214	0.8198
27	0.7127	0.5034	0.5041	0.4532	0.7748	0.8310
29	0.7127	0.5705	0.5490	0.5052	0.7997	0.8340
31	0.7127	0.5840	0.5422	0.5125	0.8148	0.8342
33	0.7127	0.6132	0.5413	0.5260	0.8242	0.8343
35	0.7127	0.6135	0.5446	0.5430	0.8332	0.8345

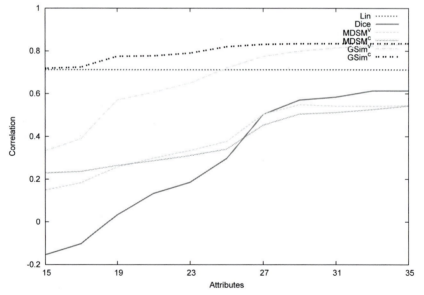

Fig. 10. Correlation of methods by varying occurrence of attributes.

Color image of this figure appears in the color plate section at the end of the book.

$MDSM^v$, $MDSM^c$. Considering that both *Dice* and *MSDM* are feature-based approaches, the *Dice* method achieves better correlation than *MDSM* by increasing the number of shared features. However, their correlations are lower than the correlations obtained by *GSim* method. Since *Lin* method is independent from features, its correlation remains invariant by varying occurrences of attributes.

In Table 7 and in Fig. 11, the correlations of the above mentioned methods with human judgment by changing the occurrences of parts are shown. We observe that the values obtained by the *GSim* method are higher

Table 7. Results about similarity methods by changing occurrences of parts.

Parts	Correlation					
	Lin	*Dice*	*MDSM^v*	*MDSM^c*	*GSim^v*	*GSim^c*
7	0.7127	0.2972	0.3552	0.4007	0.7928	0.6056
9	0.7127	0.2972	0.3585	0.3912	0.7839	0.6652
11	0.7127	0.2972	0.3621	0.3799	0.7737	0.7137
13	0.7127	0.2972	0.3659	0.3692	0.7621	0.7517
15	0.7127	0.2972	0.3697	0.3593	0.7499	0.7816
17	0.7127	0.2972	0.3737	0.3507	0.7364	0.8037
19	0.7127	0.2972	0.3774	0.3419	0.7214	0.8198
21	0.7127	0.2972	0.3821	0.3368	0.7050	0.8312
23	0.7127	0.2972	0.3864	0.3309	0.6873	0.8392
25	0.7127	0.2972	0.3908	0.3260	0.6672	0.8440
27	0.7127	0.2972	0.3948	0.3216	0.6460	0.8469
29	0.7127	0.2972	0.3984	0.3176	0.6237	0.8482
31	0.7127	0.2972	0.4014	0.3142	0.5996	0.8483

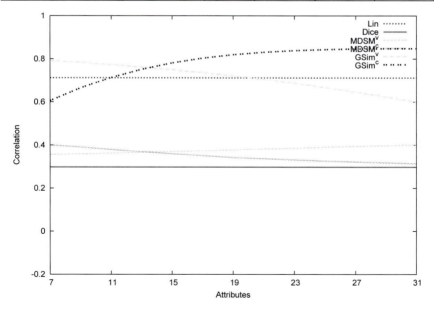

Fig. 11. Correlation of methods by varying occurrence of parts.

Color image of this figure appears in the color plate section at the end of the book.

than *MDSM* and *Dice*. We also observe that by increasing the occurrences of parts, the correlation of $MDSM^v$ increases, and the one obtained by $MDSM^c$ decreases. In particular, the values obtained by *Lin* and *Dice* are invariant, since in both cases the focus is on attributes and are independent from parts. However, we note that the correlations achieved by $GSim^c$ are higher than the ones obtained by *Lin* as the number of occurrences of parts increases from 11 to 31.

Analysis of weight effects

We analyze the effect of different weighting approaches, discussed in the Weight Assignment Methods section, on the correlations achieved by applying the similarity methods. To this end, we use the weights computed according to the frequency-based approach (w_f, see Fig. 3), the uniform probabilistic approach (w_p, see Fig. 3), the uniform probabilistic weighted arc approach (w_{pa}, see Fig. 4), and the non-uniform probabilistic weighted arc approach ($w_{\overline{pa}}$, see Fig. 5). The results are shown in Table 8, where we observe that the correlations of *Dice* and *MDSM* are invariant since in such methods the similarity of concepts are computed on the basis of common and non-common features. However, as we expect, the correlations of *Lin* and $GSim^v$, and $GSim^c$, which are hierarchy-centered methods change and the results obtained by applying the non-uniform probabilistic weighted arc approach in all three cases are better than the other two probabilistic based approaches (i.e., uniform probabilistic and uniform probabilistic weighted arc) and frequency-based approach. However, the worse results are achieved by applying the uniform probabilistic approach, indicating that it is not appropriate for hierarchies similar to that shown in Fig. 3, where most concepts have only one part. Overall, in the case of *GSim*, the results by applying the frequency approach are close to the ones obtained by the uniform and non-uniform probabilistic weighted arc and the differences of results obtained by these last two approaches are very low.

Table 8. Correlation of methods by different weighting approaches.

Approaches	Correlation					
	Lin	*Dice*	$MDSM^v$	$MDSM^c$	$GSim^v$	$GSim^c$
Frequency	0.7127	0.2972	0.3774	0.3419	0.7214	0.8198
Uniform prob.	0.5469	0.2972	0.3774	0.3419	0.6556	0.7186
Uniform prob. weigh.arc	0.7473	0.2972	0.3774	0.3419	0.8061	0.8784
Non-uniform prob. weigh.arc	0.7543	0.2972	0.3774	0.3419	0.8319	0.8863

Conclusion and Future work

In this chapter, we analyzed three different semantic similarity methods, which are defined on the basis of information content, features' commonality/non-commonality and combined approaches. The first two methods are extensively used in the domain of geographic similarity reasoning and are selected as the most representative methods in our analysis. Accordingly, *Lin, Dice, MDSM* similarity methods have been recalled and contrasted against the combined *GSim* method. This last method is an ontology–centered and information content-based method that is conceived to capture both the concept similarity within the hierarchy, and the attribute similarity (or tuple similarity) of geographic classes. Thus, the problem of weight assignment to concepts of reference ontology has been considered. To this end, the frequency and probabilistic-based methods have been analyzed. The experimental results illustrated that the *GSim* method provides more reliable measures for comparing geographic classes with respect to the selected methods.

Future research directions are the following. First, the analysis of weight assignment to DAG (Directed Acyclic Graph)-based hierarchies and the evaluation of results provided by the ontology-based similarity methods. Second, the investigation of how the methods change depending on different types of ontological relations and how it can take advantage of these relations. In this context, it is worth investigating to what extent the qualitative knowledge can be exploited in reasoning of concepts' relationships. For instance, one may need to reason about how *close* or *far* a relation is between two concepts. Similar reasoning can be made about concepts' attributes values, like, comparing the size of a *very big dig* and the size of a *small lake*. These studies effectively fall under the area of qualitative spatial reasoning (Escrig and Toledo 1998), and in particular are concerned with fuzzy spatial reasoning.

The studies on modeling fuzzy spatial relations enhance the knowledge on spatial reasoning between elements. In (Hudelot et al. 2008), the authors introduce an ontology of spatial relations as a support to recognizing structures in images, which are topological, cardinal, directional and distance relations. Topological relations between two objects are based on notions of intersection, interior, and exterior. One of the main formalism of such relationships, known as 9-intersections (Egenhofer 1991) uses a partition of space into three regions for each object (its boundary, its interior and its complement), which constitutes the basis for computing relations. Directional relationships describe the relative position of an object with respect to other ones, and require the space to be oriented, i.e., a reference system. Concerning directional relations, the most used relations are related to three axes of references: *right of, left of, above, below, in front of, behind,* regarding cardinal relations they are *north*, *east*, *west*, and *south* (Pourabbas and Rafanelli 2002). Distance relations describe the spatial arrangement of objects and the less precise forms of these types of relations are *close to*, and *far from*. A further application of similarity methods is the comparison of two or more categorical maps. This task occurs increasingly in remote sensing, geographical information analysis, spatial modelling, and landscape ecology. In many maps the definition of categories is vague, such as the ones where the categories are defined on the basis of ordinal definition, such as *high*, *medium* and *low* density incidence of a disease. Such cases also require fuzzy-based similarity approaches to identify similar categories.

Third, how multiple geo-ontologies can be exploited in order to evaluate the semantic similarity of concepts and consequently how the similarity methods can be adapted to this context for achieving accurate results. Fourth, extension of the approaches to the set of operations that can be performed on geographic data. Fifth, the analysis of the applicability of the methods to the massive corpus, like the Web. An interesting issue related to the Web is the implementation of similarity methods for geographic data based on Linked data (Heath and Bizer 2011) that significantly increments their usability in the Semantic Web. There are several proposals following the Linked Data principles for defining vocabularies and describing data (e.g., documents, people, etc.). Regarding the Linked Data initiatives, the most popular is DBpedia,[8] which aims at extracting structured content from Wikipedia and representing it in a RDF format. However, the exploitation of Linked data to implement similarity methods is a challenging problem to be investigated.

References

Baglioni, M., E. Giovannetti, M.V. Masserotti, C. Renso and L. Spinsanti. 2009. Ontology supported Querying of Geographical Databases. Transactions in GIS. 12(s1): 34–44.

Balakrishnan, N. and V.B. Nevzorov. 2005. Discrete Uniform Distribution, in A Primer on Statistical Distributions, John Wiley & Sons, Inc., Hoboken, NJ, USA.

Ballatore, A., D.C. Wilson and M. Bertolotto. 2012. Geographic Knowledge Extraction and Semantic Similarity in Open Street Map, Knowledge and Information Systems (KAIS), Springer, DOI: 10.1007/s10115-012-0571-0.

Beeri, C. 1990. A formal approach to object-oriented databases, Data & Knowledge Engineering, North-Holland. 5: 353–382.

Buccella, A., A. Cechich and P. Fillottrani. 2009. Ontology driven geographic information integration: A survey of current approaches. Computers & Geosciences. 35: 710–723.

Buccella, A., A. Cechich, D. Gendarmi, F. Lanubile, G. Semeraro and A. Colagrossi. 2011. Building a global normalized ontology for integrating geographic data sources. Computers & Geosciences. 37: 893–916.

Budanitsky, A. and G. Hirst. 2006. Evaluating wordnet-based measures of semantic distance, Comput. Ling. 32: 13–47.

---

[8] http://dbpedia.org

Castano, S., V. De Antonellis, M.G. Fugini and B. Pernici. 1998. Conceptual Schema Analysis: Techniques and Applications, ACM Trans. on Database Systems. 23(3): 286–332.

Cobb, M., M. Chung, H. Foley, F. Petry, K. Shaw and V. Miller. 1998. A Rule-Based Approach for the Conflation of Attributed Vector Data, GeoInformatica. 2(1): 7–35.

Codd, E.F. 1979. Extending the database relational model to capture more meaning, ACM Transactions on Database Systems. 4(4): 397–434.

Egenhofer, M.J. 2002. Toward the Semantic Geospatial Web. In Proceedings of the Tenth ACM International Symposium on Advances in Geographic Information Systems, McLean, Virginia.

Egenhofer, M.J. 1991. Reasoning about binary topological relations. In: 2nd International Symposium on Large Spatial Databases - SSD 1991, Lecture Notes in Computer Science, Springer-Verlag, Berlin Heidelberg New York. 525: 143–160.

Egenhofer, M.J. and R. Franzosa. 1991. Point Set Topological Spatial Relations, International Journal of Geographical Information Systems. 5: 161–174.

Egenhofer, M.J. and R. Franzosa. 1995. On the Equivalence of Topological Relations, International Journal of Geographical Information Systems. 9(2): 133–152.

Ehric, M. 2007. Ontology Alignment: Bridging the Semantic Gap (Semantic Web and Beyond). Springer Science+Business Media LLC.

Escrig, M.T. and F. Toledo. 1998. Qualitative Spatial Reasoning: Theory and Practice. Frontiers in Artificial Intelligence and Applications, IOS Press.

Fang, W.-D., L. Zhang, Y.-X. Wang and S.-B. Dong. 2005. Towards a Semantic Search Engine Based on Ontologies. In proc. of 4th Int'l Conference on Machine Learning, Guangzhou.

Fellbaum, C., J. Grabowski and S. Landes. 1997. Analysis of a hand tagging task. In proc. of ANLP-97 Workshop on Tagging Text with Lexical Semantics: Why, What, and How? Washington D.C., USA.

Francis, W.N. and H. Kucera. 1979. Brown Corpus Manual. Providence, Rhode Island. Department of Linguistics, Brown University.

Formica, A. and E. Pourabbas. 2009. Content based similarity of geographic classes organized as partition hierarchies. Knowledge and Information Systems. 20(2): 221–241.

Formica, A., M. Mazzei, E. Pourabbas and M. Rafanelli. 2012. A 16-Intersection Matrix for the Polygon-Polyline Topological Relation for Geographic Pictorial Query Languages. Multidisciplinary Research and Practice for Information Systems, Lecture Notes in Computer Science Volume 7465. 302–316.

Formica, A., E. Pourabbas and M. Rafanelli. 2013. Constraint Relaxation of the Polygon-Polyline Topological Relation for Geographic Pictorial Query Languages, Computer Science and Information Systems. 10(3): 1053–1075.

Gabrilovich, E. and S. Markovitch. 2007. Computing Semantic Relatedness using Wikipedia-based Explicit Semantic Analysis. Proc. of The 20th International Joint Conference on Artificial Intelligence (IJCAI), Hyderabad, India.

Galil, Z. 1986. Efficient algorithms for finding maximum matching in graphs, ACM Computing Surveys. 18(1): 23–38.

Gruber, T.R. 1993. A translation approach to portable ontologies. Knowledge Acquisition. 5(2): 199–220.

Guzzi, P.H., M. Mina, C. Guerra and M. Cannataro. 2012. Semantic similarity analysis of protein data: assessment with biological features and issues. Briefings in Bioinformatics. 13(5): 569–585.

Harvey, F. 1999. Designing for interoperability: Overcoming semantic differences. pp. 58–98. In: M.F. Goodchild, M.J. Egenhofer, R. Fegeas and C.A. Kottman (eds.). Interoperating Geographic Information Systems, Boston, Kluwer Academic Pub.

Heath, T. and C. Bizer. 2011. Linked Data: Evolving the Web into a Global Data Space. Synthesis Lectures on the Semantic Web: Theory and Technology, February 2011, 136 pages, (doi:10.2200/S00334ED1V01Y201102WBE001).

Hudelot, C., J. Atif and I. Bloch. 2008. Fuzzy spatial relation ontology for image interpretation, Fuzzy Sets and Systems. 159: 1929–1951.

Janowicz, K. 2008. Kinds of contexts and their impact on semantic similarity measurement. In 5th IEEE Workshop on Context Modeling and Reasoning (Co-MoRea) at the 6th IEEE International Conference on Pervasive Computing and Communication (PerCom'08), Hong Kong, IEEE Computer Society.

Janowicz, K., M. Raubal and W. Kuhn. 2011. The Semantics of Similarity in Geographic Information Retrieval, Journal of Spatial Information Science, 2: 29–57.

Jean-Marya, Y.R., E.P. Shironoshita and M.R. Kabukaa. 2009. Ontology matching with semantic verification. Web Semantics: Science, Services and Agents on the World Wide Web. 7(3): 235–25.

Jiang, J.J. and D.W. Conrath. 1997. Semantic Similarity Based on Corpus Statistics and Lexical Taxonomy. In: Proc. of the 10th International Conference on Research in Computational Linguistics (ROCLING), Taiwan. 1–15.

Jones C.B., A.I. Abdelmoty and G. Fu. 2003. Maintaining Ontologies for geographical Information Retrieval on the Web, In Proceedings of CoopIS/DOA/ODBASE, R. Meersman et al. (eds.). Lecture Notes in Computer Science 2888, Springer-Verlag Berlin, Heidelberg, pp. 934–951.

Kavouras, M., M. Kokla and E. Tomai. 2005. Comparing Categories among Geographic Ontologies, Computers & Geosciences, special issue. 31: 145–154.

Keβler, C., M. Raubal and K. Janowicz. 2007. The effect of context on semantic similarity measurement. pp. 1274–1284. In: R. Meersman, Z. Tari and P. Herrero (eds.). On the Move to Meaningful Internet Systems 2007: OTM 2007 Workshop SWWS 2007, number 4806 in Lecture Notes in Computer Science, Vilamoura, Portuga, Springer.

Kim, J.W. and K.S. Candan. 2006. CP/CV: Concept Similarity Mining without Frequency Information from Domain Describing Taxonomies. Proc. of the 15th ACM International Conference on Information and Knowledge Management-CIKM 2006, pages 483–492, New York, NY, USA, ACM Press.

Kuhn, H.W. 1955. The Hungarian method for the assignment problem. Naval Research Logistics Quarterly. 2: 83–97.

Lakoff, G. 1988. Cognitive Semantics, in Meaning and mental representations (Advances in Semiotics), U. Eco, M. Santambrogio, and P. Violi (eds.). Indiana University Press: Bloomington, Indianapolis, USA. 119–154.

Leacock, C. and M. Chodorow. 1998. Combining local context and WordNet similarity for word sense identification. pp. 265–283. In: Christiane Fellbaum (ed.). WordNet: An Electronic Lexical Database. The MIT Press, Cambridge, MA, chapter 11.

Lee, J.H., M.H. Kim and Y.J. 1993. Information Retrieval Based on Conceptual Distance in IS-A Hierarchies, Journal of Documentation. 49(2): 188–207.

Li, Y., Z.A. Bandar and D. McLean. 2003. An Approach for Measuring Semantic Similarity between Words Using Multiple Information Sources. IEEE Transactions on Knowledge and Data Engineering. 15(4): 871–882.

Lin, D. 1998. An Information-Theoretic Definition of Similarity. In proc. of 15th the International Conference on Machine Learning. Madison, Wisconsin, USA, Morgan Kaufmann, Shavlik J.W. (ed.). 296–304.

Lutz, M. and E. Klien. 2006. Ontology-based retrieval of geographic information. International Journal of Geographical Information Systems. 2(3): 233–260.

Maarek, Y.S., D.M. Berry and G.E. Kaiser. 1991. An Information Retrieval Approach For Automatically Constructing Software Libraries, IEEE Transactions on Software Engineering. 17(8): 800–813.

Medin, D.L., R.L. Goldstone and D. Gentner. 1990. Similarity involving attributes and relations: Judgments of similarity and difference are not inverses. Psychological Science. 1(1): 64–69.

Medin, D., R. Goldstone and D. Gentner. 1993. Respects for similarity. Psychological Review. 100(2): 254–278.

Miller, G. and W.G. Charles. 1991. Contextual Correlates of Semantic Similarity. Language and Cognitive Processes. 6(1): 1–28.

Miller, G.A. 1995. WordNet: A Lexical database for English. Communication of the ACM. 38(11): 39–41.

Papadias, D. and Delis. 1997. Relation-based similarity. In: Proc. of the 5th ACM Int. workshop on Advances in Geographical Information Systems, Las Vegas, Nevada, ACM Press. 1–4.

Patwardhan, S. and T. Pedersen. 2006. Using WordNet-based Context Vectors to Estimate the Semantic Relatedness of Concepts. In: Proceedings of EACL Workshop on Making Sense of Sense: Bringing Computational Linguistics and Psycholinguistics Together, Trento, Italy. pp. 1–8.

Pinto, H.S. and J.P. Martins. 2004. Ontologies: How can They be Built? Knowledge and Information Systems. 6(4): 441–464.

Pirró, G. and N. Seco. 2008. Design, Implementation and Evaluation of a New Semantic Similarity Metric Combining Features and Intrinsic Information Content. In: Proc. of OTM 2008 Confederated International Conferences CoopIS. pp. 1271–1288.

Pourabbas, E. and M. Rafanelli. 2002. A Pictorial Query Language for Querying Geographic Databases using Positional and OLAP operators. SIGMOD Record. 31(2): 22–27.

Pourabbas, E. 2003. Cooperation with Geographic Databases. pp. 393–432. In: M. Rafanelli (ed.). Multidimensional Databases. Idea Group Publishing, Hershey, PA, USA.

Pourabbas, E. and M. Rafanelli. 2003. Hierarchies. pp. 91–115. In: M. Rafanelli (ed.). Multidimensional Databases. Idea Group Publishing, Hershey, PA, USA.

Rada, L., H. Mili, E. Bicknell and M. Bletter. 1989. Development and Application of a Metric on Semantic Nets, IEEE Transactions on Systems Man and Cybernetics. 19(1): 17–30.

Rasmussen, E. 1992. Clustering Algorithms. pp. 419–442. In: W.B. Frakes and R. Baeza-Yates (eds.). Information Retrieval: Data Structures & Algorithms, Prentice Hall.

Resnik, P. 1995. Using information content to evaluate semantic similarity in a taxonomy. In Proceedings of the 14th International Joint Conference on Artificial Intelligence. IJCAI'95. 1: 448–453.

Resnik, P. 1999. Semantic Similarity in a Taxonomy: An Information-Based Measure and its Application to Problems of Ambiguity in Natural Language. J. of Artificial Intelligence Research (JAIR). 11: 95–130.

Rodriguez, A. and M. Egenhofer. 2003. Determining Semantic Similarity Among Entity Classes from Different Ontologies, IEEE Transactions on Knowledge and Data Engineering. 15(2): 442–456.

Rodriguez, A. and M. Egenhofer. 2004. Comparing Geospatial Entity Classes: An Asymmetric and Context-Dependent Similarity Measure, International Journal of Geographical Information Science. 18(3): 229–256.

Rubenstein, H. and J.B. Goodenough. 1965. Contextual correlates of synonymy. Communications of the ACM. 8(10): 627–633.

Samet, H. 2004. Indexing Issues in Supporting Similarity Searching, in Advances in Multimedia Information Processing - PCM 2004, LNCS 3332, Springer. pp. 463–470.

Sánchez, D., M. Batet, D. Isern and A. Valls. 2012. Ontology-based semantic similarity: a new feature-based approach, Expert Systems with Applications. 39(9): 7718–7728.

Sánchez, D., M. Batet, A. Valls and K. Gibert. 2010. Ontology-driven web-based semantic similarity, Journal of Intelligent Information Systems. 35: 383–41.

Schwering, A. and M. Raubal. 2005. Measuring Semantic Similarity Between Geospatial Conceptual Regions. GeoSpatial Semantics, Lecture Notes in Computer Science Volume 3799. 90–106.

Schwering, A. 2008. Approaches to Semantic Similarity Measurement for Geo-Spatial Data: A Survey. Transactions in GIS. 12(1): 5–29.

Shannon, C.E. 1948. A Mathematical Theory of Communication. The Bell System Technical Journal. 27(3): 379–423.

Shariff, A.R., M.J. Egenhofer and D.M. Mark. 1998. Natural-Language Spatial Relations Between Linear and Areal Objects: The Topology and Metric of English-Language Terms. International Journal of Geographical Information Science. 12(3): 215–246.

Sheth, A. 1999. Changing Focus on Interoperability in Information Systems: From System, Syntax, Structure to Semantics. pp. 5–30. In: M.F. Goodchild, M.J. Egenhofer, R. Fegeas and C.A. Kottman (eds.). Interoperating Geographic Information Systems, New York, Kluwer.

Storey, V.C. 1993. Understanding Semantic Relationships, VLDB Journal. 2(4): 455–488.

Tanga, Y. and J. Zheng. 2006. Linguistic modelling based on semantic similarity relation among linguistic labels. Fuzzy Sets and Systems. 157(12): 1662–1673.

Teorey, T.L., D. Yang and J.P. Fry. 1986. A Logical design methodology for relational databases using the extended entity-relational model, ACM Computing Surveys. 18(2): 197–222.

Tsichritzis, D. and F. Lochovsky. 1982. Data Models, New York, Prentice-Hall.

Tversky, A. 1977. Features of Similarity. Psycological Review. 84(4): 327–352.

Uitermark, H.T., P.J.M. van Oosterom, N.J.I. Mars and M. Molenaar. 1999. Ontology-Based Geographic Data Set Integration. M.H. Böhlen, C.S. Jensen and M.O. Scholl (eds.): STDBM'99, LNCS 1678. 60–78.

USGS. 2012. Building Ontology for The National Map. U.S. Geological Survey. Accessed January 6, 2012 at http://cegis.usgs.gov/ontology.html.

Visser, U., H. Stuckenschmidt, G. Schuster and T. Vögele. 2002. Ontologies for geographic information processing. Computers & Geosciences. 28(2002): 103–117.

Wang, J.Z., Z. Du, R. Payattakool, P.S. Yu and C.F. Chen. 2007. A new method to measure the semantic similarity of GO terms. Bioinformatics. 23: 1274–1281.

Winston, M., R. Chaffin and D. Herrmann. 1987. A Taxonomy of Part-WholeRelations. Cognitive Science. 11: 417–444.

Wu, Z. and M. Palmer. 1994. Verb semantics and lexical selection. In: Proceedings of the 32nd Annual Meeting of the Associations for Computational Linguistics. 133–138.

Yeh, A. and X. Shi. 2003. The Application of Case-based Reasoning in Development Control. pp. 223–248. In: S. Geertman and J. Stillwell (eds.). Planning Support Systems in Practice, Berlin, Springer-Verlag.

Index

Color Plate Section

Chapter 3

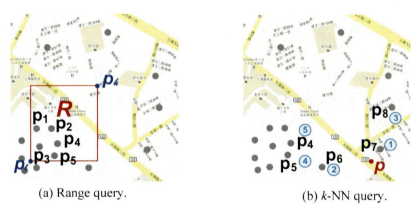

(a) Range query.

(b) *k*-NN query.

Fig. 1. Examples of multi-attribute access.

(a) An example of skewed data.

(b) R-tree with $M = 3$ and $m = 1$.

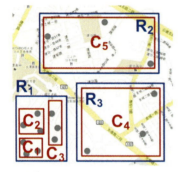

(c) R$^+$-tree with $M = 3$ and $m = 1$.

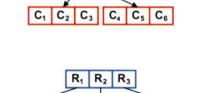

(d) The index structures for the R-tree in Fig. 2(b) and the R$^+$-tree in Fig. 2(c).

Fig. 2. Examples of R-trees and R$^+$-trees.

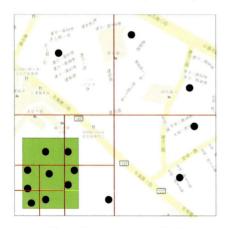

Fig. 3. Quad-tree with $M = 3$.

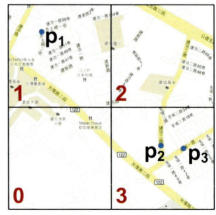

Fig. 4. A Hilbert curve on grids of a map.

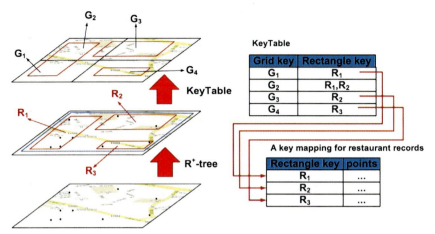

Fig. 5. Overview of the KR⁺-index.

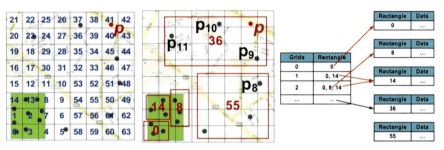

(a) A key definition for grids.

(b) A key definition for rectangles.

(c) A KeyTable for rectangles.

Fig. 6. An example of a KR⁺-index.

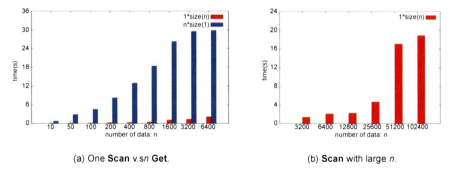

(a) One **Scan** v.s *n* **Get**.

(b) **Scan** with large *n*.

Fig. 7. The features of the CDMs.

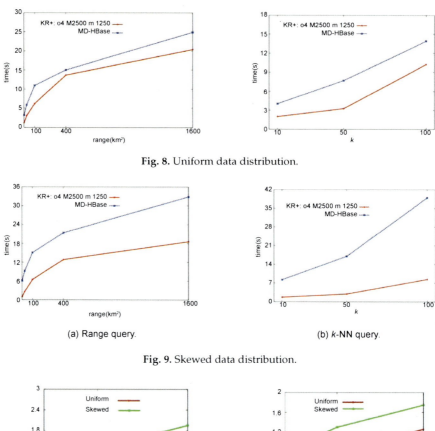

Fig. 8. Uniform data distribution.

(a) Range query.   (b) *k*-NN query.

Fig. 9. Skewed data distribution.

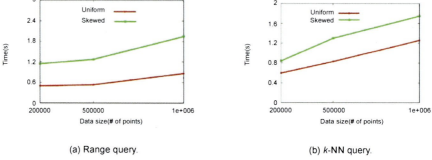

(a) Range query.   (b) *k*-NN query.

Fig. 10. Effect of data size.

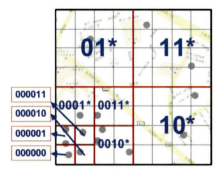

Fig. 11. A key formulation in MD-HBase.

Chapter 8

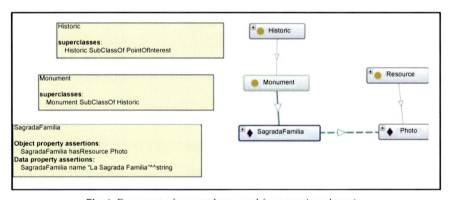

Fig. 1. Fragment of an ontology used for a tourism domain.

Fig. 7. Prototype of Android application that uses GeospatialWeb functions over ontology.

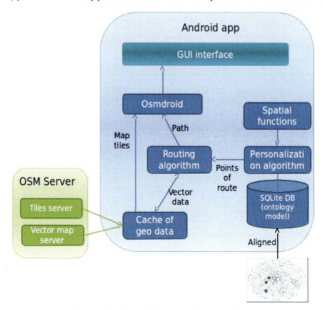

Fig. 9. Final architecture of the project.

Fig. 10. Screenshots of Android pilot application. (a) Map with an example of a route. (b) User preferences. (c) Preferred POIs types. (d) Preferred food types.

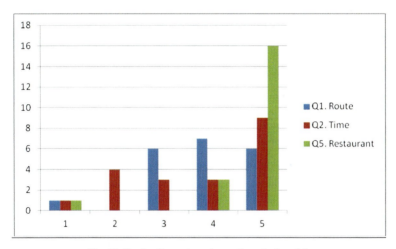

Fig. 11. Evaluation rates of questions 1, 2 and 5.

Chapter 9

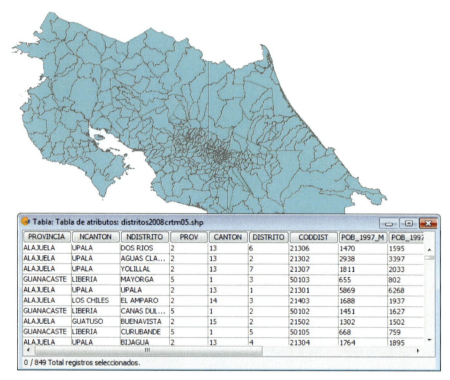

Fig. 2. An example of a shape file representing Costa Rican districts and associated populations.

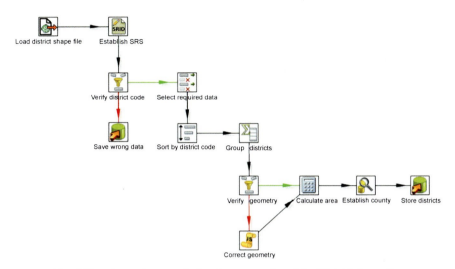

Fig. 6. Transformations applied to the geometries of the *District* shape file.

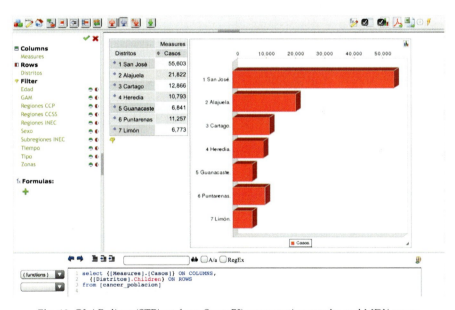

Fig. 12. OLAP client (STPivot from StrateBI) representing results and MDX query.

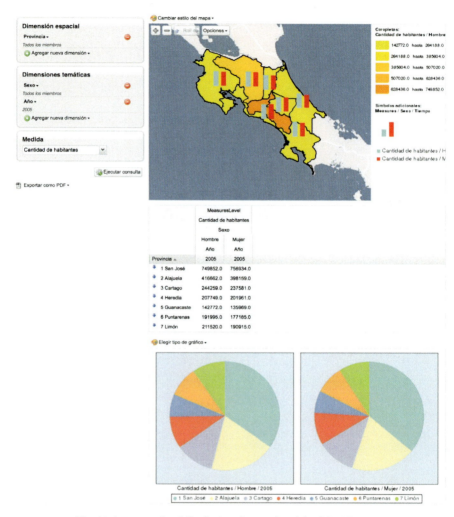

Fig. 13. An example of displaying the results of the SOLAP query.

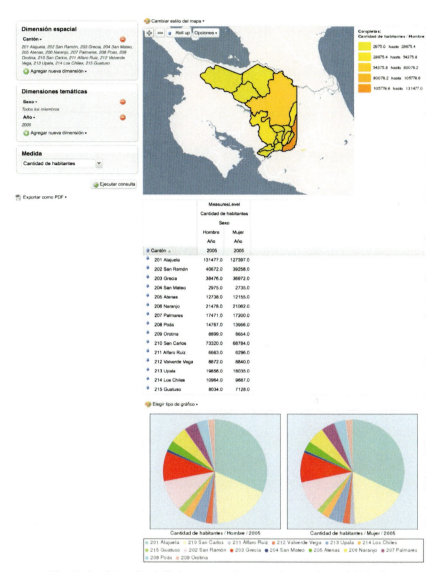

Fig. 14. Applying the drill-down operation over the province of Alajuela.

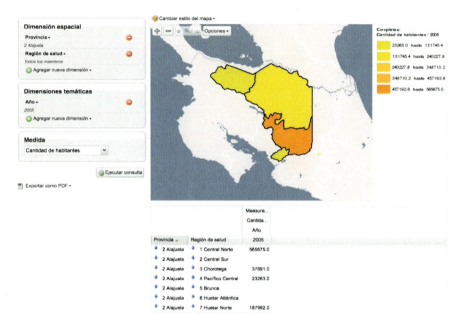

Fig. 15. Combining two spatial hierarchies of administrative and health divisions for the Alajuela province.

Chapter 10

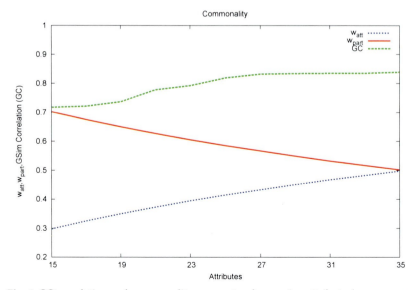

Fig. 6. GC^c correlation and commonality parameters by varying attributes' occurrences.

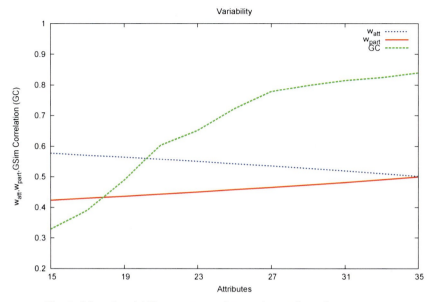

Fig. 7. GC^v and variability parameters by varying attributes' occurrences.

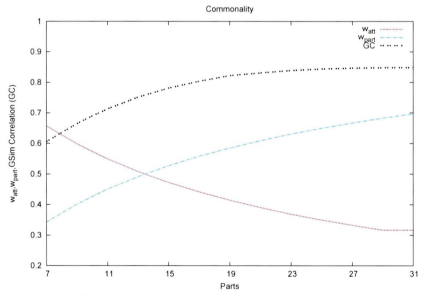

Fig. 8. GC^c and commonality parameters by varying parts' occurrences.

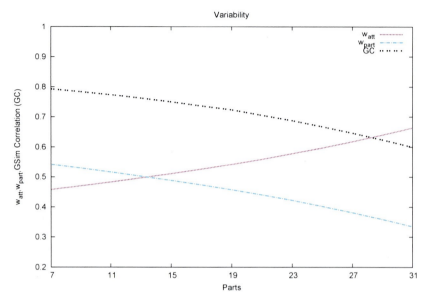

Fig. 9. GC^v and variability parameters by varying parts' occurrences

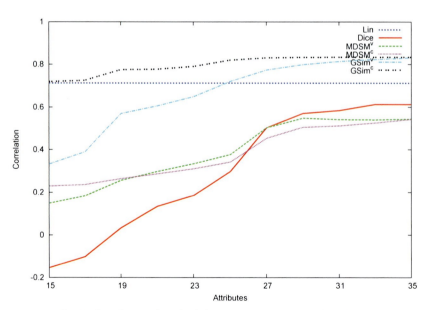

Fig. 10. Correlation of methods by varying occurrence of attributes.

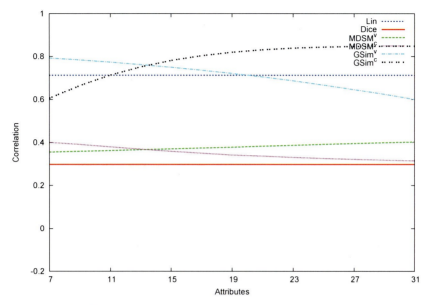

Fig. 11. Correlation of methods by varying occurrence of parts.